Introduction to Digital Mobile Communication

WILEY SERIES IN TELECOMMUNICATIONS AND SIGNAL PROCESSING

John G. Proakis, Editor
Northeastern University

Worldwide Telecommunications Guide for the Business Manager
Walter H. Vignault

Expert System Applications to Telecommunications
Jay Liebowitz

Business Earth Stations for Telecommunications
Walter L. Morgan and Denis Rouffet

Introduction to Communications Engineering, 2nd Edition
Robert M. Gagliardi

Satellite Communications: The First Quarter Century of Service
David W. E. Rees

Synchronization in Digital Communications, Volume 1
Heinrich Meyr and Gerd Ascheid

Digital Telephony, 2nd Edition
John Bellamy

Elements of Information Theory
Thomas M. Cover and Joy A. Thomas

Telecommunication Transmission Handbook, 3rd Edition
Roger L. Freeman

Digital Signal Estimation
Robert J. Mammone, Editor

Telecommunication Circuit Design
Patrick D. van der Puije

Meteor Burst Communications: Theory and Practice
Donald L. Schilling, Editor

Mobile Communications Design Fundamentals, 2nd Edition
William C. Y. Lee

Fundamentals of Telecommunication Networks
Tarek N. Saadawi, Mostafa Ammar, with Ahmed El Hakeem

Optical Communications, 2nd Edition
Robert M. Gagliardi and Sherman Karp

Wireless Information Networks
Kaveh Pahlavan and Allen H. Levesque

Practical Data Communications
Roger L. Freeman

Active Noise Control Systems: Algorithms and DSP Implementations
Sen M. Kuo and Dennis R. Morgan

Telecommunication System Engineering, 3rd Edition
Roger L. Freeman
Radio System Design for Telecommunications, 2nd Edition
Roger L. Freeman

Digital Communication Receivers: Synchronization, Channel Estimation and Signal Processing
Heinrich Meyr, Marc Moeneclaey, and Stefan Fechtel

Introduction to Digital Mobile Communication
Yoshihiko Akaiwa

Introduction to Digital Mobile Communication

Yoshihiko Akaiwa

A Wiley-Interscience Publication
JOHN WILEY & SONS, INC.
New York • Chichester • Weinheim • Brisbane • Singapore • Toronto

This text is printed on acid-free paper. ⊗

Copyright © 1997 by John Wiley & Sons, Inc. All rights reserved.

Published simultaneously in Canada.

Library of Congress Cataloging in Publication Data:

Akaiwa, Yoshihiko.
    Introduction to digital mobile communication/Yoshihiko Akaiwa.
      p. cm.—(Wiley series in telecommunications and signal processing)
    Includes index.
    ISBN 0-471-17545-5 (cloth: alk. paper)
    1. Mobile communication systems. 2. Digital communications.
  I. Series.
  TK6570.M6A39   1997
  621.3845—dc21
                                           97-4310
                                           CIP

Printed in the United States of America
10 9 8 7 6 5 4 3 2 1

CONTENTS

PREFACE

Digitization is a technical trend in telecommunications as well as in other fields; for example, digital audio systems (compact disk) and digital control of machines. A digital approach enjoys higher accuracy and stability of the system over an analog system. Driving forces for digitization are VLSI (very large-scale integrated) circuits and computers, which make it feasible to implement circuits required for the digitization of systems. In the telecommunications field, technological advances have been made on data and digital speech transmission on switched telephone networks, digital microwave communications, and digital optic fiber communications. Digitization in mobile radio communications has lagged behind in comparison with these developments. However, in recent times, explosive activity in research and development of digital mobile communications seems set to recover the delay in the digitization progress. This progress is further spurred by novel applications of digital mobile communications such as personal communication services, mobile computing, and mobile multimedia.

Some digital technologies can be applied in common to any field. However, others are not directly applicable because of differences in the requirements of a specific technical field. For mobile communications, robustness against fast fading, spectrum efficiency, power efficiency, and compactness and low price of equipment are essential. Digital modulation/demodulation is the key technology to fulfill these requirements for mobile communications.

This book is intended to introduce the digital mobile communication technologies with an emphasis on digital transmission methods.

Chapter 1 describes what a digital mobile communication system is and why mobile communications are digitized. The reader will get a rough global insight into the various technologies used in digital mobile communications.

Chapter 2 describes mathematical analyses of signals and noise. Signal analysis includes topics such as the delta function, the Fourier transform, the response of a linear system, the impulse response, the cross correlation, the autocorrelation, a representation of digital signals, the average signal power, the power spectral density, modulated signals, orthogonal signals, and the sampling theorem. The analysis of noise includes the noise figure or the noise temperature and the statistical characteristics of the noise.

Chapter 3 covers the fundamentals of a digital transmission system. It deals with the Nyquist criteria for transmission without intersymbol interference, multilevel coding, partial response systems, a matched filter receiver, and the optimum receiver.

Chapter 4 covers mobile radio channels. It includes some specific features of mobile radio channels, such as Rayleigh fading, frequency-selective fading, shadow fading, and the near–far problem.

Chapter 5 is devoted to the elements of digital modulation. The chapter includes a fundamental description of digital modulation, the power spectrum of modulated signals, demodulation, and error rate analysis.

Chapter 6 discusses digital modulation methods for mobile radio communications. Constant envelope digital modulations are described first. These are minimum shift keying, tamed frequency modulation, GMSK (Gaussian MSK), Nyquist multilevel FM, PLL-QPSK, CCPSK, duoquatenary FM, correlative PSK, and digital PM. Then linear digital modulations are described. Following the description of the significance and the problems of linear modulation when applying it to mobile radio communications, some digital modulation such as $\pi/4$ shifted QPSK, 8PSK, and 16QAM are discussed. For the circuit aspects related to linear modulation, techniques for the nonlinearity compensation in a power amplifier are introduced.

Chapter 7 describes other topics related to digital mobile radio communication. It includes antimultipath modulations, multicarrier systems, spread-spectrum communications, diversity communication systems, adaptive equalizers, error control techniques, trellis-coded modulation, adaptive interference cancellation, and voice coding.

Chapter 8 describes equipment and circuit implementation methods. Configurations of base stations and mobile stations are shown. It also includes a discussion on superheterodyne and direct conversion receivers, transmit and receive duplexing, frequency synthesizers, transmitter circuits, receiver circuits, and countermeasures against dc blocking and dc offset.

Chapter 9 describes digital mobile radio systems. Fundamental concepts such as traffic theory, the cellular concept, multiple access systems, channel assignment, and inter-base-station synchronization are discussed first. Then systems such as digital transmission in analog FM systems, paging systems, the so-called two-way digital mobile radio, mobile data service systems, digital cordless phone, digital mobile telephone systems, wireless local area networks, mobile satellite systems, personal communication, and FPLMTS are introduced.

The author's objectives in writing this book were as follows.

- Almost all important topics are covered, although the depth of their description differs.

- A beginning engineer can follow the book by starting with a study of the analyses of signals and noise and of the basic technologies for digital communications.
- Intuitive understanding of the meaning of the techniques is emphasized over a rigorous discussion.
- Circuit implementation methods are described that are useful for design engineers of digital mobile communication systems.

A certain part of this book is bound to become obsolete in the near future, when more advanced technologies for digital mobile communications are developed as the result of currently on-going active research in the field. However, the author believes that this book may be helpful to mobile radio engineers, since there is no book which comprehensively describes digital mobile communications at this time.

Engineering is a little different from science. In contrast to science, new technologies are sometimes achieved through a combination of well-known facts or by introduction of a technology that was developed in another field. It is rather easy to understand the mechanism, the importance, and the motivation of a new technique once it has been presented. But it is fairly difficult for engineers to create a new technology. A profound understanding of fundamental key technologies in conjunction with an assumed background are important. The most important things for creative engineers as well as scientists are motivation for and patience with the attainment of their technological goal. I hope this book will help mobile radio engineers in their work in this new technology field.

YOSHIHIKO AKAIWA

Kyushu University
Fukuoka, Japan
December 1996

ACKNOWLEDGMENTS

It is my sincere pleasure to thank individuals who helped me prepare this book. Some of the contents are based on my work at NEC corporation. I thank Messrs. Ichiroh Takase, Yoshinori Nagata, and Yukitsuna Furuya for their cooperation. Since I joined the university, I have enjoyed working with my research assistant and students. Some results of the research carried out with them at the university are also included in this book. Furthermore, they helped me by preparing and reviewing the manuscript. I wish to thank all of them, especially Assistant Professor Hiroshi Furukawa, Tetuhiko Miyatani, Hidehiro Andoh, Toshio Nomura, Taisuke Konishi, Masahiro Maki, Mokoto Taromaru, Takayuki Torigai, Kuninori Oosaki, Hisao Koga, Canchi Radhakrishna, Fangwei Tong, Masaya Saitou, Hitoshi Iochi, Mutsuhiko Ooishi, Takenobu Arima, Takashi Matubara, Takuya Ootani, and Takashi Seto.

I would like to thank also Mrs. Kara Maria and my wife Terumi for typing and proofreading some parts of the manuscript. My special thanks go to Professor Jeffrey Capone of Northeastern University for his generous review of the manuscript and correction of the English language. I am very grateful to the Series Editor Professor John G. Proakis of Northeastern University and the Executive Editor Mr. George Telecki of John Wiley & Sons for their efforts to publish this book.

Introduction to Digital Mobile Communication

INTRODUCTION

This chapter briefly describes the digital mobile radio communication system and the significance of digitization in mobile radio communications.

1.1 DIGITAL MOBILE RADIO COMMUNICATION SYSTEMS

A schematic block diagram of a digital mobile radio communication system is shown in Fig. 1.1. A voice signal is converted into a digital signal via a voice coder. The digitized voice signal is transmitted through a digital mobile radio channel, and is converted into a reconstructed analog voice signal at the receiver. The target of voice coding technology is to achieve a lower coding rate for an acceptable voice quality. The digital signals are processed

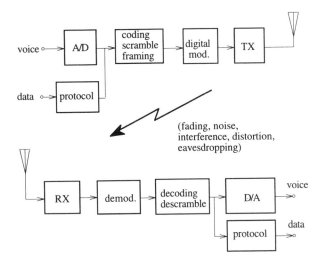

Figure 1.1. Digital mobile radio communication system.

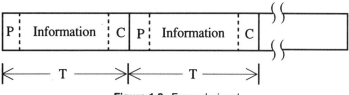

Figure 1.2. Framed signal.

by logic circuits for several system requirements: typically, channel coding, scrambling, and framing. Channel coding is a process that inserts additional bits in order to correct or to detect errors in received signals. Scrambling is designed to hide a transmitted signal from a third party by performing a complicated transformation which is known only to an authorized recipient. Framing is a process whereby information signals are grouped into blocks with other signals, as shown in Fig. 1.2. The purpose of framing is to multiplex different signals in the time domain (TDM: time division multiplexing), to introduce channel coding, and to make it possible to adopt a synchronized scrambling technique.

The next stage of a transmitter is digital modulation. There is no difference in principle between digital modulation and analog modulation: modulation is a process in which the amplitude and/or phase of an oscillating carrier signal are varied in proportion to a modulating signal. A narrow spectrum bandwidth of the modulated signal is desired. For this purpose a low-pass filter or an equivalent IF band-pass filter is employed to limit the bandwidth of the modulating signal. Transfer functions, or the impulse response of the low-pass filters, are especially designed for digital transmission as discussed in Chapter 2. After the modulation stage, frequency conversion up to a final RF frequency band, power amplification, and transmission via an antenna follow in subsequent stages.

The signal that is received through an antenna is amplified, frequency down-converted, and then band-limited with a band-pass filter. A demodulator outputs a transmitted baseband signal, which is corrupted to some extent by noise and imperfections in the function of the transmitter, channel, and/or receiver. The principle for the demodulation of a digitally modulated signal is the same as for an analog signal: that is, coherent or noncoherent demodulation can be used. In some digital transmission systems, a carrier signal, which is necessary for coherent demodulation, can be extracted from a received signal using the knowledge that the digitally modulated signals take specific phases.

The demodulated signal is fed into a decision circuit, where the received signal is sampled and discretized to one of the allowable values of the transmitted signals. A timing recovery circuit generates a timing signal for sampling. The timing signal is extracted from the received signal by detecting the change in digital signal levels at the clock frequency. The decision error rate is the main concern at the receiving side. Decision is a process particular to a digital transmission system. The principle of the decision process is based on the fact that digital signals take one of several discrete states. If there is no decision error, then noise, interference, or distortion will have no effect on the signal transmission. This fact becomes especially important for a multi-hop transmission system, where the signal is regenerated and transmitted at the repeaters: no accumulation of noise and interference occurs in the digital system, in contrast to the case with an analog system.

Output of the decision circuit takes the form of a logic signal. Channel decoding, descrambling, and deassembling of the framed signal, which are inverse operations to channel coding, scrambling, and framing, respectively, are carried out with logic circuits.

The digital signal is converted into an analog voice signal by a voice decoder. As long as a voice signal is involved, the phase of the transfer function is not important, because human ears are insensitive to the phase of the voice signal. On the other hand, decision error in the digital transmission is sensitive to both the phase and the amplitude of the transfer function. This is why pulse waveforms should be transmitted without distortion.

The transmission of data signals differs from voice signal transmission in several aspects. We should assure a very low error rate, because great value may depend on accurate data transmission, for example, in a banking service. In contrast, we can tolerate a rather higher bit error rate for speech communication, because the voice signal has a lot of redundancy, and because humans are very intelligent communication terminals: we can ask for repetition, we can confirm the meaning of words or phrases, and we can guess pronunciation and meanings.

The intelligence of the data communication system is given by protocols imbedded in the data terminal. The protocol affects the efficiency of data transmission even in the same transmission channel. In a mobile radio communication channel, in which error rate performance is not generally good, an automatic repeat request (ARQ) with error detection seems indispensable for data transmission. Real-time communication is not particularly important for data transmission, in contrast to spoken conversation.

A mobile radio communication channel is characterized as a channel without direct propagation. When mobile equipment moves quickly, a transmitted signal encounters rapid fading phenomena. The depth of the fading can reach as deep as several tens of dB. The rate of fading is proportional to the carrier frequency and the speed of motion. For example, the maximum Doppler frequency is as high as 90 Hz for a carrier frequency of 900 MHz and a car speed of 100 km/h. A receiver has to cope with those fast fading phenomena. On the other hand, in the case of hand-carried equipment, the fading speed is quite slow because of the low speed of motion. Another problem occurs: the fading duration becomes too long. In this case we can not communicate for a long time. As the digital transmission rate increases, the channel exhibits frequency-selective fading, where the fading is different at different frequencies and, as a result, the transfer function of the channel loses its distortion-free characteristic. The distortion in the frequency-selective fading channel degrades bit error performance due to intersymbol interference.

Noise is a general problem in communications. Carrier frequencies for mobile radio communications are predominantly in the VHF and UHF bands. At these frequencies man-made noise and atmospheric noise are rather high compared with the microwave communication band. For the purpose of obtaining higher receiving sensitivity, it is ineffective to use a low-noise receiver that has a noise level lower than the channel noise. Digital transmission offers a solution, an error control scheme to make communication secure in a noisy channel.

Eavesdropping on a conversation by a third party can occur in mobile radio communications if the transmitted signal is not protected: the transmitted signal propagates everywhere around the transmitter, and one can easily obtain eavesdropping equipment (i.e. a scanning receiver). Voice scramblers for analog transmission are not sufficient to assure the desired degree of protection or voice quality. Digital transmission offers high-security and high-voice-quality scrambling to combat eavesdropping.

Mobile radio communication equipment is, in some sense, comparable to other consumer products: the number of mobile terminals in the market is large in contrast to other communication systems, such as satellite communication or microwave communication equipment. A mobile terminal is so slim that it slips into pocket and is so inexpensive that consumers can easily afford it. For a digital mobile radio terminal, an increase in the price or size will not be acceptable in the market, as long as the service is equivalent.

The biggest difference between mobile radio communications and other consumer products is that the former is supported with a huge system, while the latter is not. Consider a mobile telephone system; the mobile switching centers and a number of base stations, as well as the mobile terminals, are incorporated into the system. Digitization of the mobile telephone system operates on the all of these facilities. The switching center and the wire line communications were digitized prior to the digitization of the base stations and the mobile terminals. Communications between the base station and the mobile terminals (i.e., through the air interface) is the major concern of this book. This interface includes digital modulation/demodulation, channel access methods for use of the channels by many users on demand, and other topics related to the signal transmission between the base station and the subscriber stations.

1.2 THE PURPOSE OF DIGITIZATION OF MOBILE RADIO COMMUNICATIONS

Data transmission, voice scrambling, and spectrum efficiency are the major motivations for digitization of mobile communications. The emphasis is different for each purpose.

Data Communication

Advanced mobile Data transmission plays the primary role in controlling a mobile communication system. An example of a controlling system is call set-up and termination control in a mobile telephone system. Although the data transmission is relatively slow in speed, high reliability of the transmission is required in this case. Without reliable data transmissions in an unstable mobile radio channel, no communication system could be realized.

In a digital paging system, the radio station broadcasts paging signals in the form of transmitted data. The radio paging system was the earliest example of digitization in mobile communications. The digital radio paging system can accommodate 30 000 subscribers per channel, which is three times larger than conventional tone signaling system.

Computer data transmission between mobile data terminals and host computers has been performed. For example, a sales person may send some business data to the host computer using digital mobile radio communication equipment from his station in his car, and he will be able to have the data automatically processed to yield required items, such as bills or documents, when he returns to the office. For another example, car dispatch service efficiency will be greatly enhanced by adopting a mobile radio data communication system.

Display paging service is a forerunner of mobile message communications (Fig.1.3). It is expected that message communication through digital mobile radio channels will be important as electronic mail service becomes more popular. Message communication is more advantageous than speech communication from the point of view that it does not disturb the recipient if he or she is engaged in other work, and it is more spectrum-efficient than mobile radio speech conversation.

The integrated services digital network (ISDN) has been put in operation. The application of this network technology will surely be extended in the near future to mobile radio communications. Thus, the digitization of the mobile radio channel is considered imminent in more advanced and versatile communication systems.

Voice Scrambling

Voice protection from eavesdropping in mobile radio communications is the most important requirement for security in military or police communication. Even in public mobile communications, such as by mobile telephone or cordless telephone, voice scrambling has received increased attention as those services have become widespread. Although some voice protection techniques have been developed for analog communication systems, they do not provide a sufficient degree of protection and/or voice quality. Voice scrambling in digital communication resolves those shortcomings: it is hardly

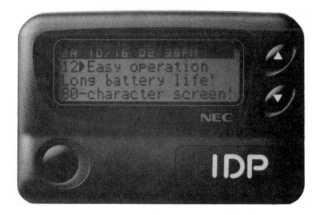

Figure 1.3. Display pager. Courtesy of NEC Inc.

recognizable by a third party whether the signal contains conversation or not, yet the voice quality is not degraded.

Spectrum Efficiency

Spectrum efficiency is the most important issue for radio communications, since the radio frequency spectrum band is considered to be a limited natural resource. In general, digital transmission needs a wider bandwidth than analog transmission: if we transmit a voice signal using the 64 kbps pulse code modulation (PCM) in two-level signaling, we need at least 32 kHz of baseband frequency. This is about eight times wider than that needed for analog transmission. The reason the development of digital technologies for mobile radio communication has been delayed so far, compared with wire line communications and fixed microwave communication, is that the requirement for spectrum efficiency is so tight and that no technology to date has been able to cope with this requirement. Recent progress in digital mobile communication has succeeded in overcoming this problem.

Digital modulation and voice coding play a dominant role in spectrum efficiency. Figure 1.4 shows spectrum efficiency in terms of the required bandwidth per channel, or the channel spacing for digital voice transmission as a function of the voice coding rate and the efficiency of the digital modulation. The efficiency of digital modulation (bps per Hz) is defined as a transmission bit rate (bps) per unit spectrum bandwidth (Hz). A 25 kHz channel spacing, which is the most conventional, can be achieved with the digital modulation of 4-level FM combined with voice coding using 16 kbps adaptive delta modulation (ADM). If voice coding at a 8 kbps coding rate or digital modulation with double the spectral efficiency, such as linear quadrature phase shift keying (QPSK) is introduced, a 12.5 kHz channel

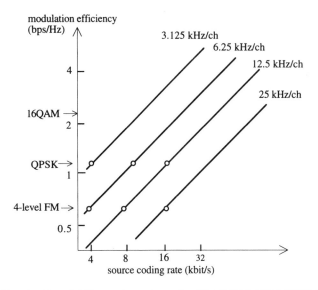

Figure 1.4. Required bandwidth per channel as a function of modulation efficiency and source coding rate.

spacing is possible. A 6.25 kHz channel spacing is possible if both the above techniques are utilized. As will be described later, the per-channel bandwidth of 3.125 kHz is possible at this moment.

For a cellular system, there is another factor that governs the spectrum efficiency, that is, frequency- or channel-reuse-distance. A shorter reuse distance is desirable. The digital system can reduce the reuse distance by using error correction techniques to mitigate the cochannel interference effect.

System Cost

Cost sometimes increases when a communication system is digitized. However, introducing an N-channel time-division multiple access (TDMA) scheme for digital transmission, the cost of the transmitter–receiver at the base station may be reduced, because the number of the transmitters and receivers is decreased to $1/N$ of those for the conventional analog system using the frequency-division multiple access (FDMA). Furthermore, by the introduction of a recently developed common amplifier for TDM systems, the base station's cost has been dramatically reduced. However, the cost of a digital mobile telephone terminal may be higher in this case than that of the conventional analog mobile telephone equipment. But it is more meaningful to consider the relative terminal cost per channel, even if the total cost increases. The increase of the total number of channels in the digital system can thus result in much lower cost per channel.

SIGNAL AND NOISE ANALYSIS

This chapter provides a mathematical foundation for the analysis of signals and noise that will be used in subsequent chapters. The reader who is familiar with these fundamental techniques may choose to move on to the following chapter.

2.1 SIGNAL ANALYSIS

2.1.1 Delta Function

DEFINITION. The Dirac delta function has many equivalent definitions [1]. In this book we use the following,

$$\int_{-\infty}^{\infty} \delta(t) f(t)\, dt = f(0) \tag{2.1a}$$

and

$$\delta(t) = 0 \qquad \text{for } t \neq 0. \tag{2.1b}$$

From the above,

$$\int_{a}^{b} \delta(t) f(t)\, dt = f(0), \qquad ab < 0, \tag{2.1c}$$

where $f(t)$ is an arbitrary function that is continuous at $t = 0$.

FORMULAS. When $f(t) = 1$,

$$\int_{-\infty}^{\infty} \delta(t)\, dt = 1. \tag{2.2}$$

In a physical sense, the delta function is an impulse to be multiplied by other functions; the duration of the impulse approaches zero while the area is unity.

Changing the variables, we have

$$\int_{-\infty}^{\infty} \delta(t - t_0)f(t)\, dt = f(t_0) \tag{2.3}$$

and

$$\int_{-\infty}^{\infty} \delta(at)f(t)\, dt = \frac{1}{|a|}f(0). \tag{2.4}$$

It can be seen that

$$\delta(t)f(t) = \delta(t)f(0). \tag{2.5}$$

2.1.2 Fourier Transform

The Fourier transform of a signal $s(t)$, is defined as

$$S(\omega) = \int_{-\infty}^{\infty} s(t)e^{-j\omega t}\, dt, \tag{2.6}$$

where $j = \sqrt{-1}$, ω is the angular frequency (rad/s) and $e^{j\omega t} \equiv \cos \omega t + j \sin \omega t$. A plot of $S(\omega)$ versus ω is called the spectrum of $s(t)$. The *energy spectral density* is given by $|S(\omega)|^2$.

The inverse Fourier transform is such that

$$s(t) = \frac{1}{2\pi}\int_{-\infty}^{\infty} S(\omega)e^{j\omega t}\, d\omega \tag{2.7a}$$

$$= \int_{-\infty}^{\infty} S(2\pi f)e^{j2\pi ft}\, df, \tag{2.7b}$$

where $\omega = 2\pi f$.

The proof for Eq. (2.7a) is as follows:

$$\frac{1}{2\pi}\int_{-\infty}^{\infty} S(\omega)e^{j\omega t}\, d\omega = \frac{1}{2\pi}\int_{-\infty}^{\infty}\int_{-\infty}^{\infty} s(\tau)e^{-j\omega\tau} e^{j\omega t}\, d\tau\, d\omega$$

$$= \frac{1}{2\pi}\int_{-\infty}^{\infty} s(\tau)\int_{-\infty}^{\infty} e^{j\omega(t-\tau)}\, d\omega\, d\tau$$

$$= \int_{-\infty}^{\infty} s(\tau)\delta(t-\tau)\, d\tau$$

$$= s(t),$$

where we have used the fact that $\int_{-\infty}^{\infty} e^{j\omega(t-\tau)} d\omega = 2\pi\delta(t-\tau)$ [1].

For a real signal $x(t)$, using Eq. (A2.8) in Appendix 2.1, we get the following:

$$X(-\omega) = X^*(\omega). \tag{2.8}$$

We denote the relation between $s(t)$ and $S(\omega)$ as $s(t) \leftrightarrow S(\omega)$. Some properties of the Fourier transform are given in Appendix 2.1 [1].

The inverse Fourier transform of a time-periodic function $f(t)$ is known as the Fourier series expansion and given by [1]

$$f(t) = \sum_{n=-\infty}^{\infty} \alpha_n e^{jn\omega_0 t}, \tag{2.9}$$

where

$$\alpha_n = \frac{1}{T} \int_{-T/2}^{T/2} f(t) e^{-jn\omega_0 t} dt \tag{2.10}$$

and $\omega_0 = 2\pi/T$ (T is the period).

2.1.3   Cross Correlation Function

The (time) cross correlation function of two signals $s_i(t)$ and $s_j(t)$ is defined as

$$R_{ij}(\tau) = \int_{-\infty}^{\infty} s_i(t) s_j(t+\tau) dt. \tag{2.11}$$

2.1.4   Autocorrelation Function

The autocorrelation function of a signal $s(t)$ is defined as

$$R(\tau) = \int_{-\infty}^{\infty} s(t) s(t+\tau) dt$$

$$= \int_{-\infty}^{\infty} s(-x) s(\tau - x) dx$$

$$= s(-\tau) * s(\tau), \tag{2.12}$$

where the symbol $*$ denotes the convolution integral that will be defined in Eqs. (2.25) or (2.26). If $R(\tau) = \delta(\tau)$, then the energy spectrum as seen from Eqs. (2.18) and (A2.12) is a constant function and is referred to as "flat."

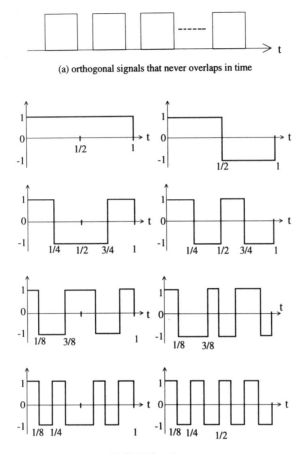

(a) orthogonal signals that never overlaps in time

(b) Walsh functions

Figure 2.1. Orthogonal signals.

2.1.5 Orthogonal Signals

The two signals $s_i(t)$ and $s_j(t)$ are orthogonal over the time range of $t_1 - t_2$, if

$$\int_{t_1}^{t_2} s_i(t)s_j(t)\, dt = 0 \qquad (i \neq j). \tag{2.13}$$

Signals that never overlap in time, as shown in Fig. 2.1(a), are *orthogonal* to each other. Another group of orthogonal signals is

$$s_{cn}(t) = \cos(n\omega_0 t) \qquad (n = 0, 1, 2, \ldots) \tag{2.14}$$

or

$$s_{sn}(t) = \sin(n\omega_0 t) \qquad (n = 1, 2, 3, \ldots), \tag{2.15}$$

where $\omega_0 = 2\pi/T$ and $T_0 = t_2 - t_1$.

Orthogonal signals that take the two values (± 1) are called Walsh functions [3] and are illustrated in Fig. 2.1(b).

Another class of orthogonal signals is discussed in Appendix 7.3 in the context of OFDM (orthogonal frequency division multiplexing) systems.

2.1.6 Energy, Power, and Power Spectral Density

The *energy* of a signal $s(t)$ is given by

$$E = \int_{-\infty}^{\infty} s^2(t)\, dt. \tag{2.16}$$

Applying the convolution integral property (Eq. A2.10) to Eq. (2.12), and using Eqs. (2.8) and (A2.3), yields

$$R(\tau) = \int_{-\infty}^{\infty} s(t)s(t+\tau)\, dt = \int_{-\infty}^{\infty} s(-t)s(\tau-t)\, dt$$

$$= \frac{1}{2\pi} \int_{-\infty}^{\infty} S(-\omega)S(\omega)e^{j\omega\tau}\, d\omega$$

$$= \frac{1}{2\pi} \int_{-\infty}^{\infty} |S(\omega)|^2 e^{j\omega\tau}\, d\omega. \tag{2.17}$$

The above equation means $R(\tau) \leftrightarrow |S(\omega)|^2$. Thus, we have

$$|S(\omega)|^2 = \int_{-\infty}^{\infty} R(\tau)e^{-j\omega\tau}\, d\tau, \tag{2.18}$$

where $|S(\omega)|^2$ is the *energy spectral density*, which is given by the Fourier transform of the autocorrelation function. Letting $\tau = 0$ in Eq. (2.17), we have

$$E = \int_{-\infty}^{\infty} s^2(t)\, dt = \frac{1}{2\pi} \int_{-\infty}^{\infty} |S(\omega)|^2\, d\omega \tag{2.19}$$

This equation is known as Parseval's theorem.

The energy may take an infinite value if the signal does not converge to zero. In order to have a signal with a finite energy, we assume

$$s_T(t) = \begin{cases} s(t), & |t| < T \\ 0, & \text{otherwise.} \end{cases}$$

Let $s_T(t) \leftrightarrow S_T(\omega)$ in the following.

The *power* is defined as

$$P_T = \lim_{T \to \infty} \frac{1}{2T} \int_{-T}^{T} s_T^2(t) \, dt. \tag{2.20}$$

Using Eq. (2.19) and taking the limit as T approaches ∞,

$$P_T = \lim_{T \to \infty} \left[\frac{1}{2T} \frac{1}{2\pi} \int_{-\infty}^{\infty} |S_T(\omega)|^2 \, d\omega \right]$$

$$= \frac{1}{2\pi} \int_{-\infty}^{\infty} G(\omega) \, d\omega, \tag{2.21}$$

where $G(\omega)$ is the *power spectral density*, defined as

$$G(\omega) = \lim_{T \to \infty} \frac{|S_T(\omega)|^2}{2T}. \tag{2.22}$$

Replacing $S(\omega)$ by $S_T(\omega)$ and $s(t)$ by $s_T(t)$, respectively, in Eqs. (2.17) and (2.18), we have

$$G(\omega) = \int_{-\infty}^{\infty} R_T(\tau) e^{-j\omega\tau} \, d\tau, \tag{2.23}$$

where

$$R_T(\tau) = \lim_{T \to \infty} \frac{1}{2T} \int_{-T}^{T} s_T(t) s_T(t + \tau) \, dt. \tag{2.24}$$

2.1.7 Response of a Linear System

A digital communication system is sometimes called as a pulsed waveform transmission system. The transmitted waveform affects the spectral characteristics as well as the bit error rate performance. Thus the response in the time domain is more used in digital systems than in analog systems, where the response is commonly analyzed in the frequency domain.

If we assume that the input signal $x(t)$ is applied to a linear system, for example, the linear filter shown in Fig. 2.2, the output signal $y(t)$, or the response of the system, can be expressed as

$$y(t) = \int_{-\infty}^{\infty} x(\tau) h(t - \tau) \, d\tau, \tag{2.25}$$

where $h(t)$ is the impulse response of the system. Making a change of

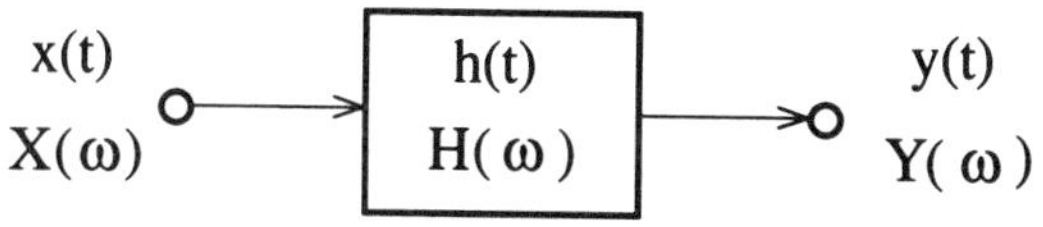

Figure 2.2. A linear system.

variables from $t - \tau$ to τ, Eq. (2.25) is equivalently expressed as

$$y(t) = \int_{-\infty}^{\infty} x(t - \tau)h(\tau)\, d\tau. \tag{2.26}$$

The above integrals are known as the convolution integrals, and are denoted as

$$y(t) = x(t) * h(t).$$

If our input $x(t)$ is an impulse, then Eq. (2.25) yields

$$y(t) = \int_{-\infty}^{\infty} \delta(\tau)h(t - \tau)\, d\tau = h(t), \tag{2.27}$$

where $\delta(\cdot)$ is the delta function. Here we see why the function $h(t)$ is called the impulse response.

Figure 2.3 shows the response of a filter with a rectangular impulse response to various inputs.

The Fourier transform of $h(t)$ is given as

$$H(\omega) = \int_{-\infty}^{\infty} h(t)e^{-j\omega t}\, dt.$$

The function $H(\omega)$ is the (voltage) transfer function. The function $|H(\omega)|^2$ is called the *power transfer function*.

Let us consider an input signal $x(t) = A_0 e^{j\omega_0 t}(-\infty \leq t \leq \infty,\ A_0 = \text{constant})$, then Eq. (2.26) yields

$$y(t) = A_0 e^{j\omega_0 t} \int_{-\infty}^{\infty} h(\tau)e^{-j\omega_0 \tau}\, d\tau$$

$$= A_0 H(\omega_0)e^{j\omega_0 t}$$

$$= H(\omega_0)x(t). \tag{2.28}$$

Here the transfer function can be interpreted as a complex factor that is multiplied with the input signal $A_0 e^{j\omega_0 t}$ to produce the output signal. In the language of integral equations, $x(t) = A_0 e^{j\omega t}$ is called the *eigenfunction* and

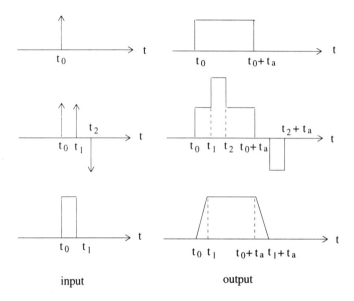

input   output

Figure 2.3. Responses of a filter with a rectangular impulse response

$$h(t) = \begin{cases} 1 & (0 < t < t_a) \\ 0 & (\text{otherwise}). \end{cases}$$

$H(\omega)$ the *eigenvalue*. The complex representation of signals is usually for the sake of mathematical simplicity. In practice, the signal is given either by the real or by the imaginary part of this complex expression.

Taking the Fourier transform on both sides of Eq. (2.25) and using Eq. (A2.10), we get

$$Y(\omega) = H(\omega)X(\omega), \tag{2.29}$$

where $Y(\omega) \leftrightarrow y(t)$ and $X(\omega) \leftrightarrow x(t)$.

Delay Line. The impulse response of a delay line with a time delay of t_0 is given by

$$h(t) = \delta(t - t_0). \tag{2.30}$$

Assuming an arbitrary input signal $x(t)$ in Eq. (2.26), the output of the delay line will be

$$y(t) = x(t) * \delta(t - t_0)$$

$$= x(t - t_0). \tag{2.31}$$

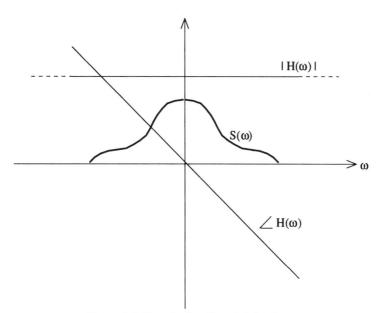

Figure 2.4. Transfer function of delay line.

No distortion occurs to the input signal when it is passed through a delay line. Taking the Fourier transform of Eq. (2.30) and using Eq. (A2.13), the transfer function becomes

$$H(\omega) = e^{-j\omega t_0} \qquad (-\infty < \omega < \infty). \tag{2.32}$$

This transfer function represents the characteristics of a distortion-free circuit. The amplitude and the phase are shown in Fig. 2.4 as a function of angular frequency. However, it is not necessary for the delay line or the distortion-free circuit to have these characteristics over an infinite frequency band; it is sufficient that the bandwidth of the circuit be wider than signal spectrum as shown in Fig. 2.4.

The time delay of a circuit is given by $\tau(\omega) = -d\angle H(\omega)/d\omega$. For a distortion-free circuit, $\tau(\omega) = \tau_0$ (constant).

Integrate-and-Dump Filter. The integrate-and-dump filter integrates an input signal $x(t)$ over the time range T to produce the output as

$$y(t) = \int_{t_0}^{t} x(\tau)\, d\tau \qquad (t_0 < t < t_0 + T), \tag{2.33}$$

where t_0 is a starting time. After the integration time T, the output signal

$y(t)$ is sampled then dumped. The start and dump times are controlled by an external timing signal. If the integration is performed continuously over the time range as

$$y(t') = \int_{t'-T}^{t'} x(\tau)\, d\tau \qquad (-\infty \leq t' \leq \infty), \tag{2.34}$$

this equation represents a moving average of $x(t)$. The output of the integrate-and-dump filter at the time $t = t_0 + T$ is the same as that of the moving average at $t' = t_0 + T$.

Equation (2.34) can be written in an alternative form as

$$y(t') = \int_{-\infty}^{\infty} x(\tau) h(t' - \tau)\, d\tau, \tag{2.35}$$

where

$$h(t) = \begin{cases} 1 & (0 < t < T) \\ 0 & (\text{otherwise}). \end{cases} \tag{2.36}$$

Thus, the moving average of $x(t)$ is achieved by applying it to the filter with the above impulse response. The transfer function of the filter is given by

$$H(\omega) = T\frac{\sin(\omega T/2)}{\omega T/2} e^{-j\omega T/2}, \tag{2.37}$$

where $e^{-j\omega T/2}$ is a signal delay term which causes no distortion on the input signal.

Causality. Causality implies that a response of a system (result) never occurs before an input signal (cause) is impressed. In other words, if the input signal $x(t)$ is zero for $t < t_1$, then the output signal $y(t)$ is also zero for $t < t_1$. Using Eq. (2.25),

$$y(t) = \int_{t_1}^{\infty} h(t - \tau) x(\tau)\, d\tau = 0 \qquad (t < t_1). \tag{2.38}$$

This equation leads to

$$h(t) = 0 \qquad (t < 0). \tag{2.39}$$

This result is a mathematical expression of the above definition of causality when the input signal is an impulse.

If we assume that an impulse response $h(t)$ is not zero for $t < 0$, then we

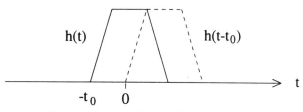

Figure 2.5. Causality and impulse response.

must shift the time origin by t_0, where t_0 is the time at which $h(t - t_0) = 0$ for $t < 0$ (Fig. 2.5). Furthermore, we often assume an impulse response $h(t)$ that is nonzero for $-\infty < t < \infty$. In that case, we can only approximate the impulse response. The approximation is better for larger t_0, but results in a larger delay in the response.

2.1.8   The Sampling Theorem

Sampling together with quantization is a fundamental process in the digital transmission of an analog signal. Sampling is also necessary for signal processing using a digital computer. The sampling theorem guarantees recovery of the analog signal from the sampled version. If the sampling theorem is satisfied, digital signal processing (transmission) of the sampled signal gives the same result as analog signal processing (transmission), provided quantization error is negligible.

The sampling theorem states that if a signal is band-limited within $0\sim f_m$ Hz, the minimum sampling frequency is $2f_m$ Hz.

In the following, the proof of the sampling theorem is shown. We define a sampling function $f_s(t)$ as

$$f_s(t) = \sum_{n=-\infty}^{\infty} \delta(t - nT_s), \tag{2.40}$$

where $1/T_s$ is the sampling frequency. The sampled version of $s(t)$ is given by

$$s_s(t) = s(t)f_s(t). \tag{2.41}$$

Taking the Fourier transform of Eq. (2.41), and using Eqs. (A2.11) and (A2.19), we get

$$S_s(\omega) = \frac{1}{2\pi} S(\omega) * F_s(\omega)$$

$$= \frac{1}{T_s} \sum_{n=-\infty}^{\infty} S(\omega - n\omega_s), \tag{2.42}$$

where $\omega_s = 2\pi/T_s$.

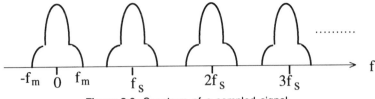

Figure 2.6. Spectrum of a sampled signal.

The spectrum $S_s(\omega)$ is the repetition of the spectrum $S(\omega)$, each element of which is separated by $1/T_s\ (=f_s)$ Hz, as illustrated in Fig. 2.6. We see intuitively that the original analog signal can be recovered by filtering the sampled signal with a low-pass filter (LPF). The LPF has gain T_s and passes only the fundamental spectrum centered at 0. The output of the LPF is given by

$$y(t) = s_s(t) * h_{LPF}(t). \tag{2.43}$$

With Eqs. (2.5), (2.27), (2.40), and (2.41), we get

$$y(t) = \sum_{n=-\infty}^{\infty} s(nT_s)h_{LPF}(t - nT_s), \tag{2.44}$$

where $h_{LPF}(t)$ is an impulse response of the LPF. $h_{LPF}(t)$ can be chosen arbitrarily as long as the LPF passes the fundamental signal without distortion but with gain of T_s, and completely attenuates the higher-order spectra. In the case when the sampling frequency takes the minimum value, i.e., $2f_m$ Hz, which is known as the Nyquist sampling rate, the LPF should take an ideal rectangular transfer function that has a pass band of $-f_m \sim f_m$ Hz; then

$$y(t) = \sum_{n=-\infty}^{\infty} s(nT_s)\, \frac{\sin 2\pi f_m(t - nT_s)}{2\pi f_m(t - nT_s)}, \tag{2.45}$$

where $T_s = 1/(2f_m)$ and $s(nT_s)$ are sampled signals.

No ideal rectangular transfer function can be realized, although approximations can be made. As the sampling frequency increases, the realization of the LPF becomes easier, since the requirements for the characteristics of the LPF can be relaxed.

2.1.9 Sampled Signal System

Consider a signal sequence sampled at $t = nT$ as

$$x_s(t) = x(t) \sum_{n=-\infty}^{\infty} \delta(t - nT); \tag{2.46}$$

using Eq. (2.5),

$$x_s(t) = \sum_{n=-\infty}^{\infty} x(nT)\delta(t - nT). \tag{2.47}$$

If $x_s(t)$ is applied to a linear system with impulse response $h(t)$, then the output signal sampled at $t = mT$ is given as

$$y(mT) = \sum_{n=-\infty}^{\infty} x(nT)h(mT - nT) \tag{2.48a}$$

or

$$y(mT) = \sum_{n=-\infty}^{\infty} x(mT - nT)h(nT). \tag{2.48b}$$

If the system is causal, then

$$y(mT) = \sum_{n=0}^{\infty} x(mT - nT)h(nT). \tag{2.49}$$

Write the sampled signal sequence as

$$y_s(t) = \sum_{m=-\infty}^{\infty} y(mT)\delta(t - mT)$$

$$= \sum_{m=-\infty}^{\infty} \sum_{n=-\infty}^{\infty} x(mT - nT)h(nT)\delta(t - mT).$$

Taking the Fourier transform of the above equation, we have

$$Y_s(\omega) = H_s(\omega)X_s(\omega), \tag{2.50}$$

where

$$H_s(\omega) = \sum_{n=-\infty}^{\infty} h(nT)e^{-jn\omega T} \tag{2.51}$$

and

$$X_s(\omega) = \sum_{n=-\infty}^{\infty} x(nT)e^{-jn\omega T}. \tag{2.52}$$

$X_s(\omega)$ and $H_s(\omega)$ are the Fourier transforms of the sampled signal sequence $x_s(t) = \sum_{n=-\infty}^{\infty} x(nT)\delta(t - nT)$ and $h_s(t) = \sum_{n=-\infty}^{\infty} h(nT)\delta(t - nT)$, respectively.

If we let $z = e^{j\omega T}$, then Eqs. (2.50)–(2.52) give

$$Y(z) = H(z)X(z), \tag{2.53}$$

$$H(z) = \sum_{n=-\infty}^{\infty} h(nT)z^{-n}, \tag{2.54}$$

$$X(z) = \sum_{n=-\infty}^{\infty} x(nT)z^{-n}, \qquad Y(z) = \sum_{n=-\infty}^{\infty} y(nT)z^{-n} \tag{2.55}$$

where the suffix s is not used. $H(z)$ and $X(z)$ can be considered as a new transform or a series expansion of $h(t)$ and $x(t)$ in terms of z. They are called z-transforms. The z-transform is a more general function than that we have derived in the above. In general, z can take any complex value. Thus our treatment is a special case of $z = e^{j\omega T}$.

Since the z-transform, $X(z)$ and $H(z)$, includes the Fourier transform, $X_s(\omega)$ and $H_s(\omega)$, as a special case of $z = e^{j\omega T}$, the properties of the Fourier transform can be applied to $X(z)$ and $H(z)$ when $z = e^{j\omega T}$.

2.1.10 Digital Signals

Expression. A sequence of digital signals is expressed as

$$s(t) = \sum_{-\infty}^{\infty} a_n h(t - nT), \tag{2.56}$$

where a_n is a symbol that is selected from a set of discrete values corresponding to the input data signal. For example, a_n takes values $+A$, or $-A$ for a binary system. $h(t)$ is the impulse response of the band-limiting filter that is used to generate a pulse waveform. T denotes the symbol duration, that is, the symbol frequency is 1/T. The fact that the digital signal is transmitted at a given fixed symbol frequency ensures carrier signal regeneration and clock timing signal regeneration at the receiver side.

If we use a non-return-to-zero (NRZ) waveform, $h(t)$ is given as

$$h(t) = \begin{cases} 1 & (0 \le t \le T) \\ 0 & \text{(otherwise)}. \end{cases}$$

Some other waveforms are discussed in Section 3.1. Assigning the waveform to the transmitting signal is called *line coding*.

Average Power. The average power of a digital signal is defined

$$P = \left\langle \lim_{N \to \infty} \frac{1}{2NT} \int_{-NT}^{NT} s^2(t) \, dt \right\rangle$$

$$= \left\langle \lim_{N \to \infty} \frac{1}{2NT} \int_{-NT}^{NT} \left| \sum_{n=-N}^{N} a_n h(t - nT) \right|^2 dt \right\rangle, \tag{2.57}$$

where $\langle \cdot \rangle$ denotes an ensemble average. Interchanging the order of operations between the limit and the ensemble average, we have

$$P = \lim_{N \to \infty} \frac{1}{2NT} \sum_{n=-N}^{N} \sum_{m=-N}^{N} \langle a_n a_m \rangle \int_{-NT}^{NT} h(t - nT)h(t - mT) \, dt. \tag{2.58}$$

If the occurrence of symbols is assumed to be independent and random, that is,

$$\langle a_n a_m \rangle = \begin{cases} 0 & (n \neq m) \\ \overline{a^2} & (n = m), \end{cases} \tag{2.59}$$

then the Eq. (2.58) yields

$$P = \lim_{N \to \infty} \frac{\overline{a^2}}{2NT} \sum_{n=-N}^{N} \int_{-NT}^{NT} h^2(t - nT) \, dt$$

$$= \frac{\overline{a^2}}{T} \int_{-\infty}^{\infty} h^2(t) \, dt. \tag{2.60}$$

Using Parseval's theorem, (Eq. 2.19), the average power is rewritten as

$$P = \frac{\overline{a^2}}{2\pi T} \int_{-\infty}^{\infty} |H(\omega)|^2 \, d\omega. \tag{2.61}$$

Assuming an NRZ waveform given by Eq. (2.36), Eq. (2.60) yields

$$P = \overline{a^2}. \tag{2.62}$$

The same result is given by Eq. (2.61) using Eq. (2.37) and the formula

$$\int_{-\infty}^{\infty} \left[\frac{\sin x}{x} \right]^2 dx = \pi. \tag{2.63}$$

The *energy per symbol* E_s is given as

$$E_s = PT. \tag{2.64}$$

For a binary transmission, the *energy per bit* E_b is the same as E_s. For a 2^n-ary $(n = 1, 2, 3, \ldots)$ transmission,

$$E_b = \frac{E_s}{n}. \tag{2.65}$$

If the symbol a_n takes $+A$ or $-A$ (binary transmission),

$$\overline{a^2} = A^2. \tag{2.66}$$

For a case of 4-level transmission, where a_n takes $\pm A$ and $\pm A/3$, then

$$\overline{a^2} = \{\mathrm{Prob}(A) + \mathrm{Prob}(-A)\}A^2 + \{\mathrm{prob}(A/3) + \mathrm{prob}(-A/3)\}\frac{A^2}{9}. \tag{2.67}$$

If the symbols, a_n, are equally likely, that is, if the probabilities are all the same (i.e., 1/4),

$$\overline{a^2} = \tfrac{5}{9}A^2. \tag{2.68}$$

Power Spectral Density. The power spectral density of a random signal $s(t)$ is defined as

$$S(\omega) = \left\langle \lim_{N \to \infty} \frac{1}{2NT} \left| \int_{-NT}^{NT} s(t)e^{-j\omega t}\,dt \right|^2 \right\rangle. \tag{2.69}$$

Using Eqs. (2.56) and Eq. (2.69), and interchanging the order of the limit and the ensemble average,

$$S(\omega) = \lim_{N \to \infty} \frac{1}{2NT} \sum_{n=-N}^{N} \sum_{m=-N}^{N} \langle a_n a_m \rangle$$
$$\times \int_{-NT}^{NT} \int_{-NT}^{NT} h(t - nT)e^{-j\omega t} h(x - mT)e^{j\omega x}\,dt\,dx. \tag{2.70}$$

Assuming that a_n is random and using Eq. (2.59),

$$S(\omega) = \lim_{N \to \infty} \frac{\overline{a^2}}{2NT} \sum_{n=-N}^{N} \left| \int_{-NT}^{NT} h(t - nT)e^{-j\omega t}\,dt \right|^2. \tag{2.71}$$

Using the Fourier transform formula given in Eq. (A2.4), the integral in Eq. (2.71) becomes

$$\left| \int_{-\infty}^{\infty} h(t - nT)e^{-j\omega t}\,dt \right|^2 = |H(\omega)e^{-j\omega nT}|^2$$
$$= |H(\omega)|^2. \tag{2.72}$$

With use of Eq. (2.72), Eq. (2.71) reduces to

$$S(\omega) = \frac{\overline{a^2}}{T} |H(\omega)|^2. \tag{2.73}$$

Thus the power spectral density of a random digital signal is found by taking the Fourier transform of a pulse waveform, for the case when the data are random.

Another method for the power spectral density is shown in the followings. Define a new variable $\tau = t - x$, then Eq. (2.70) becomes

$$S(\omega) = \lim_{N\to\infty} \frac{\overline{a^2}}{2NT} \sum_{n=-N}^{N} \int_{-NT}^{NT} \int_{-NT}^{NT} h(x + \tau - nT)h(x - nT) \, dx \, e^{-j\omega\tau} \, d\tau$$

$$= \lim_{N\to\infty} \frac{\overline{a^2}}{2NT} \sum_{n=-N}^{N} \int_{-NT}^{NT} R(\tau, nT)e^{-j\omega\tau} \, d\tau, \tag{2.74}$$

where

$$R(\tau, nT) = \int_{-\infty}^{\infty} h(x + \tau - nT)h(x - nT) \, dx. \tag{2.75}$$

It is seen that $R(\tau, nT) = R(\tau, mT)$. Thus we get,

$$S(\omega) = \frac{\overline{a^2}}{T} \int_{-\infty}^{\infty} R(\tau)e^{-j\omega\tau} \, d\tau, \tag{2.76}$$

where $R(\tau)$ is the autocorrelation function given as

$$R(\tau) = \int_{-\infty}^{\infty} h(x + \tau)h(x) \, dx. \tag{2.77}$$

For example, let us consider the case for an NRZ waveform; then,

$$H(\omega) = \int_{-\infty}^{\infty} h(t)e^{-j\omega t} \, dt$$

$$= T\frac{\sin(\omega T/2)}{\omega T/2}e^{-j\omega T/2}, \tag{2.78}$$

$$R(\tau) = \int_{-\infty}^{\infty} h(t + \tau)h(t) \, dt$$

$$= \begin{cases} T - |\tau|, & 0 < |\tau| < T \\ 0 & |\tau| > T \end{cases}, \tag{2.79}$$

$$\int_{-\infty}^{\infty} R(\tau)e^{-j\omega\tau}\,d\tau = T^2\frac{\sin^2(\omega T/2)}{(\omega T/2)^2}. \tag{2.80}$$

With the above results, Eq. (2.76) yields

$$S(\omega) = \overline{a^2}\,T\,\frac{\sin^2(\omega T/2)}{(\omega T/2)^2}. \tag{2.81}$$

On the other hand, applying the transfer function given in Eq. (2.78) to the Eq. (2.73), we obtain an identical expression. Thus, we have shown that the two methods for obtaining the power spectral density yield the same result.

2.1.11 Modulated Signals

An oscillating carrier signal is expressed as

$$c(t) = A_0\cos(\omega_c t + \theta_0)$$
$$= \mathrm{Re}(A_0 e^{j(\omega_c t + \theta_0)}), \tag{2.82}$$

where A_0 is the amplitude, ω_c is the carrier angular frequency, and θ_0 is the initial phase. The carrier signal has two degrees of freedom, namely, the amplitude and the phase or its derivative; the derivative of the phase is in other words the instantaneous frequency.

Modulation is a process whereby the amplitude and/or the phase is varied according to the modulating signal. Amplitude modulation is given by

$$A(t) = k_A m(t), \tag{2.83}$$

where $A(t)$ is the amplitude, k_A is a dimensionless constant, and $m(t)$ is the modulating signal. Phase modulation is given as

$$\theta(t) = k_P m(t), \tag{2.84}$$

and frequency modulation is expressed as

$$\omega(t) = \frac{d}{dt}\theta(t) = k_F m(t), \tag{2.85}$$

where k_P and k_F are constants that have dimensions of [radian/volt] and [radian/second/volt], respectively. The expressions for analog and digital modulation are the same. The input $m(t)$ can be an arbitrary modulating signal in either the analog or the digital format.

If the amplitude and phase are modulated simultaneously, the modulated signal is expressed as

$$s(t) = A(t) \cos[\omega_c t + \theta(t)].$$

(2.86)

The above equation is rewritten as

$$s(t) = A(t) \cos \theta(t) \cos \omega_c t - A(t) \sin \theta(t) \sin \omega_c t.$$

(2.87)

If we introduce new variables such that

$$x(t) = A(t) \cos \theta(t),$$

(2.88)

$$y(t) = A(t) \sin \theta(t),$$

(2.89)

then Eq. (2.87) is written as

$$s(t) = x(t) \cos \omega_c t - y(t) \sin \omega_c t,$$

(2.90)

where $x(t)$ and $y(t)$ are called the in-phase and quadrature-phase components, respectively. Because $\cos \omega_c t$ and $\sin \omega_c t$ are orthogonal to each other, two different modulating signals, $m_1(t)$ and $m_2(t)$, can be loaded on the same carrier signal by letting $x(t) = km_1(t)$ and $y(t) = km_2(t)$. This kind of modulation, called quadrature amplitude modulation (QAM), is a widely used digital modulation technique. This modulation is equivalent to a system where the amplitude and the phase are simultaneously modulated, such that

$$A(t) = \sqrt{[x^2(t) + y^2(t)]},$$

(2.91)

$$\theta(t) = \tan^{-1} \frac{y(t)}{x(t)}.$$

(2.92)

The amplitude and the phase are simultaneously varied in the case of single-side-band modulation (SSB), however, this is not a QAM. For an SSB signal, there is the relation $x(t) = \bar{y}(t)$, where $\bar{y}(t)$ denotes the Hilbert transform, that is, 90° phase shifting of $y(t)$.

The modulated signal can be expressed as a trajectory on a two-dimensional plane of the in-phase and the quadrature-phase components that is rotating at a speed of a carrier frequency. If we observe the signal trajectory relative to the rotating plane, we need not care about the carrier frequency. In this case, using the complex plane expression, the signal can be written as

$$s(t) = A(t)e^{j\theta(t)}$$

$$= x(t) + jy(t).$$

(2.93)

This expression is sometimes called a complex amplitude, or a zero IF (intermediate frequency) signal. The actual signal is given by $\text{Re}[s(t)e^{j\omega_c t}]$.

2.2   NOISE ANALYSIS

Noise is a random process that disturbs the signal transmission. In radio communications, most of the noise is added to the signal in the radio channel and at the receiver. The noise that occurs at the transmitter hardly deteriorates the signal transmission quality, because the signal level is sufficiently high.

2.2.1   Noise in Communication Systems

Figure 2.7 shows a system model for noise in a radio transmission channel. The received signal level and the received noise level are denoted as S_i and N_c, respectively. Noise is generated in various stages in the circuit at the receiver. The first stage is usually a low-noise amplifier. We consider a model in which noise is assumed to be generated at the input for each stage of the circuit.

The signal to noise power ratio (SNR) at the output of the receiver is given as

$$\left[\frac{S}{N}\right]_{\text{out}} = \frac{S_i}{N_c + N_1 + \dfrac{N_2}{G_1} + \dfrac{N_3}{G_1 G_2} + \cdots \dfrac{N_k}{G_1 G_2 \cdots G_{k-1}}} \tag{2.94}$$

where $N_i, G_i (i = 1, 2, \ldots, k)$ are respectively the noise level and the power gain of each stage. Assuming $G_i \gg 1$, we can see the noise level at the stages downstream hardly affect the output SNR. This is why we place a low-noise amplifier at the first stage of the receiver. It is also seen that an effort to suppress the noise levels at a receiver much lower than the channel induced noise level N_c is meaningless.

The noise power spectrum is usually much wider than that for the signal spectrum. Thus, in general, a flat (double-sided) noise power spectral density (white noise) is assumed as follows:

$$N(\omega) = \frac{N_0}{2} \qquad (-\infty < \omega < \infty), \tag{2.95}$$

where we have omitted the suffix i for the ith stage of a receiver. If we denote the power transfer function of a circuit by $G(\omega)$, then the noise power is given by

$$N = \int_{-\infty}^{\infty} N(\omega) G(\omega)\, d\omega = N_0 W_{\text{eq}}, \tag{2.96}$$

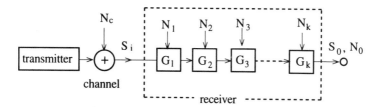

Figure 2.7. A model for noise in a transmission system.

where W_{eq} is an equivalent noise bandwidth, which is given as

$$W_{eq} = \int_0^\infty G(\omega)\, d\omega. \tag{2.97}$$

Figure 2.8 illustrates the meaning of W_{eq}.

The signal to noise power ratio always deteriorates at the output of the circuit. The noise figure NF is defined as the degradation of SNR for a single-sided input noise power density given by

$$N(\omega) = kT_a \qquad (0 < \omega < \infty), \tag{2.98}$$

where k is the Boltzmann constant ($k = 1.38 \times 10^{-23}$(W/Hz)/K) and T_a is temperature of the environment at which the measurement is performed. At room temperature $T_a = 290$ K. Thus, we have

$$NF = \frac{S_i/kT_a W_{eq}}{S_o/N_{out}}$$

$$= 1 + \frac{T_e}{T_a}, \tag{2.99}$$

where T_e is called the equivalent noise temperature and is given by

$$T_e = \frac{N_0'}{k} \tag{2.100}$$

and N_0' is the input equivalent noise power density of the circuit.

If we place an attenuator where no noise is added before a circuit with noise temperature T_e, then the total noise figure is given by

$$NF = 1 + \frac{1}{L}\frac{T_e}{T_a}, \tag{2.101}$$

where L (<1) is the power gain of the attenuator. It is seen that an attenuator degrades the noise figure.

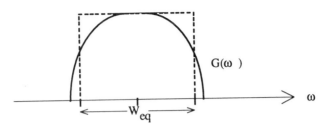

Figure 2.8. Illustration of equivalent noise bandwidth.

2.2.2 Statistics of Noise

The statistical characteristics of a noise signal $\tilde{x}$ are described by a probability distribution function, $P(x)$. The probability distribution function $P(x)$ is defined by

$$P(x) = \text{Prob}(\tilde{x} < x) \qquad (-\infty < x < \infty), \qquad (2.102)$$

where $\text{Prob}(\tilde{x} < x)$ is the probability that $\tilde{x}$ takes a value lower than x. $P(x)$ is given as

$$P(x) = \int_{-\infty}^{x} p(x)\,dx, \qquad (2.103)$$

where $p(x)$, the *probability density function*, is defined as

$$p(x) = \frac{d}{dx} P(x). \qquad (2.104)$$

The probability that $\tilde{x}$ takes a value x between x_1 and x_2 is given as

$$\text{Prob}(x_1 \le x \le x_2) = P(x_2) - P(x_1) = \int_{x_1}^{x_2} p(x)\,dx.$$

In a communication system, zero-mean additive white Gaussian noise (AWGN) is often encountered. Its probability density function is given by

$$p(x) = \frac{1}{\sqrt{2\pi}\sigma} e^{-x^2/2\sigma^2}, \qquad (2.105)$$

where σ^2 is average noise power; that is,

$$\sigma^2 = \int_{-\infty}^{\infty} x^2 p(x)\,dx. \qquad (2.106)$$

We can confirm this result by integrating by parts and using the formula

$$\int_{-\infty}^{\infty} e^{-a^2 x^2} dx = \frac{\sqrt{\pi}}{a}.$$

The mean value x_m is,

$$x_m = \int_{-\infty}^{\infty} xp(x) dx = 0. \tag{2.107}$$

In general, Gaussian noise is generated when a large number of relatively independent impulsive perturbations are applied to a low-pass filter. The effects of the impulsive perturbations are collected due to the finite impulse response time [4]. This is a good model to describe how noise is generated in electronic devices, such as transistors, where many electrons collide at random with obstacles and where the response time of the device is long enough to receive a number of collisions. From the nature of the Gaussian noise it can be seen that the filtered Gaussian noise is still Gaussian.

2.2.3 Power Spectral Density of Noise

Let $x_T(t)$ to be a noise signal by time-truncating a signal $x(t)$ within a time range $|t| < T$. The power spectral density is defined as

$$S_x(\omega) = \lim_{T \to \infty} \frac{\langle |X_T(\omega)|^2 \rangle}{2T}, \tag{2.108}$$

where $\langle \cdot \rangle$ means ensemble average and $X_T(\omega) \leftrightarrow x_T(t)$, that is,

$$X_T(\omega) = \int_{-\infty}^{\infty} x_T(t) e^{-j\omega t} dt = \int_{-T}^{T} x(t) e^{-j\omega t} dt. \tag{2.109}$$

We have

$$\langle |X_T(\omega)|^2 \rangle = \left\langle \int_{-T}^{T} x(t_1) e^{j\omega t_1} dt_1 \int_{-T}^{T} x(t_2) e^{-j\omega t_2} dt_2 \right\rangle$$

$$= \int_{-T}^{T} \int_{-T}^{T} \langle x(t_1) x(t_2) \rangle e^{(j\omega t_1 - j\omega t_2)} dt_1 dt_2. \tag{2.110}$$

If $x(t)$ is a stationary process or at least $\langle x(t_1)x(t_2) \rangle$ is not dependent on the time origin, then we can let

$$\langle x(t_1)x(t_2) \rangle = R_x(t_1 - t_2)$$

and the double integral in Eq. (2.110) is given by a single integral as [5]

$$
\int_{-T}^{T}\int_{-T}^{T} \langle x(t_1)x(t_2)\rangle e^{(j\omega t_1 - j\omega t_2)} \, dt_1 \, dt_2
$$
$$
= \int_{-2T}^{2T} (2T - |\tau|) R_x(\tau) e^{-j\omega\tau} \, d\tau, \tag{2.111}
$$

where $\tau = t_2 - t_1$.

Substituting Eq. (2.111) into Eq. (2.110) and using Eq. (2.108) we get

$$
S_x(\omega) = \lim_{T\to\infty} \int_{-2T}^{2T} \frac{1}{2T} (2T - |\tau|) R_x(\tau) e^{-j\omega\tau} \, d\tau
$$
$$
= \lim_{T\to\infty} \int_{-2T}^{2T} \left(1 - \frac{|\tau|}{2T}\right) R_x(\tau) e^{-j\omega\tau} \, d\tau. \tag{2.112}
$$

If

$$
\lim_{T\to\infty} \int_{-2T}^{2T} \frac{|\tau|}{2T} R_x(\tau) e^{-j\omega\tau} \, d\tau = 0
$$

then the power spectral density is given as

$$
S_x(\omega) = \int_{-\infty}^{\infty} R_x(\tau) e^{-j\omega\tau} \, d\tau. \tag{2.113}
$$

Thus,

$$
S_x(\omega) \leftrightarrow R_x(\tau). \tag{2.114}
$$

Power Spectral Density of Filtered Noise. If we denote filtered noise by $y(t)$, then the correlation function of $y(t)$ is given as

$$
R_y(\tau) = \langle y(t)y(t + \tau)\rangle
$$
$$
= \langle \{x(t) * h(t)\} \{x(t + \tau) * h(t + \tau)\}\rangle
$$
$$
= \int_{-\infty}^{\infty}\int_{-\infty}^{\infty} \langle x(t - t_1)x(t + \tau - t_2)\rangle h(t_1)h(t_2) \, dt_1 \, dt_2, \tag{2.115}
$$

where $x(t)$ is the input noise and $h(t)$ is the impulse response of the filter. The power spectral density is given by

$$
S_y(\omega) = \int_{-\infty}^{\infty} R_y(\tau) e^{-j\omega\tau} \, d\tau. \tag{2.116}
$$

Inserting Eq. (2.115) into Eq. (2.116) and changing variables, we get

$$S_y(\omega) = |H(\omega)|^2 S_x(\omega), \qquad (2.117)$$

where $H(\omega) \leftrightarrow h(t)$ and $S_x(\omega)$ is the power spectral density of $x(t)$, which is given as

$$S_x(\omega) = \int_{-\infty}^{\infty} R_x(\tau) e^{-j\omega\tau} d\tau, \qquad (2.118)$$

where $R_x(\tau) = \langle x(t)x(t+\tau) \rangle$.

2.2.4 Autocorrelation Function of Filtered Noise

From Eqs. (2.116) and (2.117), the autocorrelation function of filtered noise is given as

$$R_y(\tau) = \langle y(t)y(t+\tau) \rangle = \frac{1}{2\pi} \int_{-\infty}^{\infty} |H(\omega)|^2 S_x(\omega) e^{j\omega\tau} d\omega. \qquad (2.119)$$

If the input signal is white noise with power spectral density of N_0,

$$R_y(\tau) = \frac{N_0}{2\pi} \int_{-\infty}^{\infty} |H(\omega)|^2 e^{j\omega\tau} d\omega. \qquad (2.120)$$

Considering the relation given by Eq. (2.17), we have

$$R_y(\tau) = N_0 \int_{-\infty}^{\infty} h(t)h(t+\tau) \, dt. \qquad (2.121)$$

It can be shown that the filtered noise shows no correlation at different symbol time if $|H(\omega)|^2$ meets the Nyquist-I criterion [Eq. (3.76)] ($R_y(nT_s) = 0$ for $n \neq 0$, where T_s is symbol duration).

2.2.5 Band-pass Noise

The radio frequency noise $n(t)$ is expressed as

$$n(t) = n_x(t) \cos \omega_c t - n_y(t) \sin \omega_c t, \qquad (2.122)$$

where $n_x(t)$ and $n_y(t)$ are stationary and independent baseband noise processes that have zero mean ($\langle n_x(t) \rangle = \langle n_y(t) \rangle = 0$), and ω_c, is a reference center frequency. Since $n_x(t)$ and $n_y(t)$ are independent, the phase of $n(t)$ is uniformly distributed in the complex plane of $\tilde{n}(t) = n_x(t) + jn_y(t)$. The amplitude of the noise is given as

$$r(t) = \sqrt{n_x^2(t) + n_y^2(t)}. \tag{2.123}$$

If we assume zero-mean Gaussian noise of the same power for $n_x(t)$ and $n_y(t)$, the distribution of the amplitude obeys the Rayleigh distribution as is shown in the following. Using Eqs. (2.105) and (2.123), transforming the variables by $n_x(t) = r(t) \cos \theta$, $n_y(t) = r(t) \sin \theta$, and considering $p(x, y)\, dx\, dy = p(r, \theta) r\, d\theta\, dr$,

$$P(r) = \text{Prob}(r(t) \le r)$$

$$= \int_0^{2\pi} \int_0^r \frac{1}{2\pi\sigma^2} e^{-r^2/2\sigma^2} r\, dr\, d\theta$$

$$= 1 - e^{-r^2/2\sigma^2}. \tag{2.124}$$

The probability density function is given as

$$p(r) = \frac{r}{\sigma^2} e^{-r^2/2\sigma^2} \qquad (0 \le r < \infty), \tag{2.125}$$

$$p(\theta) = \frac{1}{2\pi} \qquad (-\pi \le \theta \le \pi). \tag{2.126}$$

The probability density functions for the Gaussian distribution and the Rayleigh distribution are shown in Fig. 2.9.

2.2.5.1 Power Spectral Density of Passband Noise.
The passband signal $n(t)$ (Eq. 2.122) is generated as shown in Fig. 2.10(a). The noise power spectral density $S_n(\omega)$ is given as (Eq. 2.23)

$$S_n(\omega) = \int_{-\infty}^{\infty} R_n(\tau) e^{-j\omega\tau} d\tau, \tag{2.127}$$

where R_n is the autocorrelation function of $n(t)$, that is,

$$R_n(\tau) = \left\langle \lim_{T \to \infty} \frac{1}{2T} \int_{-T}^{T} n(t + \tau) n(t)\, dt \right\rangle. \tag{2.128}$$

Inserting Eq. (2.122) into Eq. (2.128) and using the relations $\langle n_x(t) n_y(t + \tau) \rangle = \langle n_y(t) \rangle \langle n_x(t + \tau) \rangle = 0$, we have

$$R_n(\tau) = \tfrac{1}{2}[R_x(\tau) + R_y(\tau)] \cos \omega_c \tau, \tag{2.129}$$

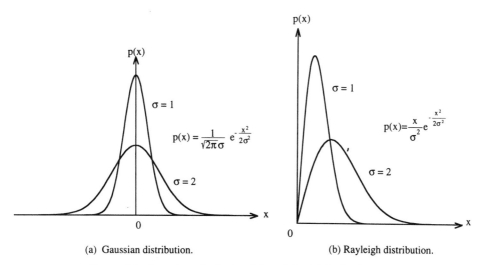

(a) Gaussian distribution.   (b) Rayleigh distribution.

Figure 2.9. Gaussian distribution (a) and Rayleigh distribution (b).

where $R_x(\tau) = \langle n_x(t)n_x(t+\tau)\rangle$ and $R_y(\tau) = \langle n_y(t)n_y(t+\tau)\rangle$ are the autocorrelation functions of $n_x(t)$ and $n_y(t)$, respectively.

Assuming that

$$R_x(\tau) = R_y(\tau) = R_0(\tau) \qquad (2.130)$$

we have

$$\langle n^2(t)\rangle = \langle n_x^2(t)\rangle = \langle n_y^2(t)\rangle. \qquad (2.131)$$

Rewriting $\cos\omega_c\tau = \frac{1}{2}(e^{j\omega_c\tau} + e^{-j\omega_c\tau})$ and from Eqs. (2.129), (2.130), and (A2.5), we have

$$S_n(\omega) = \frac{1}{2}[S_0(\omega + \omega_c) + S_0(\omega - \omega_c)], \qquad (2.132)$$

where $S_0(\omega) = \int_{-\infty}^{\infty} R_0(\tau)e^{-j\omega\tau}\,d\tau$ are the power spectral densities (PSD) of $n_x(t)$ and $n_y(t)$.

Next we wish to express the power spectral density of the baseband signals $n_x(t)$ and $n_y(t)$ with that of $n(t)$. $n_x(t)$ and $n_y(t)$ can be obtained by applying the passband signal into the circuit shown in Fig. 2.11, where the role of the low-pass filter is to pass only the baseband signal. The output signal at the multipliers becomes $m_x(t) = 2n(t)\cos\omega_c t$ and $m_y(t) = 2n(t)\sin\omega_c t$. Following the above argument for $n_x(t)$ or $n_y(t)$ instead of $n(t)$ in Eq. (2.122), we get

$$S_{n_x}(\omega) = S_{n_y}(\omega) = S_n(\omega + \omega_c) + S_n(\omega - \omega_c). \qquad (2.133)$$

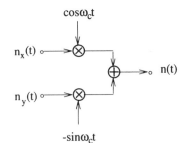

(a) Circuit configuration

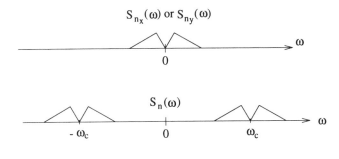

(b) Power spectrum

Figure 2.10. Generation of passband noise from quadrature-base band noise.

If we assume a narrow, flat passband spectrum,

$$S_n(\omega) = \begin{cases} N_0/2 & (|\omega \pm \omega_c| \leq \omega_B, \omega_B < \omega_c) \\ 0 & (\text{otherwise}), \end{cases}$$

where $N_0/2$ is the double-sided noise power density, then we have the baseband noise power spectrum

$$S_{n_x}(\omega) = S_{n_y}(\omega) = \begin{cases} N_0 & (|\omega| \leq \omega_B) \\ 0 & (\text{otherwise}). \end{cases} \tag{2.134}$$

In the above discussion the signal spectrum is considered in the frequency range $-\infty < \omega < \infty$. Often, the power spectral density is only defined for positive frequency. In this case, if we consider the single-sided noise power density of N_0, the same result can be obtained in the analysis of the system.

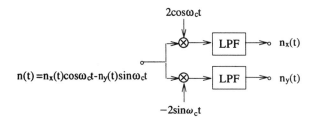

(a) Circuit configuration

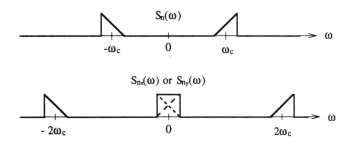

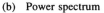

(b) Power spectrum

Figure 2.11. Conversion of passband noise into quadrature-baseband noise.

2.2.6 Envelope and Phase of a Sinusoidal Signal in Band-pass Noise

Let us consider a case when a sinusoidal signal $A \cos(\omega_c t + \varphi)$ is mixed with passband Gaussian noise $n(t)$ [5]. The mixed signal· can be written as

$$z(t) = A \cos(\omega_c t + \varphi) + n_x(t) \cos(\omega_c t + \varphi) - n_y(t) \sin(\omega_c t + \varphi)$$

$$= R(t) \cos[\omega_c t + \varphi + \theta(t)], \tag{2.135}$$

where the envelope $R(t)$ and phase $\theta(t)$ are

$$R(t) = \sqrt{\{[A + n_x(t)]^2 + n_y^2(t)\}}, \tag{2.136}$$

$$\theta(t) = \tan^{-1} \frac{n_y(t)}{A + n_x(t)}, \qquad |\theta| < \pi, \tag{2.137}$$

and where $n_x(t)$ and $n_y(t)$ are zero-mean Gaussian processes with the variance σ^2. We observe

$$n_x^2 + n_y^2 = R^2 - A^2 - 2An_x$$
$$= R^2 - 2A(A + n_x) + A^2$$
$$= R^2 - 2AR\cos\theta + A^2. \tag{2.138}$$

Then we have

$$p(R, \theta) = \frac{R}{2\pi\sigma^2}\exp[-(R^2 - 2AR\cos\theta + A^2)/2\sigma^2]. \tag{2.139}$$

The probability density function of the envelope is given as [5],

$$p(R) = \int_{-\pi}^{\pi} p(R, \theta)\,d\theta$$
$$= \frac{R}{\sigma^2}\exp[-(R^2 + A^2)/2\sigma^2]I_0\left(\frac{AR}{\sigma^2}\right), \tag{2.140}$$

where $I_0(\cdot)$ is the modified Bessel function of the first kind.

The probability density function of the phase is given by

$$p(\theta) = \int_0^{\infty} p(R, \theta)\,dR, \qquad |\theta| < \pi$$
$$= \frac{1}{2\pi}e^{-A^2/2\sigma^2}\left\{1 + \frac{A}{\sigma}\sqrt{2\pi}\cos\theta\,e^{A^2\cos^2\theta/2\sigma^2}\left[1 - Q\left(\frac{A\cos\theta}{\sigma}\right)\right]\right\}, \tag{2.141}$$

where

$$Q(y) = \frac{1}{\sqrt{2\pi}}\int_y^{\infty} e^{-x^2/2}\,dx.$$

APPENDIX 2.1 FOURIER TRANSFORM PROPERTIES

(a) Linearity

$$a_1 f_1(t) + a_2 f_2(t) \leftrightarrow a_1 F_1(\omega) + a_1 F_2(\omega), \tag{A2.1}$$

where a_1, a_2 are constants.

(b) Symmetry. If

$$f(t) \leftrightarrow F(\omega)$$

then

$$F(t) \leftrightarrow 2\pi f(-\omega). \tag{A2.2}$$

(c) Time scale expansion

$$f(at) \leftrightarrow \frac{1}{|a|} F\left(\frac{\omega}{a}\right). \tag{A2.3}$$

(d) Time shift

$$f(t - t_0) \leftrightarrow F(\omega)e^{-j\omega t_0}. \tag{A2.4}$$

(e) Frequency shift

$$f(t)e^{j\omega_0 t} \leftrightarrow F(\omega - \omega_0). \tag{A2.5}$$

(f) Time derivative

$$\frac{d^n f(t)}{dt^n} \leftrightarrow (j\omega)^n F(\omega). \tag{A2.6}$$

(g) Frequency derivative

$$\frac{d^n F(\omega)}{d\omega^n} \leftrightarrow (-jt)^n f(t). \tag{A2.7}$$

(h) Complex conjugate

$$f^*(t) \leftrightarrow F^*(-\omega). \tag{A2.8}$$

(i) Integral

$$\int_{-\infty}^{t} f(\tau)\, d\tau \leftrightarrow \frac{1}{j\omega} F(\omega) + \pi F(0)\, \delta(\omega) \tag{A2.9}$$

(j) Convolution integral

$$\int_{-\infty}^{\infty} f_1(\tau) f_2(t - \tau)\, d\tau \leftrightarrow F_1(\omega) F_2(\omega). \tag{A2.10}$$

$$f_1(t) f_2(t) \leftrightarrow \frac{1}{2\pi} \int_{-\infty}^{\infty} F_1(y) F_2(\omega - y)\, dy. \tag{A2.11}$$

(k) Some common Fourier transform pairs including the delta function

$$\delta(t) \leftrightarrow 1. \tag{A2.12}$$

$$\delta(t - t_0) \leftrightarrow e^{-j\omega t_0}. \tag{A2.13}$$

$$e^{j\omega_0 t} \leftrightarrow 2\pi\delta(\omega - \omega_0). \tag{A2.14}$$

$$\cos(\omega_0 t) \leftrightarrow \pi[\delta(\omega - \omega_0) + \delta(\omega + \omega_0)]. \tag{A2.15}$$

$$\sin(\omega_0 t) \leftrightarrow -j\pi[\delta(\omega - \omega_0) - \delta(\omega + \omega_0)]. \tag{A2.16}$$

$$\text{sgn}(t) = \frac{t}{|t|} \leftrightarrow \frac{2}{j\omega}. \tag{A2.17}$$

$$U(t) \leftrightarrow \pi\delta(\omega) + \frac{1}{j\omega}, \tag{A2.18}$$

where

$$U(t) = \tfrac{1}{2} + \tfrac{1}{2}\text{sgn}(t).$$

$$\sum_{n=-\infty}^{\infty} \delta(t - nT_0) \leftrightarrow \omega_0 \sum_{n=-\infty}^{\infty} \delta(\omega - n\omega_0), \tag{A2.19}$$

where

$$\omega_0 = \frac{2\pi}{T_0}.$$

REFERENCES

1. A. Papoulis, *The Fourier Integral and Its Application*, McGraw-Hill, New York, 1962.
2. A. Papoulis, *Probability, Random Variables and Stochastic Process*, McGraw-Hill, New York, 1965.
3. H. F. Harmuth, *Transmission of Information by Orthogonal Functions*, Springer-Verlag, New York, 1970.
4. J. M. Wozencraft and I. M. Jacobs, *Principles of Communication Engineering*, Wiley, New York, 1965.
5. B. P. Lathi, *Modern Digital and Analog Communication Systems*, in HRW Series in Electrical and Computer Engineering, Holt, Rinehart and Winston, New York, 1983.

3

THE ELEMENTS OF DIGITAL COMMUNICATION SYSTEMS

This chapter briefly covers some general topics in digital signal transmission and detection theory. The reader may refer to references [1–9] for a more comprehensive treatment of the material.

Figure 3.1 shows the block diagram of a baseband digital transmission system. The transmitted binary data symbols are applied at the pulse repetition frequency to an encoding circuit that performs logical manipulations such as differential encoding and two-to-multilevel conversion. Corresponding to the logical data symbols, the pulse shaping circuit sends out digitally encoded waveforms into the channel.

Pulse shaping is intended to produce a narrower signal spectrum or to match the spectrum to the transfer characteristics of the channel, for example, a dc blocked channel. The manipulation of data by the encoder or by the pulse shaping circuits is called *line coding*.

The transmitted signal is detected at the receiver. Typically the transmitted signal is corrupted by noise and interference in the channel. The detection process includes noise band limitation and decision-making on the digital signal. Achieving a lower error rate for the decisions is the main concern in the receiver. The determined logical symbol is decoded appropriately according to the encoding.

3.1 PULSE SHAPING

Pulse shaping with limited bandwidth results in an impulse response of the pulse shaping circuit/filter with a long time period. For example, let us

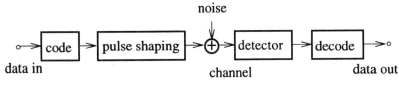

Figure 3.1. Digital transmission system.

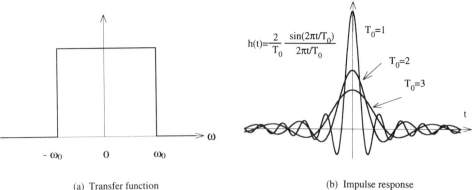

(a) Transfer function   (b) Impulse response

Figure 3.2. Characteristics of an ideal low-pass filter. $T_0 = 2\pi/\omega_0$.

consider an ideal low-pass filter with a rectangular transfer function $H(\omega)$ as shown in Fig. 3.2(a). The impulse response is given by the inverse Fourier transform as

$$h(t) = \frac{1}{2\pi} \int_{-\infty}^{\infty} H(\omega)e^{j\omega t}\, d\omega$$

$$= \frac{\omega_0}{\pi} \frac{\sin(\omega_0 t)}{\omega_0 t} = \frac{2}{T_0} \frac{\sin(2\pi t/T_0)}{2\pi t/T_0}, \tag{3.1}$$

where $T_0 = 2\pi/\omega_0$.

Figure 3.2(b) shows $h(t)$. The part of $h(t)$ that spreads into other symbol periods causes interference with signals, that is, *intersymbol interference*. The intersymbol interference degrades decision error rate performance at the receiver. Figure 3.3 (these are termed eye-diagrams) shows an overlapped drawing over two symbol periods of a randomly generated binary data sequence with pulse shaping by Eq. (3.1). At each symbol time, the signal is sampled and is subjected to decision-making. The minimum distance between the sampled signal levels corresponding to different binary signals is called eye-opening. The ideal eye-opening is given when the intersymbol interference is null at the sampling instant, as is shown in Fig. 3.3(a).

3.1.1 Nyquist's First Criterion

Nyquist's first criterion guarantees that no intersymbol interference occurs. It requires that the impulse response $h(t)$ of a pulse shaping filter satisfy

$$h(t) = \begin{cases} 1 & (t = 0) \\ 0 & (t = nT, n \neq 0), \end{cases} \tag{3.2}$$

where $1/T$ is the pulse repetition frequency. Decision is made at $t = nT$.

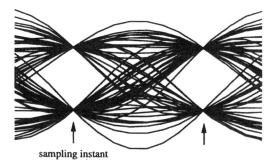

sampling instant

(a) T / T_0 = 0.5

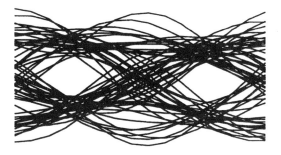

(b) T / T_0 = 0.475

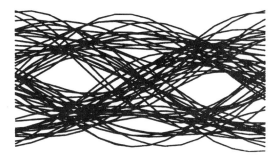

(c) T / T_0 = 0.45

Figure 3.3. Eye-diagrams of different pulse shapings. _T_ is pulse duration.

Considering the above requirement for no intersymbol interference, the sampled impulse response $h_s(t)$ is given as

$$h_s(t) = h(t) \sum_{n=-\infty}^{\infty} \delta(t - nT) = \delta(t). \tag{3.3}$$

Let $h(t) \leftrightarrow H(\omega)$. Taking the Fourier transform of the above equation, we obtain, in similar process to Eq. (2.42),

$$\frac{1}{T} \sum_{n=-\infty}^{\infty} H(\omega - n\omega_0) = 1 \tag{3.4}$$

or

$$\sum_{n=-\infty}^{\infty} H(\omega - n\omega_0) = T, \tag{3.5}$$

where $\omega_0 = 2\pi/T$.

Let us assume that $H(\omega)$ is band-limited within $0 < |\omega| < \omega_0$. If we consider a frequency range $0 \sim \omega_0$, then Eq. (3.5) becomes

$$H(\omega) + H(\omega - \omega_0) = T \ (0 < \omega < \omega_0). \tag{3.6}$$

Take a new variable $x = \omega - \omega_0/2$, then

$$H\left(\frac{\omega_0}{2} + x\right) + H\left(x - \frac{\omega_0}{2}\right) = T \quad \left(|x| < \frac{\omega_0}{2}\right). \tag{3.7}$$

Applying Eq. (2.8) to the above we have

$$H\left(\frac{\omega_0}{2} + x\right) + H^*\left(\frac{\omega_0}{2} - x\right) = T \quad \left(|x| < \frac{\omega_0}{2}\right). \tag{3.8}$$

Taking the real part and the imaginary part of Eq. (3.8),

$$\text{Re}\left[H\left(\frac{\omega_0}{2} + x\right)\right] + \text{Re}\left[H\left(\frac{\omega_0}{2} - x\right)\right] = T \quad \left(|x| < \frac{\omega_0}{2}\right), \tag{3.9}$$

$$\text{Im}\left[H\left(\frac{\omega_0}{2} + x\right)\right] - \text{Im}\left[H\left(\frac{\omega_0}{2} - x\right)\right] = 0 \quad \left(|x| < \frac{\omega_0}{2}\right). \tag{3.10}$$

These relations are illustrated in Fig. 3.4. There are a large number of functions that will fulfill these conditions. If we consider a real function $G(\omega)$, then Eq. (3.10) is automatically satisfied. However, a transfer function that

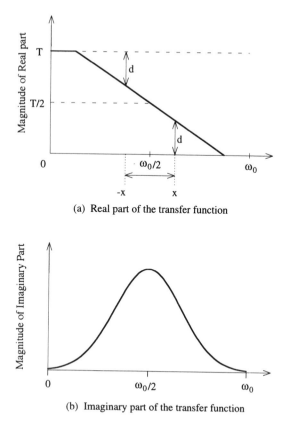

(a) Real part of the transfer function

(b) Imaginary part of the transfer function

Figure 3.4. Transfer characteristics to meet Nyquist's first criterion.

has no imaginary part cannot be realized. Thus, a realizable transfer function must be in the form

$$H(\omega) = G(\omega)e^{-j\omega t_0}, \tag{3.11}$$

where $G(\omega)$ is a real function meeting Eq. (3.9), and t_0 is a time delay constant. Because the term $e^{-j\omega t_0}$ presents a transfer function of a delay line, it does not distort the transmitted waveform. Since $G(\omega)$ is a real function, the impulse response $(g(t) \leftrightarrow G(\omega))$ becomes an even function, that is, $g(-t) = g(t)$. Then from Eqs. (3.11) and (A2.4) we have $h(t) = g(t - t_0)$. Since $G(\omega)$ is band-limited within the finite bandwidth $|\omega| < \omega_0$, $g(t)$ shows an infinite response. Therefore the time delay t_0 must be infinite to get an ideal Nyquist-I transfer function. Practically, approximations are made for reasonable values of t_0.

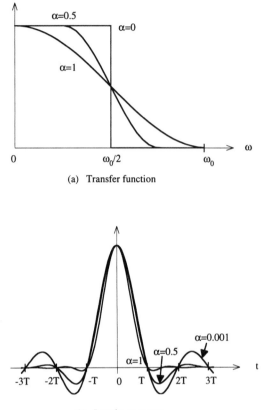

(a) Transfer function

(b) Impulse response

Figure 3.5. Transfer characteristics and impulse response of a raised cosine roll-off filter for different roll-off factors.

A widely used filter that meets Nyquist's first criterion is the raised cosine roll-off filter, whose transfer function is given as

$$
G(\omega) = \begin{cases}
1 & (|\omega| < (1-\alpha)\omega_0/2) \\
\frac{1}{2}\left\{1 + \cos\left[\left(\frac{|\omega| - \omega_0/2}{\alpha\omega_0/2} + 1\right)\frac{\pi}{2}\right]\right\} & ((1-\alpha)\omega_0/2 < |\omega| < (1+\alpha)\omega_0/2) \\
0 & (|\omega| > (1+\alpha)\omega_0/2),
\end{cases} \quad (3.12)
$$

where $\alpha(\leq 1)$ is a roll-off factor.

The impulse response of the filter is given by

$$
g(t) = \frac{1}{T}\frac{\sin \pi t/T}{\pi t/T}\frac{\cos \alpha\pi t/T}{1 - 4\alpha^2 t^2/T^2}. \tag{3.13}
$$

Figures 3.5 and 3.6 show $G(\omega)$, $h(t)$, and eye-diagrams of the raised cosine roll-off filter.

$\alpha = 0$

$\alpha = 0.5$

$\alpha = 1$

Figure 3.6. Eye-diagram of a raised cosine roll-off filter for different roll-off factors.

3.1.2 Nyquist's Second Criterion

This criterion guarantees an eye-opening at the midpoint of symbol times. It requires an impulse response

$$
h(t) = \begin{cases} 1, & t = \pm T/2 \\ 0, & t = \pm(n + \frac{1}{2})T \quad (n \neq 0). \end{cases}
$$
(3.14)

Multiplying the sampling function by the above, we get

$$
h(t) \sum_{n=-\infty}^{\infty} \delta\left[t - \left(n + \frac{1}{2}\right)T\right] = \delta\left(t + \frac{T}{2}\right) + \delta\left(t - \frac{T}{2}\right).
$$
(3.15)

Taking the Fourier transform, we have

$$
\sum_{n=-\infty}^{\infty} H(\omega - n\omega_0)e^{-jn\pi} = 2T\cos\left(\frac{\omega T}{2}\right),
$$
(3.16)

where $\omega_0 = 2\pi/T$.

If we restrict the bandwidth of $H(\omega)$ within a frequency range of $|\omega| \leq \omega_0/2$, then from the above equation we have

$$
H(\omega) = \begin{cases} 2T\cos\left(\frac{\omega T}{2}\right) & \left(|\omega| \leq \frac{\omega_0}{2}\right) \\ 0 & \text{(otherwise)}. \end{cases}
$$
(3.17)

If we restrict the bandwidth of $H(\omega)$ within $|\omega| \leq \omega_0$ and only consider the frequency over $0 < \omega < \omega_0$, then we get

$$
H(\omega) - H(\omega - \omega_0) = 2T\cos\left(\frac{\omega T}{2}\right) \quad (0 < \omega < \omega_0).
$$
(3.18)

The following transfer function satisfies the above condition:

$$
H(\omega) = \begin{cases} T\left[1 + \cos\left(\frac{\omega T}{2}\right)\right] & (0 < |\omega| < \omega_0) \\ 0 & \text{(otherwise)}. \end{cases}
$$
(3.19)

We can see that this is a special case of the transfer function in Eq. (3.12) where the roll-off factor $\alpha = 1$. Thus, the raised cosine roll-off filter with a roll-off factor of $\alpha = 1$ satisfies Nyquist's first and second criteria.

Figure 3.7 shows eye-diagrams of the above two transfer functions given by Eqs. (3.17) and (3.19). As shown later, a Nyquist's second criterion (Nyquist-II) filter is used for a duobinary transmission system.

Figure 3.7. Eye-diagrams of pulse shapings with Nyquist's second criterion. Top: $H(\omega)$; Eq. (3.17). Bottom: $H(\omega)$; Eq. (3.19).

3.1.3 Nyquist's Third Criterion

This criterion insures that the integral of an impulse response over interfering symbol periods is zero and is a constant over the considered time period. It is expressed as

$$y(nT) = \int_{nT-T/2}^{nT+T/2} h(t)\,dt = \begin{cases} \text{constant} & (n = 0) \\ 0 & (n \neq 0). \end{cases} \tag{3.20}$$

The transfer function is given as

$$H_{\text{III}}(\omega) = \frac{\omega T/2}{\sin(\omega T/2)} H_{\text{I}}(\omega), \tag{3.21}$$

where $H_{\text{I}}(\omega)$ meets Nyquist's first criterion.

By referring to the diagram in Fig. 3.8, let us examine whether the above

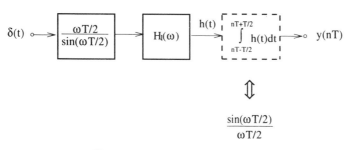

Figure 3.8. Nyquist's third criterion filter.

transfer function meets Nyquist's third criterion. We have shown in Section 2.1.7 that the integral of an input signal for a time range over T is equivalent to the sampled output of the filter which has the transfer function $H(\omega) = T\sin(\omega T/2)/(\omega T/2)$. Thus, the first term on the right side of Eq. (3.21) is canceled with $H(\omega)$, as far as the sampled signal $y(nT)$ is concerned. The transfer function $H_{\mathrm{I}}(\omega)$ ensures that $y(nT) = 0$ ($n \neq 0$). Thus Eq. (3.20) is satisfied.

A Nyquist's third criterion filter is used in digital FM systems to generate a modulated signal that takes fixed phases.

3.1.4 Other Pulse Shaping Methods

For a rectangular pulse, the ratio of the pulse time-width to the pulse repetition duration is called the pulse duty ratio. A pulse with a pulse duty ratio of 100% is called an NRZ (non-return-to-zero) signal, and other pulses are called RZ (return-to-zero) signals. An NRZ signal has a rectangular pulse shape such that

$$h(t) = \begin{cases} 1 & (-T/2 \leq t \leq T/2) \\ 0 & (\text{otherwise}), \end{cases} \tag{3.22}$$

and has a transfer function

$$H(\omega) = T\frac{\sin(\omega T/2)}{\omega T/2}. \tag{3.23}$$

Another frequently used pulse shape has the form of a Gaussian function,

$$h(t) = \frac{1}{\sqrt{2\pi}t_0}e^{-t^2/2t_0^2}, \tag{3.24}$$

where t_0 is a constant that represents the width of the pulse. The transfer function also has Gaussian shape,

$$H(\omega) = e^{-\omega^2/2B_0^2}, \tag{3.25}$$

where $B_0 = 1/t_0$ is the bandwidth of the pulse shaping filter.

Another pulse shape has a raised cosine waveform and is given as

$$h(t) = \begin{cases} \dfrac{1}{2LT}\left[1 - \cos\left(\dfrac{2\pi t}{LT}\right)\right] & (0 \leq t \leq LT) \\ 0 & (\text{otherwise}), \end{cases} \tag{3.26}$$

where L is an integer. The above two waveforms are shown in Fig. 3.9.

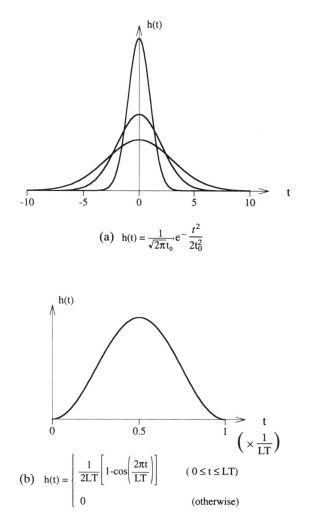

(a) $h(t) = \dfrac{1}{\sqrt{2\pi}t_0} e^{-\frac{t^2}{2t_0^2}}$

(b) $h(t) = \begin{cases} \dfrac{1}{2LT}\left[1-\cos\left(\dfrac{2\pi t}{LT}\right)\right] & (0 \le t \le LT) \\ 0 & (\text{otherwise}) \end{cases}$

Figure 3.9. Pulse waveforms of (a) a Gaussian, and (b) the raised cosine impulse response.

3.2 LINE CODING

Line coding includes encoding and pulse shaping. Pulse shaping has been described in Section 3.1.

3.2.1 Unipolar (On–Off) Code and Polar Codes

A unipolar code assigns a pulse $p(t)$ to a signal "1" and no pulse to "0". On the other hand, a polar code assigns $p(t)$ to a signal "1" and $-p(t)$ to "0". Figures 3.10(a) and 3.10(b) illustrate a unipolar and a polar code,

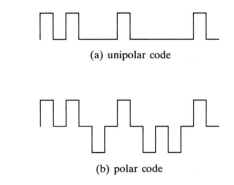

(a) unipolar code

(b) polar code

Figure 3.10. Unipolar code and polar code (duty ratio is 50%).

respectively, with a pulse duty ratio of 50%. An NRZ unipolar code consists of an NRZ polar code and a dc signal, whose level is one-half of the amplitude of the polar code.

3.2.2   Multilevel Codes

A sequence of N bits signal can be represented by 2^N different states (symbols). A line code with 2^N-levels ($N > 1$) is called a *multilevel code*. The symbol frequency f_s will be $1/N$ times the bit frequency f_b. Thus, with the use of 2^N-level codes, the bandwidth of the baseband signal is reduced by a factor of $1/N$ when compared to the 2-level coded system. For example, using 4-, 8-, and 16-level codes, the bandwidth decreases to 1/2, 1/3, or 1/4 of that of the 2-level code.

Under the condition that the average transmit power is kept the same, the minimum distance between the code levels decreases as the number of levels increases. Suppose that the levels are symmetrically spaced with respect to zero voltage (see Fig. 3.11) and that the levels are equally likely to occur, then the minimum distance $2d_m$ between signal levels will be given as

$$d_m \propto \frac{1}{\sqrt{\dfrac{2}{2^N} \displaystyle\sum_{n=1}^{2^{N-1}} (2n-1)^2}}, \tag{3.27}$$

where a 2^N-level code is assumed

3.2.3   The Gray Codes

The one-to-one mapping of N-bit sequences to 2^N levels is arbitrary. The most straightforward mapping is to use a natural binary code. On the other hand, the mapping where the bits assigned to the two neighboring levels differ

-3d$_m$	×
d$_m$	×
-d$_m$	×
-3d$_m$	×

Figure 3.11. 4-Level code.

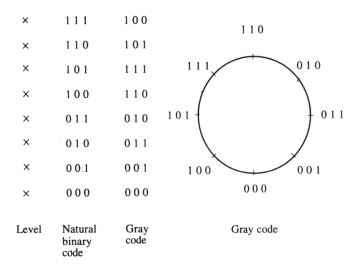

Level	Natural binary code	Gray code
×	1 1 1	1 0 0
×	1 1 0	1 0 1
×	1 0 1	1 1 1
×	1 0 0	1 1 0
×	0 1 1	0 1 0
×	0 1 0	0 1 1
×	0 0 1	0 0 1
×	0 0 0	0 0 0

Figure 3.12. Natural binary code and Gray code (8-level).

only by one bit is called *Gray coding*. This property is advantageous in a multilevel transmission system where decision errors occur mostly between two neighboring levels; a symbol error results in a one-bit error in this case, while a symbol error may cause a two-bit error or more for the case when a natural binary code is used. Figure 3.12 shows the Gray code and the natural code for an eight-level system. The property of Gray codes that only one bit differs from the neighboring levels is maintained even when the levels are arranged on a circle.

Consider a 2^N-level ($N \geq 2$) system and express the bit patterns as $(a_{n-1},$

$a_{n-2}, a_{n-3}, \ldots, a_0$) for the natural binary code and ($b_{n-1}, b_{n-2}, b_{n-3}, \ldots, b_0$) for the Gray code. Gray code is generated such that

$$b_{n-1} = a_{n-1}, \tag{3.28a}$$

$$b_i = a_{i+1} \oplus a_i \quad (i = n-2, n-3, n-4, \ldots, 0), \tag{3.28b}$$

where $\oplus$ denotes addition modulo 2.

3.2.4 Manchester (Split-Phase) Code

The waveforms of this code are shown in Fig. 3.13. The spectrum is given as

$$P(\omega) = \int_{-\infty}^{\infty} p(t) e^{-j\omega t} \, dt$$

$$= jT \frac{\sin^2(\omega T/4)}{\omega T/4}. \tag{3.29}$$

In a symbol duration, the plus and the minus areas of each waveform are balanced and therefore the spectrum of the signal is null at zero frequency (dc). This property of Manchester codes is advantageous for application to systems where the dc frequency is blocked. Actually, Manchester codes are used in analog (voice) FM systems, in order to transmit the slow-speed data signal in the analog voice channel, which is dc-blocked (i.e., ac-coupled): the voice signal contains no dc component. The ac-coupled circuits are easier to implement, because they are free from, dc-offset effects.

The other advantage of Manchester codes is ease of the clock signal extraction due to the property that polarity of the signal changes in every symbol.

A Manchester code can be seen as binary phase shift keying where the carrier signal takes a rectangular waveform with a frequency of the bit

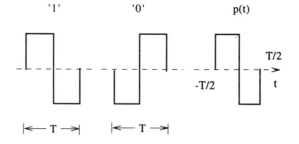

Figure 3.13. Manchester (split-phase) code.

frequency and is synchronized to the bit clock signal. This property of Manchester codes ensures the use of delay detection, coherent detection as well as matched filter detection.

We can band-limit Manchester codes without losing the properties that the spectrum has no dc-component and that no intersymbol interference occurs. The idea is that two filters are combined that have impulse responses $h_1(t)$ and $h_2(t)$, respectively,

$$h_1(t) = \delta(t + T/4) - \delta(t - T/4), \tag{3.30}$$

and $h_2(t)$ meets the first Nyquist criterion for the symbol duration of $T/2$. Thus the transfer function of the combined filter is

$$H(\omega) = H_1(\omega)H_2(\omega), \tag{3.31}$$

where

$$h_1(t) \leftrightarrow H_1(\omega) = j2\sin\left(\frac{\omega T}{4}\right), \tag{3.32}$$

$$h_2(t) \leftrightarrow H_2(\omega), \tag{3.33}$$

and $H_2(\omega)$ has, for example, the raised cosine roll-off characteristic for the symbol duration of $T/2$. The spectrum of a band-limited Manchester code, as above, has a maximum frequency of $(1 + \alpha)1/T$, where α is the roll-off factor. Consequently, the bandwidth of Manchester codes is two times the bandwidth of the polar codes. Figure 3.14 shows the spectra or transfer functions given by Eqs. (3.29) and (3.31).

3.2.5 Synchronized Frequency Shift Keying Code

As in the case of the Manchester codes, the baseband signal of synchronized frequency shift keying codes can be presented as a frequency-modulated signal where the carrier signal has a frequency that is comparable with the bit rate frequency and is synchronized to the clock signal. An example of the codes is shown in Fig. 3.15. These codes are produced using a rectangular carrier signal with frequency of $5/4T$ and frequency modulation index of 0.5. If we assume a sinusoidal waveform carrier instead of rectangular waveform, the code becomes the minimum shift keying, which is used for data transmission in analog voice FM systems. The modulated signal is matched to systems that have dc-blocked baseband channels.

The mark and the space frequencies that correspond to the digital signal '1' and '0', respectively, should be chosen so that the carrier signal is

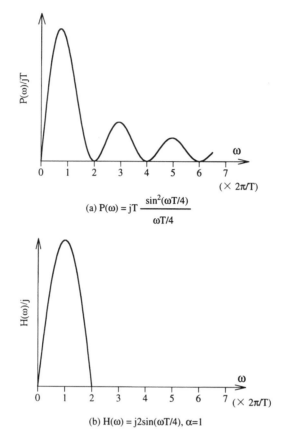

Figure 3.14. Spectra of the Manchester code: (a) without band limitation; (b) with band limitation.

synchronized with the clock signal. They are given as

$$\left.\begin{array}{c} f_M = N_1 f_B \\ f_S = N_2 f_B \end{array}\right\} \quad \left(N_1, N_2 = \frac{N}{2},\ N = 1, 2, 3, \ldots\right), \quad (3.34)$$

where f_B is bit rate frequency. In the case of Fig. 3.15 $N_1 = 1$ and $N_2 = 3/2$.

3.2.6 Correlative Coding

Correlative coding is a coding scheme in which the level of the output (coded) signal is intentionally correlated by manipulating the input digital data. The purpose of correlative coding is to shape the spectrum of the coded signal: for example to get a narrower or a dc-null spectrum.

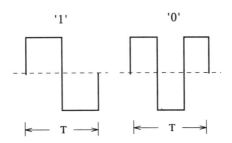

Figure 3.15. Synchronized frequency shift keying code.

The pulse shaping, such as the raised cosine roll-off Nyquist filtering, can be performed independently of the correlative coding. Actually, Nyquist filtering with a roll-off factor of zero is used extensively. Correlative codings fall into two categories. One category is where a nonlinear operation is applied to the input digital data. A number of coding methods of this category are used in the PCM (pulse code modulation) multiplexed transmission system. The other is a category where the operation applied to the input signal is linear. Linear coding schemes are discussed in the following.

3.2.6.1 Partial Response Codes.

This category of codes is generated by passing a polar code into a linear filter with the impulse response

$$h(t) = \sum_{n=0}^{N-1} a_n \delta(t - nT), \tag{3.35}$$

where a_n are integer constants and T is the bit duration. The partial response code where a_n takes real numbers is called the *generalized partial response system*. If the digitally encoded signal is sampled with frequency $1/T$, and we denote the operation for a time-delay of nT by z^{-n} (z-transform), then the partial response filter has a z-transform

$$H(z) = \sum_{n=0}^{N-1} a_n z^{-n} \tag{3.36}$$

and the output signal spectrum is

$$H(\omega) = \sum_{n=0}^{N-1} a_n e^{-jn\omega T}. \tag{3.37}$$

An input pulse produces the output pulse that lasts over a duration of N symbol periods. Memory length is NT. In practice, the waveform spreads further owing to pulse shaping as in the case of Nyquist roll-off filtering.

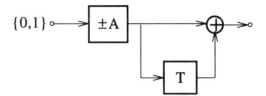

Figure 3.16. Duobinary coding.

Duobinary (Class I Partial Response) Code. The case of $N = 2$ and $a_0 = a_1 = 1$ is called duobinary or class I partial response code. The coder is shown in Fig. 3.16. It has three levels, $2A$, 0, and $-2A$. The transfer function is

$$H(\omega) = 2 \cos\left(\frac{\omega T}{2}\right) e^{-j\omega T/2}. \tag{3.38}$$

If we use the Nyquist-I filter with a roll-off factor of zero for pulse shaping, the transmit signal spectrum becomes

$$G(\omega) = \begin{cases} 2 \cos\left(\dfrac{\omega T}{2}\right) e^{-j\omega T/2} & \left(0 < |\omega| < \dfrac{\pi}{T}\right) \\ 0 & \text{(otherwise)}. \end{cases} \tag{3.39}$$

The spectrum $G(\omega)$ is drawn in Fig. 3.17(a). The eye-pattern is shown in Fig. 3.18.

The transmission system with duobinary code is shown in Fig. 3.19. The received signal is full-wave rectified to produce the two-level signal of $2A$ and 0. Level $2A$ corresponds to '0' and level 0 to '1'. The signal determined by the decision-making process is applied to a feedback circuit, where modulo 2 operation is assumed.

Figure 3.20 illustrates the transmission process using an assumed data sequence. Without transmission errors the output signal $\hat{d}_k$ is equal to the transmitted signal d_k. If an error occurs in the decision process, $\hat{c}_k$, the output signal $\hat{d}_k$ is inverted after this error and the inversion continues until the next error occurs.

Precoding is introduced to prevent the above error propagation in the partial response system (Fig. 3.21). For the duobinary system, the precoding is expressed as

$$d_k = d'_k \oplus d'_{k-1} \qquad \text{(mod 2)} \tag{3.40}$$

or

$$d'_k = d_k \oplus d'_{k-1} \qquad \text{(mod 2)}. \tag{3.41}$$

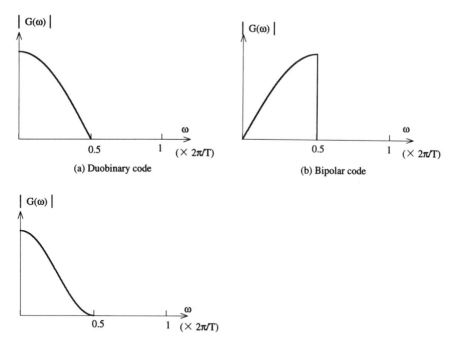

Figure 3.17. Spectra of duobinary code, bipolar code, and class II partial response code. Nyquist-I filtering with zero roll-off factor is assumed.

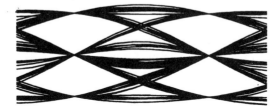

Figure 3.18. Eye-diagram of the duobinary code.

The z-transform expression of the precoder is

$$H_p(z) = \frac{1}{1 + z^{-1}} \quad (\text{mod } 2). \tag{3.42}$$

On the other hand, the duobinary coding is expressed as

$$H_c(z) = 1 + z^{-1}, \tag{3.43}$$

where arithmetic addition instead of mod 2 addition is assumed.

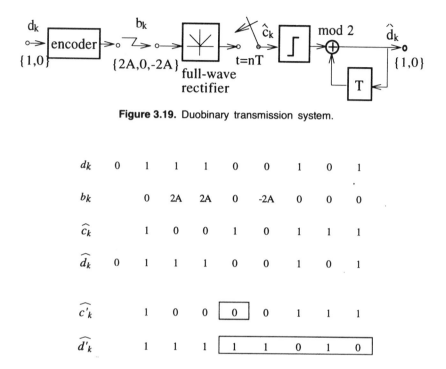

Figure 3.19. Duobinary transmission system.

d_k	0	1	1	1	0	0	1	0	1
b_k		0	2A	2A	0	-2A	0	0	0
$\widehat{c_k}$		1	0	0	1	0	1	1	1
$\widehat{d_k}$	0	1	1	1	0	0	1	0	1
$\widehat{c'_k}$		1	0	0	0	0	1	1	1
$\widehat{d'_k}$		1	1	1	1	1	0	1	0

Figure 3.20. An example of the duobinary transmission process.

Since the levels $2A$ and $-2A$ correspond to "0" and the level 0 to "1," the decision process can be seen as performing the mod 2 operation on the duobinary coded signal:

$$H_c(z) = 1 + z^{-1} \text{ (mod 2).} \tag{3.44}$$

Thus, the duobinary transmission system with precoding is interpreted as performing the logical manipulation

$$H_p(z)(\text{mod } 2)H_c(z)(\text{mod } 2) = 1$$

on the transmitted signal d_k.

Comparing Figs. 3.19 and 3.21, the receiver feedback circuit is moved to the transmitter as the precoding circuit.

For the L-level input signal, the precoder is expressed as

$$H_p(z) = \frac{1}{H_c(z)} \quad (\text{mod } L), \tag{3.45}$$

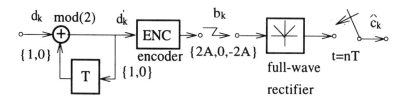

precoder

Figure 3.21. Duobinary transmission system with precoding.

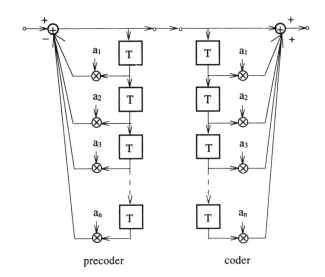

precoder   coder

Figure 3.22. Precoder and coder for partial response system.

where $H_c(z)$ is given corresponding to the partial response coding used. The precoder and the partial response coder are realized as shown in Fig. 3.22. We can see that $H_p(z)H_c(z) = 1$ holds regardless of the type of the input signal and of the modulo operation.

Another method for receiving the duobinary coded signal is shown in Fig. 3.23. This detection circuit can be seen as a decision feedback equalizer (Section 7.5.3), where the intersymbol interference caused intentionally by partial response coding is equalized.

Bipolar Code. Bipolar code is also called *AMI* (*alternative-mark-inversion*) *code*. The code assigns pulses with levels A and $-A$ to signal "1" and no pulse to signal "0." The level $\pm A$ alternates in every "1" as illustrated in Fig. 3.24. Thus, the coded waveform is balanced in polarity and has no dc

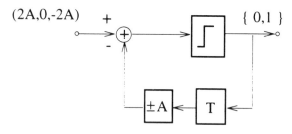

Figure 3.23. Decision feedback detection of duobinary coded signals.

1 0 1 1 0 0 0 1

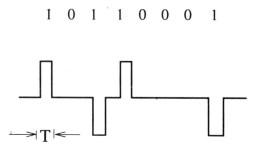

Figure 3.24. Bipolar (AMI) coding.

component. The spectrum is shown in Fig. 3.17(b). The balance is the result of correlating the two successive "1"s. The rule that the levels take alternating polarity can be used for detection of errors by monitoring the violation of the rule.

The encoding process has infinite-length memory; a signal "1" affects the entire output after the signal. A bipolar code can be detected by full-wave rectification followed by two-level decision. A bipolar code can be generated using a partial response coder

$$H_c(z) = 1 - z^{-1} \tag{3.46}$$

and the precoder $H_p(z) = 1/H_c(z)$ (mod 2). The spectrum of the coded signal becomes

$$H(\omega) = j2 \sin\left(\frac{\omega T}{2}\right) e^{-j\omega T/2} \qquad (-\infty < \omega < \infty). \tag{3.47}$$

If we assume the Nyquist-I filter with roll-off factor of zero, the spectrum is limited over the frequency range, $|\omega| < \pi/T$ (see Fig. 3.17(b)).

Duoquatenary Code. This is a 4-level signal input partial response code with $H_c(z) = 1 + z^{-1}$. A duoquatenary transmission system is shown in Fig. 3.25. The transmitted signal has 7 levels as shown in Fig. 3.26(b). The 7-level decision is made at the receiver followed by the modulo 4 operation.

The spectrum of the duoquatenary encoded signal has one-half of the bandwidth of the duobinary signal provided that the bit rate is the same.

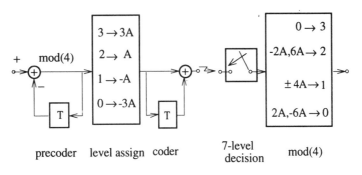

Figure 3.25. Duoquatenary transmission system.

(a) input 4-level signal

(b) duoquaternary signal

Figure 3.26. Eye-diagrams of the duoquatenary coded signal.

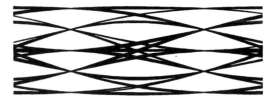

Figure 3.27. Eye-diagram of the class II partial response coded signal.

Class II Partial Response Code. This code is characterized with $H_c(z) = 1 + 2z^{-1} + z^{-2}$. The coded signal takes 5 levels when a 2-level signal input is applied as shown in Fig. 3.27. The spectrum is

$$H(\omega) = 2(1 + \cos \omega T)e^{-j\omega T}. \tag{3.48}$$

If we assume the Nyquist-I filter with roll-off factor of zero, the spectrum covers the frequency range $|\omega| < \pi/T$ (see Fig. 3.17(c)).

3.2.7 Differential Encoding

In a digital transmission system the information is sometimes carried by relative changes of the signal states: for example, in differential detection the relative phase shift between successive signals is determined. In coherent detection, differential encoding prevents error propagation caused by a carrier phase slip (Section 5.5.1). For this purpose, differential encoding is employed at the transmitter.

The differential encoding circuit and the relevant decoding circuit are shown in Fig. 3.28. The encoder and the decoder are characterized with the z-transform

$$H_{\text{enc}}(z) = \frac{1}{1 - z^{-n}} \quad (\text{mod } L), \tag{3.49a}$$

$$H_{\text{dec}}(z) = 1 - z^{-n} \quad (\text{mod } L), \tag{3.49b}$$

respectively.

Typically $n = 1$ and, in this case, the differential encoder is the same as the precoder for the duobinary coding. For $n = 2$, the two-symbol delay differential encoding ($n = 2$) is equivalent to repeating the one-symbol delay differential encoding ($n = 1$) twice. This can be seen from the fact that $1 - z^{-2} = (1 - z^{-1})^2 \pmod 2$.

It can be seen from Fig. 3.28 that a single error in the channel causes two successive errors in the decoding process.

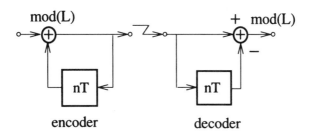

Figure 3.28. Differential encoder and decoder.

3.3 SIGNAL DETECTION

The bit error rate is a measure of the quality of the receiver. The error occurs at the decision process due to the noise and interference. Decision-making on the digital signal at the receiver is the crucial difference between digital communication and analog communication. This section describes some fundamental techniques in digital signal detection.

3.3.1 *C/N, S/N*, and *E_b/N_0*

In order to compare the performances of different modulation and detection methods, a statement of the general condition is desired that is independent of the actual received power, the bit rate frequency, and the noise figure of the receiver. The energy per bit to noise power density ratio, E_b/N_0 meets this purpose. In the following, E_b/N_0 is expressed by the carrier to noise power ratio *C/N* for a modulated signal and by the signal to noise power ratio *S/N* for a baseband signal.

For the modulated signal, the energy per bit is defined as

$$E_b = CT_b, \qquad (3.50)$$

where *C* is the average power of the modulated signal measured at the input of the demodulator and T_b is the bit duration. (The nonmodulated carrier power is the same as the modulated signal power with constant envelope modulation and it is different in the nonconstant envelope modulation.) When the signal power is measured at the output of the band-pass filter at the receiver, the measured value must take into consideration the signal spectrum and the band-pass filter transfer characteristics. The noise power *N* measured at the output of the band-pass filter is related to the noise power density N_0 as

$$N = W_e N_0, \qquad (3.51)$$

where W_e is the equivalent noise bandwidth of the band-pass filter. Thus, we have

$$\frac{E_b}{N_0} = \frac{CT_b}{N_0} = \frac{W_e}{f_b}\frac{C}{N},\tag{3.52}$$

where $f_b = 1/T_b$ is the bit rate frequency.

For a baseband signal, the above argument can be applied by considering a low-pass filter instead of the band-pass filter.

3.3.2 Bit Error Rate

A decision error occurs at the sampling instant when noise perturbs the received signal level over the threshold, as shown in Fig. 3.29(c).

If we consider a two-level polar signaling system with the additive Gaussian noise and without any intersymbol interference, the bit error rate is given as

$$P_e = \tfrac{1}{2}\text{Prob}(|x| > d_m)$$

$$= \tfrac{1}{2}\text{erfc}\left(\frac{d_m}{\sqrt{2}\sigma}\right),\tag{3.53}$$

where x is the noise level, $2d_m$ is the signal distance between the two levels at the sampling instants, $\sigma^2(\equiv\langle x^2\rangle)$ is the average noise power, and

$$\text{erfc}(x) = 1 - \text{erf}(x)$$

$$= \frac{2}{\sqrt{\pi}}\int_x^\infty e^{-t^2}\,dt.\tag{3.54}$$

In the literature the following expression $Q(x)$ is sometimes used instead of $\text{erfc}(x)$:

$$Q(x) = \tfrac{1}{2}\text{erfc}\left(\frac{x}{\sqrt{2}}\right)$$

$$= \frac{1}{\sqrt{2\pi}}\int_x^\infty e^{-t^2/2}\,dt.\tag{3.55}$$

The quantity d_m^2/σ^2 is the signal to noise power ratio $[S/N]_s$ at the sampling instants. Thus,

$$P_e = \tfrac{1}{2}\text{erfc}\left(\sqrt{\frac{1}{2}\left[\frac{S}{N}\right]_s}\right).$$

The bit error rate is shown in Fig. 3.30.

(a) $E_b/N_0 = \infty$

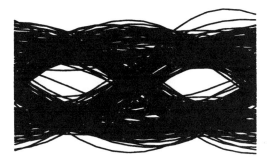

(b) $E_b/N_0 = 8$ dB

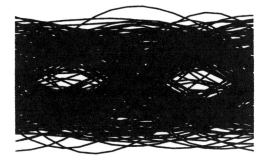

(c) $E_b/N_0 = 4$ dB

Figure 3.29. Eye-diagram of the 2-level polar signal with additive Gaussian noise.

In the case of an N-level $(N > 2)$ transmission system, considering that the error rates for the highest or the lowest levels are half of those of the other levels, we have symbol error rates of

$$P_e = [1 - \tfrac{1}{2}P(\pm M)]\mathrm{erfc}\left(\frac{d_m}{\sqrt{2}\sigma}\right), \qquad (3.56)$$

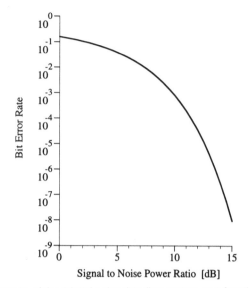

Figure 3.30. Bit error rate of the 2-level polar signaling system as a function of sampled signal to noise power ratio.

where $P(\pm M)$ is the probability of the occurrence of the highest or the lowest level.

The relation between the sampled signal to noise power ratio $[S/N]_s$ and the energy per bit to noise power density ratio E_b/N_0 depends on the pulse waveform and the receiving filter characteristics. We will investigate this matter by considering two systems.

3.3.3 NRZ Signaling with Integrate-and-Dump Filter Detection

Assume an NRZ signal with amplitude of $\pm A$ and bit duration of T, shown in Fig. 3.31 (a). The eye diagrams are shown in Fig. 3.31(b). The energy per bit becomes $E_b = A^2 T$. The signal with added white Gaussian noise, whose double-sided noise power density is $N_0/2$, is received by the integrate-and-dump (I&D) filter. The signal output of the I&D filter takes levels $\pm AT$: the sampled signal power S is $(AT)^2$.

It is shown that at the sampling instant the signal output of the I&D filter is equivalent to that of the filter with transfer function $H(\omega)$ given by Eq. (2.37). Using this fact and the relation between the average power and the power spectral density given by Eq. (2.21), the average noise power N at the output of the I&D filter becomes (Eq. 2.63)

$$N = \frac{1}{2\pi} \int_{-\infty}^{\infty} \frac{N_0}{2} |H(\omega)|^2 \, d\omega = \frac{N_0 T}{2}. \tag{3.57}$$

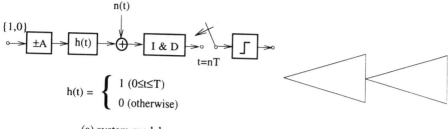

$$h(t) = \begin{cases} 1 \ (0 \le t \le T) \\ 0 \ (\text{otherwise}) \end{cases}$$

(a) system model   (b) eye diagram

Figure 3.31. NRZ signaling with integrate-and-dump filter detection.

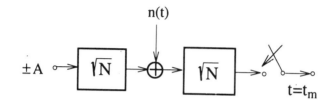

Figure 3.32. A 2-level Nyquist-I signaling system. *N* represents Nyquist's first criterion filter transfer function.

Then we have

$$\left[\frac{S}{N}\right]_s = \frac{(AT)^2}{N_0 T/2} = 2\frac{E_b}{N_0}. \tag{3.58}$$

3.3.4 Nyquist-I Signaling System

Consider a case where the filter characteristic, which meets Nyquist's first criterion, is divided equally to the transmit and the receive filters (Fig. 3.32): for example, the transmit and receive filters have the transfer functions $H_T(\omega)$ and $H_R(\omega)$, respectively, which are the square roots of Eq. (3.12).

Using Eqs. (2.61) and (3.12) and Eqs. (2.64)–(2.66), the energy per bit is given as

$$E_b = \frac{A^2}{2\pi} \int_{-\infty}^{\infty} |H_T(\omega)|^2 \, d\omega$$

$$= \frac{A^2}{2\pi} \int_{-\infty}^{\infty} |H_I(\omega)| \, d\omega$$

$$= \frac{A^2}{T}. \tag{3.59}$$

With the fact that $H_T(\omega)H_R(\omega) = H_I(\omega)$ and Eq. (3.13), the signal at the output of the receive filter takes $\pm A/T(= \pm Ah(0))$: the sampled signal power S is $(A/T)^2$. The noise power at the output of the receive filter becomes

$$N = \frac{1}{2\pi} \int_{-\infty}^{\infty} \frac{N_0}{2} |H_R(\omega)|^2 \, d\omega$$

$$= \frac{1}{2\pi} \int_{-\infty}^{\infty} \frac{N_0}{2} |H_I(\omega)| \, d\omega$$

$$= \frac{N_0}{2} \frac{1}{T}. \tag{3.60}$$

The signal to noise power ratio at the sampling instant becomes $2A^2/(TN_0) = 2E_b/N_0$. Thus we have the same results as in the case of NRZ signaling with the I&D filter detection system.

3.3.5 The Matched Filter

The filter that maximizes the signal to noise power ratio at the sampling instant is known as the *matched filter*. Consider the case that a signal $p(t)$ with added noise $n_i(t)$ is applied to a filter (Fig. 3.33). The power spectral density of the input noise $n_i(t)$ is denoted by $G(\omega)$. The signal output from the filter at time t_m is given as

$$r(t_m) = p(t) * h(t)|_{t=t_m}$$

$$= \frac{1}{2\pi} \int_{-\infty}^{\infty} P(\omega)H(\omega)e^{j\omega t_m} \, d\omega, \tag{3.61}$$

where $p(t) \leftrightarrow P(\omega)$, $h(t) \leftrightarrow H(\omega)$.

The average filtered noise power is

$$\langle n^2(t_m) \rangle = \langle |n_i(t) * h(t)|_{t=t_m}^2 \rangle$$

$$= \frac{1}{2\pi} \int_{-\infty}^{\infty} G(\omega)|H(\omega)|^2 \, d\omega. \tag{3.62}$$

The signal to noise power ratio at the sampling instant $t = t_m$ becomes

$$\frac{S}{N} = \frac{r^2(t_m)}{\langle n^2(t_m) \rangle}$$

$$= \frac{|(1/2\pi) \int_{-\infty}^{\infty} P(\omega)H(\omega)e^{j\omega t_m} \, d\omega|^2}{(1/2\pi) \int_{-\infty}^{\infty} G(\omega)|H(\omega)|^2 \, d\omega}. \tag{3.63}$$

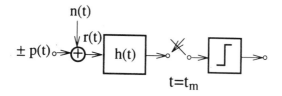

$$h(t) = p(t_m - t)$$

Figure 3.33. Matched filter receiver.

To get the optimum transfer function $H(\omega)$, which maximizes this value, we use the Schwartz inequality, viz.,

$$\int_{-\infty}^{\infty} |X(\omega)|^2 \, d\omega \int_{-\infty}^{\infty} |Y(\omega)|^2 \, d\omega \geq \left| \int_{-\infty}^{\infty} X(\omega) Y(\omega) \, d\omega \right|^2, \quad (3.64)$$

where the equality holds when

$$X(\omega) = kY^*(\omega) \qquad (k = \text{constant}; \,^* \text{ denotes complex conjugate}). \ (3.65)$$

With $X(\omega) = \sqrt{G(\omega)} H(\omega) e^{j\omega t_m}$ and $Y(\omega) = P(\omega)/\sqrt{G(\omega)}$ we get

$$\frac{S}{N} \leq \frac{1}{2\pi} \frac{\displaystyle\int_{-\infty}^{\infty} \frac{|P(\omega)|^2}{G(\omega)} \, d\omega \int_{-\infty}^{\infty} |\sqrt{G(\omega)} H(\omega) e^{j\omega t_m}|^2 \, dt}{\displaystyle\int_{-\infty}^{\infty} G(\omega) |H(\omega)|^2 \, d\omega}$$

$$= \frac{1}{2\pi} \int_{-\infty}^{\infty} \frac{|P(\omega)|^2}{G(\omega)} \, d\omega \quad (3.66)$$

with

$$H_m(\omega) = k \frac{P^*(\omega)}{G(\omega)} e^{-j\omega t_m}. \quad (3.67)$$

For simplicity of argument, assume white noise with $G(\omega) = N_0/2$. The maximum $[S/N]_m$ becomes

$$\left[\frac{S}{N} \right]_m = \frac{2E_b}{N_0}, \quad (3.68)$$

where E_b denotes the pulse energy. We have also

$$H_m(\omega) = P^*(\omega) e^{-j\omega t_m}, \quad (3.69)$$

where $k = N_0/2$ is assumed.

Taking the inverse Fourier transform of Eq. (3.69) and using Eqs. (2.8) and (A2.4) and noting $p(t) \leftrightarrow P(\omega)$, we have

$$h_m(t) = p(t_m - t), \tag{3.70}$$

where $h_m(t) \leftrightarrow H_m(\omega)$.

Thus, the impulse response of the matched filter is given by flipping the time axis and delaying the pulse waveform by t_m. The signal output $y(t)$ of the matched filter becomes

$$
\begin{aligned}
y(t) &= \int_{-\infty}^{\infty} r(\tau) h_m(t - \tau)\, d\tau \\
&= \int_{-\infty}^{\infty} r(\tau) p(t_m - t + \tau)\, d\tau.
\end{aligned}
\tag{3.71}
$$

Then, at the sampling instant $t = t_m$,

$$y(t_m) = \int_{-\infty}^{\infty} r(\tau) p(\tau)\, d\tau. \tag{3.72}$$

If there is no noise,

$$y(t_m) = \int_{-\infty}^{\infty} p^2(\tau)\, d\tau. \tag{3.73}$$

For an example of a matched filter response, consider the input signal

$$p(t) = A\,\delta(t) + B\,\delta(t - t_b). \tag{3.74}$$

The system under consideration can be seen in Fig. 3.34(a). We have

$$y(t) = AB\,\delta(t - t_m + t_b) + (A^2 + B^2)\,\delta(t - t_m) + AB\,\delta(t - t_m - t_b) \tag{3.75}$$

as in Fig. 3.34(b). We get the pulse peak at the instant t_m, which can be set arbitrarily as long as $t_m > t_b$. The pulse peak never depends on the sign of the input signal amplitude and is generated by collecting the spread input signals.

The previously discussed receive filters—I&D filter for NRZ signals and the receive filter for a Nyquist-I transmission system—are both matched filters. When applying a matched filter to a digital transmission system, we must take the intersymbol interference into consideration. Both of the above filters meet the requirements of the matched filter and of intersymbol interference-free transmission. This is not always true.

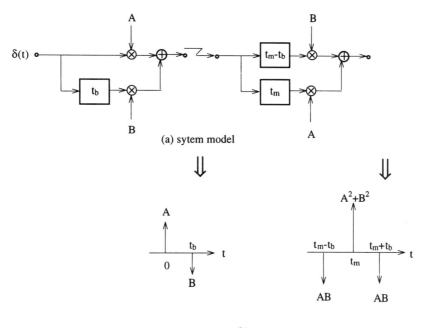

(a) sytem model

(b) response

Figure 3.34. An example of a matched filter system.

A matched filter never causes intersymbol interference if the input pulse is limited in a time range less than the symbol duration. In this case, although the output signal from the matched filter is spread over twice the time range of a pulse duration, it becomes null at the next sampling instant, as illustrated in Fig. 3.35. Even if the input pulse is not limited in the time range of the bit duration, the intersymbol interference can be avoided if

$$y(t_m - nT) = \int_{-\infty}^{\infty} p(\tau)h(t_m - nT - \tau)\,d\tau$$

$$= \int_{-\infty}^{\infty} p(\tau)p(\tau + nT)\,d\tau$$

$$= \begin{cases} \int_{-\infty}^{\infty} p^2(\tau)\,d\tau & (n = 0) \\ 0 & (n \neq 0). \end{cases} \qquad (3.76)$$

As previously discussed, this is the case where the receive matched filter with the transmit filter as a whole meets the Nyquist-I criterion. When symbol-by-symbol instantaneous decisions are assumed, a matched filter receiver without intersymbol interference is optimum for attaining the lowest bit error

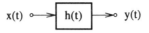

$$x(t)=p(t)-p(t-T), \quad h(t)=p(-t)$$

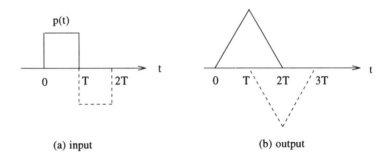

(a) input   (b) output

Figure 3.35. Matched filter response for a pulse waveform limited within a symbol duration.

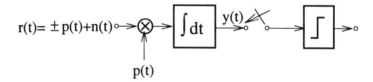

Figure 3.36. Correlation receiver.

rate, as will be discussed later. It is worth noting that the pulse shape can be arbitrary as long as the pulse energy is the same (Eq. 3.68).

Equation (3.72) suggests another implementation of the receiver, which is known as a *correlation receiver* (Fig. 3.36). In this system, the locally generated signal $p(t)$ is multiplied with and is synchronized with the incoming signal $r(t)$. The output signal at the sampling instant t_m is the same as that of the matched filter. An NRZ signaling system with I&D filter detection is a correlation receiver. The matched filter receiver is superior to the correlation receiver in an engineering sense, since it requires no local signal that is synchronized to the incoming signal.

3.3.6 Joint Optimization of the Transmit and Receive Filters

The above arguments deal with the receive filtering assuming that the transmit filtering is given. This section discusses the optimization of the transmit and the receive filter together. The results of the optimization can

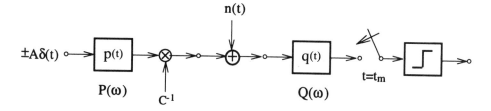

Figure 3.37. Joint optimization of transmit and receive filters.

be different depending on the assumed conditions. We assume that (i) the transmitted signal power is kept constant and (ii) the intersymbol interference is absent at the sampling instant. Under these conditions, the transmit and the receive filter combination that maximizes the signal to noise power ratio at the sampling instants will be given. Figure 3.37 shows the system to be optimized. From assumption (ii), the transfer functions $P(\omega)$ and $Q(\omega)$ are related as

$$P(\omega)Q(\omega) = H_I(\omega), \tag{3.77}$$

where $H_I(\omega)$ fulfills the Nyquist-I criterion, and $(h_I(t) \leftrightarrow H_I(\omega))$ takes the peak value at $t = t_m$; it follows that,

$$H_I(\omega) = |H_I(\omega)|e^{-j\omega t_m}. \tag{3.78}$$

The coefficient C^{-1} is introduced to keep the transmitted signal power constant irrespective of the transmit filter characteristics.

The transmitted power P_t is given as (Eq. 2.61.)

$$P_t = \frac{A^2}{2\pi T} \int_{-\infty}^{\infty} |P(\omega)|^2 \, d\omega, \tag{3.79}$$

where A is the pulse amplitude and T is the symbol duration. The coefficient C^{-1} is chosen as

$$C^{-1} \equiv \frac{1}{C} = \frac{A}{\sqrt{P_t}}. \tag{3.80}$$

At the sampling instant $(t = t_m)$, the receive filter output signal power becomes

$$S = \left| \frac{1}{2\pi} \int_{-\infty}^{\infty} A P(\omega) C^{-1} Q(\omega) e^{j\omega t_m} \, d\omega \right|^2. \tag{3.81}$$

With Eqs. (3.77) and (3.78), this is reduced to

$$S = \left| \frac{1}{2\pi} \int_{-\infty}^{\infty} AC^{-1} |H_I(\omega)| \, d\omega \right|^2. \tag{3.82}$$

The output noise power becomes

$$N = \frac{1}{2\pi} \int_{-\infty}^{\infty} G(\omega) |Q(\omega)|^2 \, d\omega, \tag{3.83}$$

where $G(\omega)$ is the noise power spectral density. With Eqs. (3.79) and (3.80) the signal to noise power ratio is given as

$$\frac{S}{N} = \frac{\left| \dfrac{A}{2\pi} \displaystyle\int_{-\infty}^{\infty} |H_I(\omega)| \, d\omega \right|^2}{\dfrac{1}{2\pi T} \displaystyle\int_{-\infty}^{\infty} |P(\omega)|^2 \, d\omega \dfrac{1}{2\pi} \displaystyle\int_{-\infty}^{\infty} G(\omega)|Q(\omega)|^2 \, d\omega}. \tag{3.84}$$

S/N is maximized when the denominator takes the minimum value. Using the Schwartz inequality (Eq. 3.64) with $X(\omega) = P(\omega)$, $Y(\omega) = \sqrt{G(\omega)} Q(\omega) e^{j\omega t_m}$, the minimum value is obtained when

$$P(\omega) = G(\omega)^{1/2} Q^*(\omega) e^{-j\omega t_m}. \tag{3.85}$$

Using Eqs. (3.77), (3.78), and (3.85) we get

$$|P(\omega)|^2 = G(\omega)^{1/2} |H_I(\omega)|, \tag{3.86a}$$

$$|Q(\omega)|^2 = G(\omega)^{-1/2} |H_I(\omega)|. \tag{3.86b}$$

The maximum S/N becomes

$$\left[\frac{S}{N} \right]_m = \frac{A^2 \left| \displaystyle\int_{-\infty}^{\infty} |H_I(\omega)| \, d\omega \right|^2}{\dfrac{1}{T} \left| \displaystyle\int_{-\infty}^{\infty} G(\omega)^{1/2} H_I(\omega) e^{j\omega t_m} \, d\omega \right|^2}. \tag{3.87}$$

Assuming a white noise with $G(\omega) = N_0/2 \, (-\infty < \omega < \infty)$, we have

$$|P(\omega)|, |Q(\omega)| \propto \sqrt{|H_I(\omega)|}$$

and

$$\left[\frac{S}{N}\right]_m = \frac{2A^2 T}{N_0} = \frac{2E_b}{N_0},$$

where E_b is the transmitted energy per bit. These results are the same as those for the previously discussed Nyquist-I signaling system.

3.3.7 The Optimum Receiver

The fact that the digital signal has a finite number of states makes possible a decision process with finite number of calculations. Finding the most probable message is possible, even if the received message is corrupted by noise and/or interference.

Let us consider the message expressed as $m_i = (m_{i1}, m_{i2}, m_{i3} \ldots, m_{iN})$, where m_{in} is the digital signal to be transmitted. The transmitted signal is given as

$$s_i(t) = \sum_{n=1}^{N} a_{in} h(t - nT), \qquad 0 < t < T_m, \ i = 1, 2, \ldots, I_m, \qquad (3.88)$$

where a_{in} takes one of the L levels corresponding to m_{in}, $h(t)$ is the impulse response of the transmit filter, T is the symbol duration, N is the message length, I_m is the number of different messages, and T_m is the time length of the message. If the message length is N, then $I_m = L^N$. In this case we can choose the most probable message from these possible candidates by L^N calculations of the objective function for decision.

The received message signal can be expressed as

$$r(t) = u_i(t) + n(t), \qquad 0 < t < T_0 \equiv T_m + \tau_0, \qquad (3.89)$$

where

$$u_i(t) = \sum_{n=1}^{N} a_{in} g(t - nT_s). \qquad (3.90)$$

$g(t)$ denotes the received pulse waveform, which is produced by applying the pulse waveform $h(t)$ to the channel; that is, $g(t) = h(t) * c(t)$, where $c(t)$ is the impulse response of the channel, assumed to last for time τ_0, and $n(t)$ denotes the noise added in the channel.

Given the received signal $r(t)$, our goal is to find the optimum receiver, which yields the lowest error probability for the received candidate messages $\hat{m}_i$. Figure 3.38 shows the system under consideration.

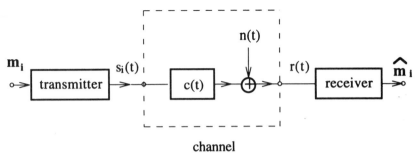

channel

Figure 3.38. Digital transmission system.

We proceed further with our discussion by referring to [4] and [5]. Complex signals are treated in this section. Here we assume that the channel impulse response is known. (Estimating the channel impulse response is another problem, which will be discussed later.) Further, we assume that the probability of the occurrence of each message is known and that the noise is Gaussian with zero mean.

When a message m_i is sent, the received signal $r(t)$ is a Gaussian variable with the mean value of $u_i(t)$. In order to analyze the stochastic process it is convenient to represent the received signal with independent variables. The method of treatment is the Karhunen–Loève series expansion [10].

We expand $n(t)$ in a series of the form

$$n(t) = \lim_{N \to \infty} \sum_{n=1}^{N} n_n f_n(t), \qquad (3.91)$$

where n_n is the expansion coefficient and $\{f_n\}$ is a set of orthonormal functions on the interval $(0, T_0 \equiv T_m + \tau_0)$; that is,

$$\int_0^{T_0} f_n(t) f_m^*(t) \, dt = \begin{cases} 1, & n = m \\ 0, & n \neq m. \end{cases} \qquad (3.92)$$

Multiplying both sides of Eq. (3.91) by $f_m^*(t)$ and using the above relation, we have

$$n_m = \int_0^{T_0} n(t) f_m^*(t) \, dt. \qquad (3.93)$$

We impose the condition that the coefficients $\{n_m\}$ are mutually uncorrelated; that is,

$$\langle n_m n_n^* \rangle = \begin{cases} |\sigma_n|^2, & m = n \\ 0, & m \neq n. \end{cases} \qquad (3.94)$$

Consequently the coefficients n_n are independent Gaussian variables with variance σ_n^2. This condition can be fulfilled if the orthonormal functions $f_n(t)$ satisfy the following integral equation [10]:

$$\int_0^{T_0} R(t,s) f_n(s)\, ds = |\sigma_n|^2 f_n(t), \tag{3.95}$$

where $R(t,s)$ is the correlation function defined as

$$R(t,s) = \langle n(t) n^*(s) \rangle = \left\langle \sum_{m=1}^{\infty} n_m f_m(t) \sum_{n=1}^{\infty} n_n^* f_n^*(s) \right\rangle. \tag{3.96}$$

We can confirm that the random variables n_n are uncorrelated, as follows. Under the condition that Eq. (3.95) holds, using Eq. (3.93) we have

$$\langle n_n n_m^* \rangle = \left\langle \int_0^{T_0} n(t) f_n^*(t)\, dt \int_0^{T_0} n^*(s) f_m(s)\, ds \right\rangle$$

$$= \int_0^{T_0} \int_0^{T_0} R(t,s) f_n^*(t) f_m(s)\, dt\, ds$$

$$= \int_0^{T_0} f_n^*(t) |\sigma_m|^2 f_m(t)\, dt$$

$$= \begin{cases} |\sigma_n|^2, & n = m \\ 0, & n \neq m. \end{cases} \tag{3.97}$$

In the language of integral equations, the functions $f_k(s)$ are eigen (or characteristic) functions and the values $|\sigma_k|^2$ are eigen (or characteristic) values.

In general, it is difficult to solve Eq. (3.95) to get the eigenfunctions $f_k(t)$ and the eigenvalues $|\sigma_k|^2$, unless $n(t)$ is white noise. By assuming white noise, the generality of the argument holds through the introduction of the concept of a noise-whitening filter, described later.

For the white noise, we have

$$R(t,s) = \frac{N_0}{2} \delta(t-s), \tag{3.98}$$

where $\delta(\cdot)$ is the delta function and N_0 is the noise power density. Applying this equation to Eq. (3.95), we get the result $(N_0/2) f_k(t) = |\sigma_k|^2 f_k(t)$. This means that any orthonormal function $f_k(t)$ can be used with

$$|\sigma_k|^2 = \frac{N_0}{2} \equiv \sigma_0^2.$$

as its eigenvalues.

Now we expand the signals $u_i(t)$ in terms of orthonormal functions $f_k(t)$. The dimension of the signals $u_i(t)$ is N. Thus, the signals can be expanded in terms of the orthonormal functions as [4]

$$u_i(t) = \sum_{n=1}^{N} u_{in} f_n(t), \qquad i = 1, 2, \ldots, I_m, \qquad (3.99)$$

where the expansion coefficient u_{in} is given as

$$u_{in} = \int_0^{T_0} u_i(t) f_n^*(t)\, dt \qquad (3.100)$$

The noise can be expressed as

$$n(t) = n_N(t) + n_0(t), \qquad (3.101)$$

where

$$n_N(t) = \sum_{n=1}^{N} n_n f_n(t), \qquad (3.102)$$

$$n_0(t) = \sum_{n=N+1}^{\infty} n_n f_n(t), \qquad (3.103)$$

We can see that the noise term $n_0(t)$ is orthogonal to the signals $u_i(t)$, that is

$$\int_0^{T_0} n_0(t) u_i^*(t)\, dt = 0.$$

Thus the noise term $n_0(t)$ is irrelevant to the decision-making at the receiver, and can be discarded.

We can express the received signals as

$$r(t) = \sum_{n=1}^{N} r_n f_n(t), \qquad (3.104)$$

where

$$r_n = \int_0^{T_0} r(t) f_n^*(t)\, dt$$

$$= u_{in} + n_n, \qquad i = 1, 2, \ldots, I_m,\ n = 1, 2, \ldots, N. \qquad (3.105)$$

The r_n are independent Gaussian variables with the mean of u_{in}. Since there is one-to-one correspondence between the received signal $r(t)$ and the coefficients r_n, we express the received signal as a vector by $r = (r_1, r_2, \ldots, r_N)$.

Given that $r = r'$, the conditional probability of the correct decision of m_i becomes

$$P(C|r = r') = P(m_i|r = r'), \tag{3.106}$$

where $P(C|r = r')$ is the conditional probability of making the correct decision given that $r = r'$. $P(m_i|r = r')$ is the conditional probability that the message m_i was sent, given that $r = r'$ was received. Our task is to maximize $P(m_i|r = r')$ or to find the message m_i for which $P(m_i|r = r') > P(m_j|r = r')$ for all $j \neq i$. The probability $P(m_i|r = r')$ is called the a posteriori probability of m_i. Our receiver is the maximum a posteriori probability detector.

We use Bayes' mixed rule to get the a posteriori probabilities as

$$P(m_i|r = r') = \frac{P(m_i)P(r = r'|m_i)}{P(r = r')}. \tag{3.107}$$

$P(r = r'|m_i)$ is the conditional probability that $r = r'$, given that the message m_i was sent. The probability $P(r = r')$ is common to all the decision candidates, so it can be ignored. Our task is to find a candidate message $\hat{m}_i$, which maximizes the numerator of the right-hand side of Eq. (3.107). We denote this term by

$$J_i = P(\hat{m}_i)P(r = r'|\hat{m}_i), \quad i = 1, 2, \ldots, I_m. \tag{3.108}$$

Since $n_n(= r_n - u_{in})$ are Gaussian variables,

$$P(r_n = r'_n|\hat{m}_i) = \frac{1}{(2\pi\sigma_0^2)^{1/2}} \exp\left(-\frac{|r'_n - \hat{u}_{in}|^2}{2\sigma_0^2}\right),$$

and since they are independent, the joint probability function for r is

$$P(r|\hat{m}_i) = \frac{1}{\displaystyle\prod_{n=1}^{N} (2\pi\sigma_0^2)^{1/2}} \exp\left(\sum_{n=1}^{N} -\frac{|r_n - \hat{u}_{in}|^2}{2\sigma_0^2}\right), \tag{3.109}$$

where $\hat{u}_{in}$ corresponds to $\hat{m}_i$.

Since

$$|r_n - \hat{u}_{in}|^2 = |r_n|^2 - 2\mathrm{Re}\{r_n\hat{u}_{in}^*\} + |\hat{u}_{in}|^2, \tag{3.110}$$

then using Eqs. (3.92), (3.99), and (3.104) we have

$$\sum_{n=1}^{N} |r_n|^2 = \int_0^{T_0} |r(t)|^2 \, dt, \tag{3.111}$$

$$\sum_{n=1}^{N} r_n \hat{u}_{in}^* = \int_0^{T_0} r(t)\hat{u}_i^*(t) \, dt, \tag{3.112}$$

and

$$\sum_{n=1}^{N} |\hat{u}_{in}^*|^2 = \int_0^{T_0} |\hat{u}_i(t)|^2 \, dt. \tag{3.113}$$

Thus we have

$$\sum_{n=1}^{N} |r_n - \hat{u}_{in}|^2 = \int_0^{T_0} |r(t) - \hat{u}_i(t)|^2 \, dt. \tag{3.114}$$

If $P(m_i)$ are the same, our task can be reduced to finding a candidate message which minimizes the squared error, $\int_0^{T_0} |r(t) - \hat{u}_i(t)|^2 \, dt$.

Since the logarithmic function is monotonic, maximizing J_i is equivalent to maximizing $\ln(J_i)$, where $\ln(\cdot)$ is the natural logarithmic function. Using this fact, we get

$$\ln(J_i) = Q_i + a, \qquad i = 1, 2, \ldots, I_m, \tag{3.115}$$

where

$$Q_i = \ln\{P(m_i)\} - \frac{1}{2\sigma_0^2} \int_0^{T_0} |r(t) - \hat{u}_i(t)|^2 \, dt, \tag{3.116a}$$

$$a = -\frac{N}{2} \ln(2\pi\sigma_0^2). \tag{3.116b}$$

The constant term is common to all the candidate messages and can be ignored in decision-making. Multiplying both sides of Eq. (3.116a) by $2\sigma_0^2(=N_0,)$, we have

$$U_i = N_0 \ln\{P(m_i)\} - \int_0^{T_0} |r(t) - \hat{u}_i(t)|^2 \, dt, \qquad i = 1, 2, \ldots, I_m. \tag{3.117}$$

where $U_i = 2\sigma_0^2 Q_i$. We can rewrite Eq. (3.117) as

$$U_i = 2\mathrm{Re}\left[\int_0^{T_0} r(t)\hat{u}_i^*(t) \, dt\right] - e_i - c + N_0 \ln\{P(m_i)\}, \tag{3.118}$$

where

$$e_i = \int_0^{T_0} |\hat{u}_i(t)|^2 \, dt,$$

$$c = \int_0^{T_0} |r(t)|^2 \, dt.$$

The term e_i represents the energy of the candidate signal $\hat{u}_i(t)$; the term c is common to all the candidates and can be discarded.

Thus, the optimum receiver is configured with a bank of correlators as shown in Fig. 3.39(a), where the term b_i equals $\{N_0 \ln[P(m_i)] - e_i\}/2$. Since the sampled output of the matched filter is equivalent to the output of the correlator, we have another configuration of the optimum receiver as shown in Fig. 3.39(b).

Let us recall that $u_i(t)$ is expanded in terms of the orthonormal functions as in Eq. (3.99). Then we have

$$\int_0^{T_0} r(t)\hat{u}_i^*(t) \, dt = \sum_{n=1}^{N} \hat{u}_{in}^* \int_0^{T_0} r(t)f_n^*(t) \, dt$$

$$= \sum_{n=1}^{N} \hat{u}_{in}^* r_n, \qquad i = 1, 2, \ldots, I_m. \tag{3.119}$$

This relation suggests another configuration of the optimum receiver as shown in Fig. 3.40.

3.3.8 The Maximum-likelihood Receiver and the Viterbi Algorithm

The number of calculations to find the optimum candidate message $\hat{m}_i$ increases exponentially with the message length, N, and is proportional to $I_m = L^N$, where L is the number of digital signal states. A brute force method is ineffective for long messages. Instead, the Viterbi algorithm or dynamic programming is used to decrease drastically the number of calculations.

In practice, it is difficult to know $P(m_i)$, the probability of the occurrence of the message m_i, so we may assign equal probabilities to all the messages. The optimum receiver under this condition is called the maximum-likelihood receiver.

The likelihood is maximized by finding the candidate message $\hat{m}_i$, or equivalently the candidate signal $\hat{u}_i(t)$, that minimizes the integral

$$V_i = \int_0^{T_0} |r(t) - \hat{u}_i(t)|^2 \, dt. \tag{3.120}$$

Minimizing this integral corresponds to maximizing Eq. (3.117). Expanding

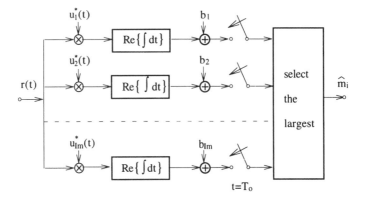

(a) correlator receiver

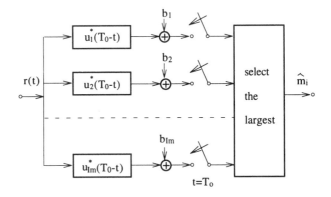

(b) matched filter receiver

Figure 3.39. Optimum receiver implementation.

the function under the integral sign, the objective function to be maximized can be expressed as

$$V_i = -2\text{Re}\left[\int_0^{T_0} r(t)\hat{u}_i^*(t)\,dt\right] + \int_0^{T_0} |\hat{u}_i(t)|^2\,dt, \qquad (3.121)$$

where the term $\int_0^{T_0}|r(t)|^2\,dt$ is ignored since it is irrelevant to the decision-making. Using Eq. (3.90), Eq. (3.121) can be rewritten as

$$V_i = -2\text{Re}\left[\sum_{n=1}^{N} \hat{a}_{in}^* r_n\right] + \sum_{n=1}^{N} \sum_{m=1}^{N} \hat{a}_{in}\hat{a}_{im}^* q_{n-m}, \qquad (3.122)$$

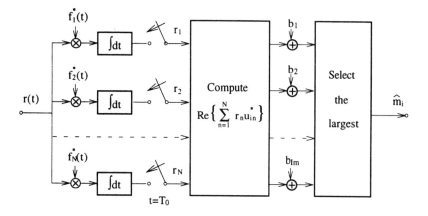

(a) correlator receiver

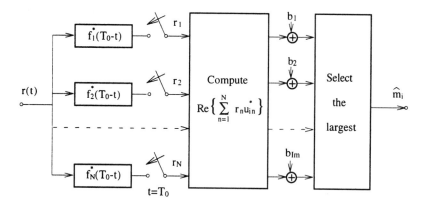

(b) matched filter receiver

Figure 3.40. Other configuration of the optimum receiver.

where

$$r_n = \int_0^{T_0} r(t) g^*(t - nT_s) \, dt, \qquad n = 1, 2, \ldots, N \qquad (3.123)$$

and

$$q_{n-m} = \int_0^{T_0} g(t - nT_s) g^*(t - mT_s) \, dt. \qquad (3.124)$$

For simplicity, we omit the suffix i that represents the ith message.

We assume

$$q_{n-m} = 0, \qquad |n - m| > M. \tag{3.125}$$

This assumption means that the symbols a_m and a_n never interfere with each other for $|n - m| > M$. Equation (3.122) can be rewritten (omitting the suffix i) as

$$V = \sum_{n=1}^{j-1} K_n + K_j + L_{N-j} \qquad (j = 1, 2, 3, \ldots, N), \tag{3.126}$$

where

$$K_n = -2\text{Re}[\hat{a}_n r_n] + 2\text{Re}\left[\hat{a}_n \sum_{m=n+1}^{M+n} \hat{a}_m^* q_{n-m}\right] + |\hat{a}_n|^2 q_0, \tag{3.127}$$

$$L_{N-j} = -2\text{Re}\left[\sum_{n=j+1}^{N} \hat{a}_n r_n\right] + \sum_{n=j+1}^{N} \sum_{m=j+1}^{N} \hat{a}_n \hat{a}_m^* q_{n-m}. \tag{3.128}$$

We have applied Eq. (3.125) to the second term of Eq. (3.127) and $\hat{a}_i = 0$ for $i > N$ is understood. The reason why we express the objective function V (metric) as Eq. (3.126) will be understood through description in the following.

Upon receiving the signal r_1, we start the calculation of the objective function V (metric): the values of K_1 are calculated for L^{M+1} candidate message sequences $\{\hat{a}_1, \hat{a}_2, \ldots, \hat{a}_{M+1}\}$. We can divide these L^{M+1} sequences into L^M groups, corresponding to the possible choices of $(\hat{a}_2, \hat{a}_3, \ldots, \hat{a}_{M+1})$. Each group consists of L sequences that differ only in the value of $\hat{a}_1$. For each group we choose the state of $\hat{a}_1$ which minimizes K_1 and the other $L - 1$ states of $\hat{a}_1$ are discarded. The chosen $\hat{a}_1$ and the corresponding minimum K_1 value, denoted as $K_{1\,\text{min}}$, are stored for each sequence group. Let $\hat{\sigma}_n$ denote the candidate message sequence $(\hat{a}_{n+1}, \hat{a}_{n+2}, \ldots, \hat{a}_{n+M})$. Thus, $K_{1\,\text{min}} = \min_{\hat{a}_1} K_1 (r_1; \hat{a}_1, \hat{\sigma}_1)$. Discarding the $L - 1$ states of $\hat{a}_1$ for each $\hat{\sigma}_1$ is justified by the fact that they cannot be a candidate later. This is true since their contributions to the metric for the signal sequences are limited within $\hat{\sigma}_1$.

At this stage, if all the sequence groups $\hat{\sigma}_1$ have the same value, then we decide on $\hat{a}_1$. Otherwise, the decision on $\hat{a}_1$ is deferred to a later time.

Receiving the signal r_2, the K_2 values (called branch metrics) are calculated for the L candidates $\hat{a}_2$. The L^M sequences $\hat{\sigma}_2$ are calculated in the same way as $\hat{a}_1$ and $\hat{\sigma}_1$ in the previous step of the algorithm. The path metric, that is $K_{2\,\text{min}} = \min_{\hat{a}_2}[K_{1\,\text{min}} + K_2(r_2; \hat{a}_2, \hat{\sigma}_2)]$ is selected for each $\hat{\sigma}_2$ with the corresponding $\hat{a}_2$ and the others are discarded. The selected $\hat{a}_2$ and the path metrics $K_{2\,\text{min}}$ are stored for each $\hat{\sigma}_2$. If all the sequences $\hat{\sigma}_2$ have the same $\hat{a}_1$ or $(\hat{a}_1, \hat{a}_2)$ value, a decision is made on these values. Otherwise the decision-making is deferred to a later time.

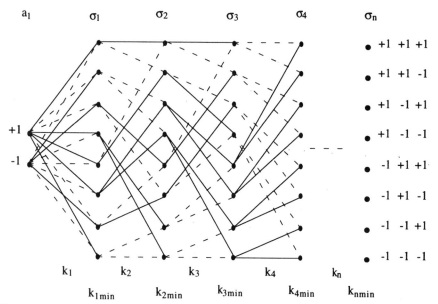

Figure 3.41. Lattice diagram of the Viterbi algorithm. Solid lines and broken lines show the surviving and the discarded path, respectively.

Receiving a new signal r_n, we proceed with the calculation and the selection in the same way: we have L^{M+1} possible message sequences $(\hat{a}_n, \hat{a}_{n+1}, \ldots, \hat{a}_{n+M})$ which are the continuations of the L^M surviving sequences of the previous stage of the process. We divide these sequences into L^M groups, each group containing L sequences with the same $\hat{\sigma}_n$ value, differing only in $\hat{a}_n$, and calculate the branch metric K_n. From each group we select the $\hat{a}_n$ which minimizes $[K_{(n-1)\min} + K_n(r_n; \hat{a}_n, \hat{\sigma}_n)]$, discarding the other $L-1$ sequences. The selected $\hat{a}_n$ together with the precursors $(\hat{a}_1, \hat{a}_2, \ldots, \hat{a}_{n-1})$ and the branch metric $K_{n\min}$ are stored. At the end of the procedure we can make decision on a sequence which shows the minimum path metric.

In each stage we compute K_j for L^{M+1} sequences. This means NL^{M+1} computations for detecting N symbols.

The lattice diagram of the Viterbi algorithm is shown in Fig. 3.41 for an example ($M = 3$). The succession of the states $\hat{\sigma}_n$ has one-to-one correspondence to the symbol sequence of $\{a_n\}$. Solid lines show the surviving path and dotted lines represent the discarded path. When all the surviving paths derive from the same state $\hat{\sigma}_n$, that is, when merging occurs, we can make a decision on that symbol $\hat{a}_n$ and its precursors. The merging occurs stochastically depending on the signal sequence and noise. If merging does not occur, the decision-making is made at the end of the received signal. The total number of calculations on the metrics is NL^{M+1}, in contrast to L^N with the brute force method, for $N \gg 1$. Selecting one candidate from the L states of a symbol for a sequence $\hat{\sigma}_n$ at every stage of process and discarding the others become the key of the Viterbi algorithm in reducing the number of calculations.

3.3.9 The Optimum Receiver for Signals without Intersymbol Interference

Consider signals for which the matched filter output shows no intersymbol interference. Since the output of a matched filter is equivalent to that of a correlation receiver at the sampling instants, the signal $g(t)$ fulfills $q_{n-m} = 0$ for $n \neq m$ (Eqs. 3.124 and 3.76). This means that there is no correlation between the sampled signals r_n.

The objective function V_i given by Eq. (3.122) for the optimum receiver becomes

$$V_i = -2\text{Re}\left[\sum_{n=1}^{N} \hat{a}_{in}^* r_n\right] + \sum_{n=1}^{N} |\hat{a}_{in}|^2 q_0, \qquad i = 1, 2, \ldots, I_m, \qquad (3.129)$$

where we have returned to the use of suffix i. This result shows that decision-making on a symbol $\hat{a}_{in}$ never affects decision-making on other symbols. Thus, the symbol-by-symbol decision becomes the optimum (equal $P(m_i)$ is assumed) because there is no correlation between the different symbols at the sampled output of the matched filter or correlator. NRZ signaling and transmit pulse shaping with a filter that has a transfer function of the square root of the Nyquist-I characteristics fall into this case.

Instead of examining the objective function of Eq. (3.129), we can adopt threshold detection as follows. Threshold detection is made by classifying the received signals r_n into one of the regions whose boundaries are the threshold values (Fig. 3.42). The detection algorithm is equivalent to determining a level that is the nearest to r_n, as

$$d_m = \min_{\hat{a}_{in}} \left| \int_0^{T_0} [r(t)g^*(t-nT) - \hat{a}_{in}|g(t-nT)|^2] \, dt \right|$$

$$= \min_{\hat{a}_{in}} |r_n - \hat{a}_{in}q_0|. \qquad (3.130)$$

Taking the square of d_m, we have

$$d_m^2 = \min_{\hat{a}_{in}} \{|r_n|^2 - 2\text{Re}[\hat{a}_{in}^* r_n q_0] + |\hat{a}_{in}|^2 q_0^2\}. \qquad (3.131)$$

The term $|r_n|^2$ is irrelevant to decision-making. The second and third terms are the same as those in Eq. (3.129) except for the constant q_0.

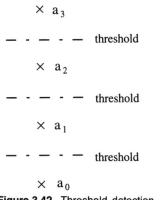

Figure 3.42. Threshold detection.

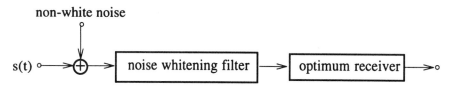

Figure 3.43. Optimum receiver for nonwhite noise.

The thresholds must be drawn at the midpoints of the neighboring signal levels $a_{in}q_0$.

3.3.10 Noise Whitening Filter

When the input noise is non-white noise, we use a noise whitening filter to achieve a flat noise spectrum and we can apply the optimum receiver for the signal and noise at the output of the filter (Fig. 3.43).

3.4 SYNCHRONIZATION

Receive symbol timing and frame timing must be synchronized with the transmitted signal. This section describes how to establish this synchronization.

3.4.1 Symbol Timing Recovery

We do not send the symbol timing signal along with the information signals permanently, since this would decrease the capacity of transmission of the information signals. Instead, we send a short signal only at the start of the transmission for the purpose of aiding the receiver in generating the timing signal. After receiving the timing signal, there are two strategies for maintaining the timing signal: one is to use a sufficiently stable oscillator, and the other is to continue extracting the timing signal from the information signals. The former method can be adopted rather easily for low-speed transmission, since a timing error owing to the frequency instability of the oscillator becomes small relative to the symbol duration.

The latter method is used widely. The principle of this method is based on detecting the change in the signal level or polarity, which occurs according to the change in data signals. Although the change in signal levels takes place at random, its average timing is precise since it is synchronized to the transmit timing generator. Figure 3.44 shows a block diagram of the symbol timing regenerator. The demodulated signal is applied to a filter that emphasizes the signal level change. The filtered signal is full-wave rectified. The rectifier is necessary to regenerate the timing signal components, since the level change occurs at random in opposite directions, resulting in the cancellation of the timing signal.

The tank circuit is a resonator that is tuned to the clock signal frequency. It is intended to suppress the jitter of the regenerated clock timing signal. The Q-value of the resonator is chosen by a compromise between the performance for low jitter and short rise-up time; a high Q-value decreases the jitter but increases the rise-up time.

When the noise is absent, we can design the prerectification filter so that jitter never occurs. Jitter never occurs if the filtered waveform crosses the zero level at the exact midpoint of the symbol times. That is, the impulse response $h(t)$ of the system consisting of the filter with other filters in the channel must meet the condition

$$h(t) \sum_{n=-\infty}^{\infty} \delta[t - (n + \tfrac{1}{2})T] = 0, \tag{3.132}$$

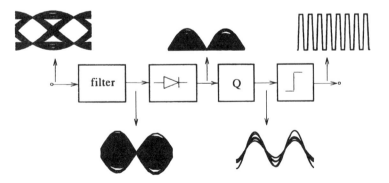

Figure 3.44. Symbol timing regeneration circuit and its waveforms.

where T denotes symbol duration. Taking the Fourier transform of Eq. (3.132), and proceeding in a similar manner to Eqs. (3.15)–(3.19), we get

$$H\left(\frac{\omega_0}{2} + x\right) = H^*\left(\frac{\omega_0}{2} - x\right), \tag{3.133}$$

where $\omega_0 = 2\pi/T$ and we have assumed that the bandwidth is limited within $0 \sim \omega_0$. The left and right sides of Eq. (3.133) must be conjugate mirror images of each other with respect to half of the symbol frequency. The computer-simulated waveforms in the system with prerectification filter satisfying Eq. (3.133) are shown in Fig. 3.44.

We cannot conclude that the prerectification filter with Eq. (3.133) is the best in the presence of noise. The optimum clock signal regeneration may depend on the signal to noise power ratio at the demodulator output. There are many other clock signal regeneration systems but these are beyond the scope of this book.

3.4.2 Frame Synchronization

Digital signals are transmitted collectively in a unit of frame, as shown in Fig. 3.45. A framed signal consists of the information signals and some extra signals, one of which is the frame synchronization (SYNCH) word. The SYNCH word is used for a receiver to determine the starting point of the information signals. Supplied with the received data signal and the regenerated symbol timing signal, the receiver uses a correlator or a matched filter to detect the SYNCH word. For this purpose, a code with a sharp correlation function is desirable. A failure in the SYNCH word detection due to symbol errors makes it impossible to receive all the information signals. There are two modes of failure: (i) the receiver misses the detection of the SYNCH word and (ii) the receiver detects an erroneous SYNCH word. We can control the failures by allowing some degree of mismatch between the received and local SYNCH word. By allowing higher degree of mismatch, the former failures can be decreased. On the other hand, this increases the latter failures. Thus, we must have a trade-off between the two failures by taking symbol error probability into consideration. A longer SYNCH word can decrease both failures. However, it decreases the efficiency of the channel.

For a system where the framed signals are transmitted periodically, we

Sync.	Information

Figure 3.45. Framed signal format.

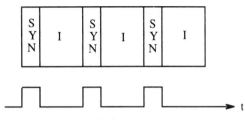

window

Figure 3.46. Framed signal format with periodic synchronization word.

can use a shorter SYNCH word. This is because the second type of failure can be decreased by introducing a periodic window for detecting the SYNCH word in synchronism with the reception of the framed signals, as illustrated in Fig. 3.46. Since the SYNCH word is short, it is dangerous to decide the SYNCH timing by only one-time detection of the SYNCH word. Therefore, the receiver makes a decision on the timing only when the SYNCH words are detected successively several times, and after this the receiver is in a locked mode. The randomness of data signals justifies the approach of the method. We must pay attention to maintaining the frame timing and fast detection of the loss of synchronization. The receiver recognizes the lack of synchronization after failing to detect the SYNCH word several times successively. Then the receiver changes into a hunting mode to reestablish the synchronization. The frame synchronizing circuit opens the window for the SYNCH word correlator to output the frame timing signals. This operation was explained previously. The principle of this frame synchronizing method is based on the "flywheel effect" which is produced by successive detections of the SYNCH word.

3.5 SCRAMBLING

In contrast to analog signal scrambling, the scrambling of digital signals is easily carried out using logic circuits or a processor. The purpose of scrambling is to secure the digital transmission. For example, consider a dc cut-off channel case; the digital signal consisting of a constant sequence of "0" or "1" data cannot be transmitted, since the signal becomes a dc signal. Scrambling prevents such a situation by randomizing the transmitted data. A long sequence of "0" or "1" data often occurs in data terminals. Another case is a channel with an automatic equalizer that requires random data for a stable operation. Standard scrambling is recommended in data communication systems.

Another purpose of scrambling is to protect the signal transmission from eavesdropping by a third party. The degree of protection can be made sufficiently high using sophisticated digital scrambling methods. Figure 3.47

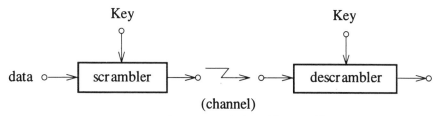

Figure 3.47. Signal transmission with scrambling.

shows the block diagram of a communication system with scrambling. The scrambling algorithm may or may not be open to everyone. Even if the scrambling algorithm and hence the descrambling algorithm is open, the key must be known for descrambling. Thus, the number of keys must be large enough for protection against a trial-and-error or a systematic attack. The data encryption standard (DES), described later, offers 2^{56} ($\approx 7 \times 10^{16}$) different keys. Key management, such as key distribution, becomes an important issue.

There are several different ways of scrambling: addition of pseudo-noise (PN), permutation, and transformation. The addition of pseudo-noise literally means that random data is added to the transmitted signal. In permutation scrambling the unit of the permutation is a bit or a block of bits (word) and it is performed within a frame or over several frames. Transformation is performed in a unit of a word.

For synchronization between scrambler and descrambler there are synchronized and self-synchronized systems. Figures 3.48(a) and 3.48(b) show the synchronized and the self-synchronized scrambling system, respectively.

The two methods of frame synchronization described earlier can be applied to synchronize the descrambler to the scrambler. The method where the SYNCH word is transmitted only at the start of the transmission can enhance the security of communication since a third party who missed the SYNCH word cannot synchronize his descrambler to the scrambled signal.

The operation of the self-organized scrambling system is explained as follows. The input data is scrambled (ciphered) by the output of a RAM (random access memory). The ciphered data is transmitted into the channel and fed at the same time into a shift register. The data contained in the shift register determine a RAM address. The RAM outputs random data corresponding to this address value. The output of the RAM is possibly a highly nonlinear function of the input data. Thus, the scrambler can be considered a nonlinear feedback circuit. The received data is fed into the shift register and descrambled by the output of the RAM. The contents of the RAM in the scrambler and in the descrambler are the same.

The synchronization of the descrambler to the scrambler is automatically

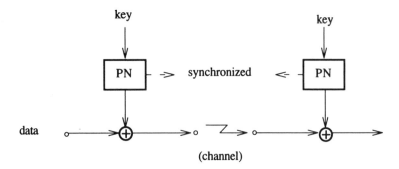

(a) synchronized system

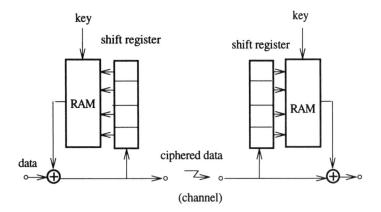

(b) self-synchronized system.

Figure 3.48. Scrambling/descrambling system.

established after the first transmitted signal reaches the end of the shift register: the addresses are always the same for the scrambler and the descrambler RAMs, as long as transmission error is absent. The penalty we must pay for using the self-synchronized scrambling system is the error propagation effect. A transmission error in the data affects the descrambler while the data errors are in the shift register. Taking into consideration that the RAM outputs "0" or "1" with equal probability for random data, we have on average number of error of $N/2$ at the scrambler output, where N is the length of the shift register. In the synchronized scrambling system, we never have such an error propagation effect. This effect results in deterioration of the sensitivity of the receiver.

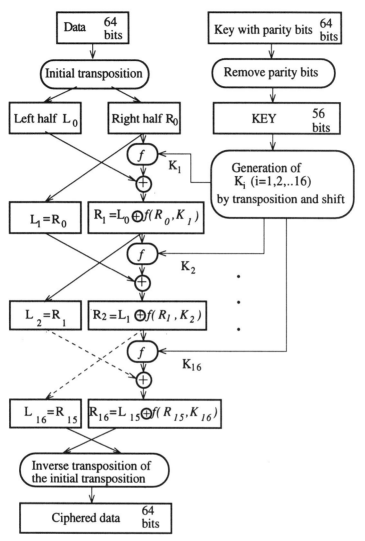

Figure 3.49. DES scrambler.

The well-known scrambling algorithm DES [11] is shown in Fig. 3.49. The data are processed in a unit of 64 bits. The input data is subjected to an initial transposition. The transposed data are divided into left- and right-hand sides. The right-hand side data become left-hand side data of the next stage and are also substituted using the subkey of 48 bits. The substituted data are added to the left-hand side data to produce the right-hand side data. This process is repeated sixteen times with the subkeys K_i ($i = 1, 2, \ldots, 16$). The final data are transposed in the opposite way to the initial transposition. The key consists of 56 bits and 8 parity bits. The subkeys are generated by substituting and shifting the key.

REFERENCES

1. W. R. Bennett and J. R. Davey, *Data Transmission*, McGraw-Hill, New York, 1965.
2. M. Schwartz, W. R. Bennett, and S. Stein, *Communication Systems and Techniques*, McGraw-Hill, New York, 1966.
3. R. W. Lucky, J. Saltz, and E. J. Weldon, Jr., *Principles of Data Communication*, McGraw-Hill, New York, 1968.
4. B. P. Lathi, *Modern Digital and Analog Communication Systems*, in HRW Series in Electrical and Computer Engineering, Holt, Rinehart and Winston, New York, 1983.
5. J. G. Proakis, *Digital Communications*, 3rd ed., McGraw-Hill, New York, 1995.
6. J. G. Proakis and M. Salehi, *Communication Systems Engineering*, Prentice-Hall, Englewood Cliffs, N.J., 1994.
7. E. A. Lee and D. G. Messerschmitt, *Digital Communication*, 2nd ed., Kluwer Academic Publishers, Norwell, Mass., 1994.
8. S. Haykin, *Communication Systems*, 3rd ed., Wiley, New York, 1994.
9. J. M. Wozencraft and I. M. Jacobs, *Principles of Communication Engineering*, Wiley, New York, 1965.
10. W. B. Davenport, Jr. and W. L. Root, *An Introduction to the Theory of Random Signals and Noise*, IEEE Press, New York, 1987.
11. A. S. Tannenbaum, *Computer Networks*, Prentice-Hall, Englewood Cliffs, N.J., 1981.

4

MOBILE RADIO CHANNELS

A land mobile radio channel is characterized by out-of-sight communication to/from a moving terminal; it becomes a multipath propagation channel with fast fading. The characteristics of the mobile radio channel introduce new challenges for the designer of mobile communication systems: for example, requirements for isotropic directivity antenna, choice of appropriate carrier frequency, and techniques for stable transmission under the condition of fast fading.

Wave propagation in the multipath channel depends on the actual environment, including factors such as antenna height, profile of buildings, roads, and terrain. Therefore, we must describe mobile radio channels in a statistical way.

Received signal power P_r is expressed as

$$P_r = L_C G P_t, \tag{4.1}$$

where P_t is the transmitted signal power, G denotes antenna gain, and L_C represents the propagation loss in the channel.

Wave propagation in a mobile radio channel is characterized by three aspects: path loss, shadowing, and fast fading.

$$L_C = L_P L_S L_F, \tag{4.2}$$

where L_P, L_S, and L_F denote path loss, shadowing loss, and fading loss, respectively (Fig. 4.1). The path loss L_P is an average propagation loss over wide areas. It is determined by macroscopic parameters, such as the distance between the transmitter and receiver, the carrier frequency, and land profile. The shadowing loss L_S represents variation of the propagation loss in a local area (several tens of meters). The shadowing is caused by variation of propagation conditions due to buildings, roads, and other obstacles in a relatively small area. The fast fading L_F is due to the motion of the terminal in a standing wave that consists of many diffracted waves; it represents the microscopic aspect of the channel on the order of the wavelength.

97

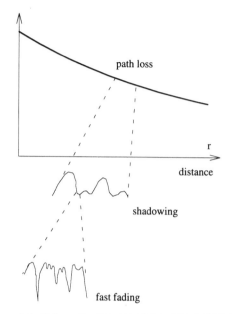

Figure 4.1. Schematic diagram of the propagation loss.

4.1 PATH LOSS

The simplest formula for path loss is

$$L_P = Ar^{-\alpha}, \tag{4.3}$$

where A and α are propagation constants and r is the distance between the transmitter and the receiver. α takes a value of $3 \sim 4$ in a typical urban area. For prediction of the propagation constants, many nomographs are obtained through propagation measurements [1–3]. The Okumura curves [1] are famous for practical use. Based on the Okumura curves, Hata [2] (Copyright © IEEE 1980) presented an empirical formula for prediction of path loss. The results are cited here.

(a)   URBAN AREA

$$L_{PU}(\text{dB}) = 69.55 + 26.16 \log f_c - 13.82 \log h_b - a(h_m)$$
$$+ (44.9 - 6.55 \log h_b) \log R, \tag{4.4}$$

where $L_P(\text{dB}) = -10 \log L_P$ and $a(h_m)$ is a correction factor for vehicular antenna height, given as follows:

Small and medium-size city:

$$a(h_m) = (1.1 \log f_c - 0.7)h_m - (1.56 \log f_c - 0.8).$$

Large city:

$$a(h_m) = \begin{cases} 8.29(\log 1.54h_m)^2 - 1.1; & f_c \le 200 \text{ MHz} \\ 3.2(\log 11.75h_m)^2 - 4.97; & f_c \ge 400 \text{ MHz}. \end{cases}$$

(b) SUBURBAN AREA

$$L_{PS}(\text{dB}) = L_P(\text{dB; urban area}) - 2[\log(f_c/28)]^2 - 5.4. \tag{4.5}$$

(c) OPEN AREA

$$L_{PO}(\text{dB}) = L_P(\text{dB; urban area}) - 4.78(\log f_c)^2 + 18.33 \log f_c - 40.94, \tag{4.6}$$

where f_c = frequency (MHz) $(150 \sim 1500 \text{ MHz})$
    h_b = base station effective antenna height (m) $(30 \sim 200 \text{ m})$
    h_m = vehicular station antenna height (m) $(1 \sim 10 \text{ m})$
    R = distance (km) $(1 \sim 20 \text{ km})$

and $\log x$ denotes the logarithmic function $\log_{10} x$.

Path loss characteristics in urban areas for small and medium size cities, and for open areas, are shown in Figs. 4.2 and 4.3, respectively. For a large city, the path loss is the same as that for small and medium size city under $h_m = 1.5$ m.

4.2 SHADOWING

Many experiments have shown that the shadowing loss obeys the log-normal distribution. In this case, the probability density function of the received signal level in decibels becomes

$$p(x) = \frac{1}{\sqrt{2\pi}\sigma} e^{-(x-x_0)^2/2\sigma^2}, \tag{4.7}$$

where x is the received power in decibels, $x_0 = \langle x \rangle$ and α^2 is the variance in decibels. The averaging of x is defined over a distance that is long enough for averaging microscopic variation (several wavelengths). The variance takes values of $4 \sim 12$ dB depending on the propagation environment.

In the following we develop a hypothetical argument supporting the log-normal distribution for the received signal level. It is probable that the

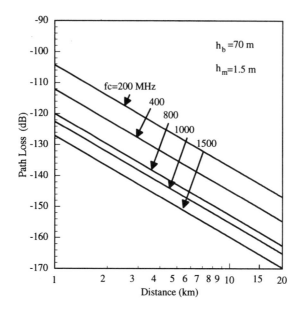

Figure 4.2. Path loss vs. diatance (small and medium size city).

received signal is subjected to a number of random reflections while propagating from the transmitter to the receiver. We can then express the received signal level as

$$P_r = P_0 \prod_{n=1}^{\infty} \Gamma_n, \tag{4.8}$$

where P_0 denotes the transmitted power and Γ_n are the random power losses due to reflections and propagation. If we express Eq. (4.8) in decibels, we have

$$10 \log_{10} P_r = 10 \log_{10} P_0 + \sum_{n=1}^{\infty} 10 \log_{10} \Gamma_n. \tag{4.9}$$

The central limit theorem states that the sum of many random values, $10 \log_{10} \Gamma_n$ in our case, obeys the normal distribution. Thus the loss, $10 \log_{10} P_r - 10 \log_{10} P_0$, in decibels, obeys the normal distribution.

In actual mobile radio channels, the signal propagates over multipaths instead of one path to the receiver. In this case, the log-normal distribution is also supported, since the addition of several signals with the log-normal distribution is well approximated by a log-normal distribution [6].

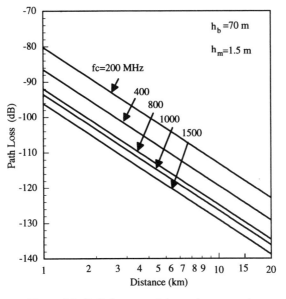

Figure 4.3. Path loss vs. distance (open area).

4.3 FAST FADING

Many random signals that propagate through different signal paths from the transmitter are superposed at the receiver to produce standing waves. When a receiver and/or a transmitter moves in the standing waves, the receiver experiences random variation in the signal level and in the phase, and also a Doppler shift. As a consequence of the standing wave characteristic, the minimum distance between signal level drops is one half of the wavelength; for example, the half-wavelength becomes as short as 15 cm at a frequency of 1000 MHz. This microscopic deviation of the received signal is called fast fading.

The fast fading phenomena are analyzed in [3] and [4] with a propagation model in which many independent signals reach a moving receiver uniformly from different directions (Fig. 4.4). Here we restrict our argument to citing some important results, referring only to [3] (Copyright © IEEE 1994).

We assume vertically polarized plane waves. The vertical electric field is expressed as

$$E_z = T_c(t) \cos \omega_c t - T_s(t) \sin \omega_c t, \qquad (4.10)$$

where ω_c is the carrier frequency. $T_c(t)$ and $T_s(t)$ are

$$T_c(t) = E_0 \sum_{n=1}^{N} C_n \cos(\omega_n t + \varphi_n), \qquad (4.11)$$

Incident wave

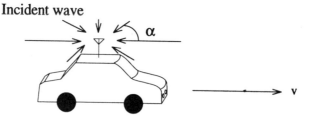

Figure 4.4. Analytical model for fast fading.

$$T_s(t) = E_0 \sum_{n=1}^{N} C_n \sin(\omega_n t + \varphi_n), \qquad (4.12)$$

where N is the number of the incident waves, $E_0 C_n$ is the real amplitude of the electric field (C_n are normalized, i.e. $\sum_{n=1}^{N} C_n^2 = 1$), ϕ_n are random phases uniformly distributed from 0 to 2π. ω_n are the Doppler frequencies given as

$$\omega_n = \beta v \cos \alpha_n, \qquad (4.13)$$

where v is the velocity, α_n is the incident angle to the direction of the movement, and β is the propagation constant expressed as

$$\beta = \frac{2\pi}{\lambda} = \frac{\omega_c}{c}, \qquad (4.14)$$

and λ is the wavelength and c is the velocity of light ($c = 3 \times 10^8$ m/s). The Doppler frequencies vary within $\pm \beta v (= \omega_c v/c)$.

$T_c(t)$ and $T_s(t)$ are the in-phase and quadrature components, respectively, of E_z. For a modulated signal, they also include the modulating signal. However, a nonmodulated carrier signal is treated here for simplicity of argument. $T_c(t)$ and $T_s(t)$ become Gaussian random processes for a large N, as a consequence of the central limit theorem. We denote the random variables $T_c(t)$ and $T_s(t)$ at the time t by T_c and T_s, respectively. They have zero mean, zero cross correlation and variance:

$$\langle T_c \rangle = \langle T_s \rangle = 0, \qquad (4.15)$$

$$\langle T_c^2 \rangle = \langle T_s^2 \rangle = \langle |E_z|^2 \rangle = E_0^2/2, \qquad (4.16)$$

$$\langle T_c T_s \rangle = 0, \qquad (4.17)$$

where $\langle \cdot \rangle$ means ensemble average with respect to α_n, ϕ_n, and C_n. If we denote T_c or T_s by x, the probability density function of x is given by

$$p(x) = \frac{1}{\sqrt{2\pi b}} e^{-x^2/2b} \qquad (4.18)$$

where $b = E_0^2/2$ is the mean power of E_z. The envelope of E_z is given by

$$r = (T_c^2 + T_s^2)^{1/2}. \tag{4.19}$$

Referring to the argument in Section 2.2.5, the envelope obeys the Rayleigh distribution. The probability density function becomes

$$p(r) = \frac{r}{b} e^{-r^2/2b}. \tag{4.20}$$

The phase of E_z is given by

$$\theta = \tan^{-1} \frac{T_s}{T_c}. \tag{4.21}$$

Therefore, θ is distributed uniformly; that is, the probability density function becomes

$$p(\theta) = \frac{1}{2\pi}. \tag{4.22}$$

4.3.1 RF Power Spectra

The received signal spectrum spreads due to the Doppler effect. The frequency can be expressed as a function of the incident angle:

$$f(\alpha) = f_m \cos \alpha + f_c, \tag{4.23}$$

where $f_m = \beta v/2\pi = v\lambda$ is the maximum Doppler shift.

We denote incident power included in an interval $d\alpha$ by $bp(\alpha)\,d\alpha$ and the receiving antenna directivity in the horizontal plane by $G(\alpha)$. Then the received power within $d\alpha$ becomes $bG(\alpha)p(\alpha)\,d\alpha$. Using Eq. (4.23) this value is equated to the differential variation of the power with frequency as

$$S(f)|df| = b[p(\alpha)G(\alpha) + p(-\alpha)G(-\alpha)]|d\alpha|. \tag{4.24}$$

On the other hand,

$$|df| = |-f_m \sin \alpha|\,|d\alpha| = [f_m^2 - (f - f_c)^2]^{1/2}|d\alpha|. \tag{4.25}$$

Thus,

$$S(f) = \begin{cases} b[p(\alpha)G(\alpha) + p(-\alpha)G(-\alpha)][f_m^2 - (f - f_c)^2]^{-1/2}, & 0 < |f - f_c| < f_m \\ 0, & \text{otherwise,} \end{cases} \tag{4.26}$$

where

$$\alpha = \cos^{-1}\frac{f-f_c}{f_m}.$$

The power spectrum depends on the antenna directivity $G(\alpha)$ and the distribution of arrival angles $p(\alpha)$. We assume a uniform distribution of arrival angles,

$$p(\alpha) = \frac{1}{2\pi}. \tag{4.27}$$

The antenna directivity depends solely on the kind of the antenna to be used. We assume a vertical whip antenna to sense the E_z component. The directivity in the horizontal plane becomes

$$G(\alpha) = 1.5. \tag{4.28}$$

Thus the received signal power spectrum density is given as

$$S(f) = \begin{cases} \dfrac{3b}{\omega_m}\left[1-\left(\dfrac{f-f_c}{f_m}\right)^2\right]^{-1/2}, & 0 \le |f-f_c| \le f_m \\ 0, & \text{otherwise,} \end{cases} \tag{4.29}$$

where $\omega_m = 2\pi f_m$.

4.3.2 Correlations between the In-phase and Quadrature Components

The correlation of the in-phase and quadrature components is given as in [3]:

$$\langle T_{c1} T_{c2}\rangle = \langle T_{s1} T_{s2}\rangle = g(\tau), \tag{4.30}$$

$$\langle T_{c1} T_{s2}\rangle = -\langle T_{s1} T_{c2}\rangle = h(\tau), \tag{4.31}$$

where subscripts 1 and 2 refer to the times t and $t+\tau$, respectively, and

$$g(\tau) = \int_{f_c-f_m}^{f_c+f_m} S(f)\cos 2\pi(f-f_c)\tau\, df, \tag{4.32}$$

$$h(\tau) = \int_{f_c-f_m}^{f_c+f_m} S(f)\sin 2\pi(f-f_c)\tau\, df, \tag{4.33}$$

and $S(f)$ is the received signal power spectrum. Using Eq. (4.29), $g(\tau)$ and $h(\tau)$ become

$$g(\tau) = b_0 J_0(\omega_m \tau), \qquad b_0 = 1.5b, \tag{4.34}$$

$$h(\tau) = 0, \tag{4.35}$$

where $J_0(\cdot)$ is the zeroth-order Bessel function of the first kind. The relation $b_0 = 1.5b$ appears due to the assumed antenna gain (Eq. 4.28).

4.3.3   Correlation of the Envelope

The autocorrelation of the envelope r is given as

$$R_r(\tau) \equiv \langle r(t)r(t+\tau) \rangle$$

$$= \frac{\pi}{2}b_0[1 + \tfrac{1}{4}\rho^2(\tau) + \tfrac{1}{64}\rho^4 + \cdots], \tag{4.36}$$

where

$$\rho^2(\tau) = \frac{1}{b_0^2}[g^2(\tau) + h^2(\tau)]$$

$$= J_0^2(\omega_m \tau). \tag{4.37}$$

Dropping terms beyond second degree,

$$R_r(\tau) \approx \frac{\pi}{2}b_0[1 + \tfrac{1}{4}J_0^2(\omega_m \tau)]. \tag{4.38}$$

The autocovariance function of r is defined for a stationary process as

$$L_e(\tau) \equiv \langle [r(t) - \langle r \rangle][r(t+\tau) - \langle r \rangle] \rangle$$

$$= R_r(\tau) - \langle r \rangle^2. \tag{4.39}$$

In our case, considering the antenna gain factor $G(\alpha) = 1.5$, $p(r) = (r/b_0)e^{-r^2/2b_0}$. Thus

$$\langle r \rangle = \frac{1}{b_0} \int_0^\infty r^2 e^{-r^2/2b_0}\, dr = \sqrt{\left(\frac{\pi}{2}b_0\right)}. \tag{4.40}$$

Substituting Eqs. (4.38) and (4.40) into Eq. (4.39), we have

$$L_e(\tau) \approx \frac{\pi}{8}b_0 J_0^2(\omega_m \tau). \tag{4.41a}$$

4.3.4 Spatial Correlation of the Envelope

Let us consider the correlation of the envelope at two places separated by distance d under the conditions assumed so far. Since we assume that the distribution of the arrival angle is uniform and that the receiver is moving at a speed v, the correlation function in terms of the time difference τ is equivalent to that in terms of the distance $d = v\tau$. Inserting the relations $d = v\tau$ and $\omega_m = 2\pi v/\lambda$ into Eq. (4.41a), we have

$$L_e(d) \approx \frac{\pi}{8} b_0 J_0^2(2\pi d/\lambda).$$

The normalized autocovariance function becomes

$$L_{en}(d) = \frac{L_e(d)}{L_e(0)} = J_0^2(2\pi d/\lambda). \tag{4.41b}$$

This relation is useful for determining the correlation function for a dual space-diversity system at a receiver. The correlation decreases with increase of d, and becomes zero for the spacing $d = 0.38\lambda$, 0.88λ, ...

The above results usually cannot be applied for a base station, since the uniform distribution of the arrival angle is hardly satisfied. Because of the height of the base station antenna, there are few scattering objects around the base station. The assumption that the receiver is moving can be equivalently satisfied by considering that the base station is moving instead of the mobile station. The spatial correlation at a base station is analyzed in [3] and [4] with a system model shown in Fig. 4.5. Since the arrival angle is not uniformly distributed, the spread of the Doppler frequencies becomes small and the spatial correlation at a base station becomes high. A spatial separation on the order of ten times the wavelength is required for a base station diversity system.

4.3.5 Random Frequency Modulation

The fast fading causes random variation in the phase θ or its derivative $\dot{\theta}$, the instantaneous frequency, as well as in the envelope r. The random variation of $\dot{\theta}$ is equivalent to frequency modulation by the random signal.

The joint density function is given as [3]

$$p(r, \dot{r}, \theta, \dot{\theta}) = \frac{r^2}{4\pi^2 b_0 b_2} \exp\left[-\frac{1}{2}\left(\frac{r^2}{b_0} + \frac{\dot{r}^2}{b_2} + \frac{r^2\dot{\theta}^2}{b_2}\right)\right], \tag{4.42}$$

where the dot denotes the time derivative and $b_1 = 0$ is assumed. The b_n

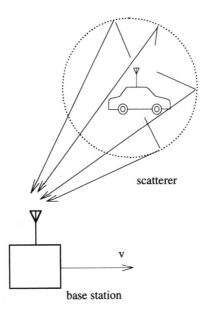

Figure 4.5. Analytical model for spatial correlation at a base station.

are the moments defined as

$$b_n = (2\pi)^n \int_{f_c-f_m}^{f_c+f_m} S(f)(f-f_c)^n \, df. \tag{4.43}$$

The b_n are given for the electric field as

$$b_n = \begin{cases} b_0\omega_m^n \dfrac{1\cdot3\cdot5\cdot\,\cdots\,\cdot(n-1)}{2\cdot4\cdot6\cdot\,\cdots\,\cdot n}, & n \text{ even} \\ 0, & n \text{ odd.} \end{cases} \tag{4.44}$$

Integrating $p(r,\dot{r},\theta,\dot{\theta})$ over r, $\dot{r}$, and θ, we have

$$
\begin{aligned}
p(\dot{\theta}) &= \frac{1}{4\pi^2 b_0 b_2} \int_0^\infty dr \int_{-\infty}^\infty d\dot{r} \int_0^{2\pi} d\theta\, r^2 \exp\left[-\frac{1}{2}\left(\frac{r^2}{b_0} + \frac{\dot{r}^2}{b_2} + \frac{r^2\dot{\theta}^2}{b_2} \right) \right] \\
&= \frac{1}{2}\sqrt{\frac{b_0}{b_2}}\left(1 + \frac{b_0}{b_2}\dot{\theta}^2 \right)^{-3/2}.
\end{aligned} \tag{4.45}
$$

For the electric field $b_2/b_0 = \omega_m^2/2$, we have

$$p(\dot{\theta}) = \frac{1}{\omega_m\sqrt{2}}\left[1 + 2\left(\frac{\dot{\theta}}{\omega_m} \right)^2 \right]^{-3/2}. \tag{4.46}$$

4.4 DELAY SPREAD AND FREQUENCY-SELECTIVE FADING

The results for fast fading discussed so far are derived from the assumption that there is no difference between the arrival times of the multipath signal [Eqs. (4.11) and (4.12)]. In fact, differences exist in the multipath delays. This fact causes no significant effect on signal transmission as long as we consider a signal with bandwidth much less than the inverse of the magnitude of the difference in the delay times. Otherwise the effects cannot be neglected and the channel causes linear distortion of the received signal. The transfer function is such that the amplitude and the phase show dependence on frequencies over the signal spectrum.

The multipath channel is characterized by the impulse response

$$h(t) = \sum_{n=1}^{N} c_n \, \delta(t - \tau_n), \tag{4.47}$$

where c_n are complex coefficients that express the (complex) amplitude of the paths and τ_n are the time delays. Since the phase of c_n is not correlated with τ_n, the power (i.e. $|c_n|^2$) versus τ_n is considered.

The power delay profile becomes different depending on the environment. In a mountainous area we often encounter a large delay difference because echoes from far mountains reach the receiver with large delay times. In a city area, where buildings are not so tall, the delay difference becomes smaller. Many measurements for the delay profile have been carried out [3], [4].

Some models for the delay profile are presented to evaluate signal transmission performance. Figure 4.6 shows the models used for the Pan-European digital mobile telephone (GSM) system. The two-path model is also often used for analytical treatments. The impulse response for this channel is

$$h(t) = c_1 \, \delta(t - \tau_1) + c_2 \, \delta(t - \tau_2), \tag{4.48}$$

where τ_1 and τ_2 are the delay times and the complex coefficients c_1 and c_2 are random variables whose amplitude obeys the Rayleigh distribution and whose phase has a uniform distribution. The power delay profile becomes

$$P(\tau) = |c_1|^2 \, \delta(\tau - \tau_1) + |c_2|^2 \, \delta(\tau - \tau_2). \tag{4.49}$$

Another assumption for the power delay profile is an exponential distribution. If we assume a continuous time delay, then

$$P(\tau) = \frac{1}{\sigma} e^{-\tau/\sigma} \qquad (\tau \geq 0), \tag{4.50}$$

where σ is a coefficient denoting the degree of difference in time delay.

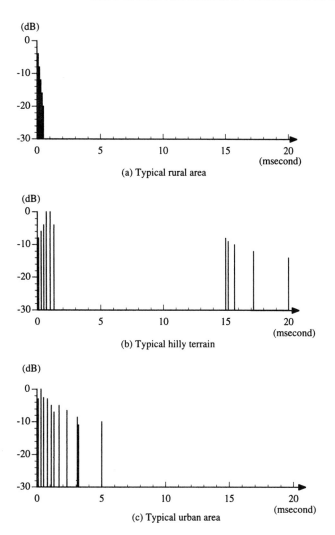

Figure 4.6. Propagation model for the Pan-European digital cellular (GMS) system.

In order to represent the degree of difference in delay, the delay spread Δ is defined as

$$\Delta \equiv \left[\frac{\displaystyle\int_{0}^{\infty} (\tau - d_m)^2 P(\tau)\, d\tau}{\displaystyle\int_{-\infty}^{\infty} P(\tau)\, d\tau} \right]^{1/2}, \tag{4.51}$$

where d_m is an average delay defined as

$$d_m = \frac{\displaystyle\int_0^\infty \tau P(\tau)\, d\tau}{\displaystyle\int_{-\infty}^\infty P(\tau)\, d\tau}. \tag{4.52}$$

The delay spread takes a value of around 3 microseconds for a city area and up to 10 microseconds in hilly terrain. The delay spread can be calculated for the two-path model as

$$\Delta = \left[\frac{|c_1|^2(\tau_1 - d_m)^2 + |c_2|^2(\tau_2 - d_m)^2}{|c_1|^2 + |c_2|^2}\right]^{1/2}, \tag{4.53}$$

where

$$d_m = \frac{\tau_1|c_1|^2 + \tau_2|c_2|^2}{|c_1|^2 + |c_2|^2}.$$

When $|c_1|^2 = |c_2|^2$,

$$\Delta = \frac{|\tau_1 - \tau_2|}{2}. \tag{4.54}$$

For the exponential distribution we have

$$\Delta = \sigma. \tag{4.55}$$

4.4.1 Coherence Bandwidth

Let us consider correlation between envelopes at the same time ($\tau = 0$ in section 1.5.1 of [3]) and at different frequencies. The correlation coefficient is defined as

$$\rho(r_1, r_2) = \frac{\langle r_1 r_2\rangle - \langle r_1\rangle\langle r_2\rangle}{\{[\langle r_1^2\rangle - \langle r_1\rangle^2][\langle r_2^2\rangle - \langle r_2\rangle^2]\}^{1/2}}, \tag{4.56}$$

where r_1 and r_2 are the amplitudes of the signals at frequencies ω_1 and ω_2, respectively. Introducing the time delay to Eqs. (4.11) and (4.12), the correlation function under the exponential power-delay profile of Eq. (4.50) is given as [3]

$$\rho(r_1, r_2) = \frac{1}{1 + s^2\sigma^2}, \tag{4.57}$$

where $s = |\omega_1 - \omega_2|$ is the difference of the frequencies. The coherence bandwidth W_c is defined as the bandwidth where $\rho(s) = 0.5$. Thus, $W_c = 1/(2\pi\sigma)$.

For the two-path model, we normalize the power-delay profile (Eq. 4.49) as

$$p(\tau) = \frac{|c_1|^2\,\delta(\tau - \tau_1) + |c_2|^2\,\delta(\tau - \tau_2)}{|c_1|^2 + |c_2|^2}. \tag{4.58}$$

Using Eq. (4.52) and following the derivation process in [3]—calculating λ^2 in equations (1.5–17) and using $\rho(r_1, r_2) \approx \lambda^2$ in equations (1.5–26) in [3]—we have

$$\rho(r_1, r_2) \approx \frac{|c_1|^4 + |c_2|^4 + 2|c_1|^2|c_2|^2 \cos[s(\tau_1 - \tau_2)]}{[|c_1|^2 + |c_2|^2]^2}. \tag{4.59}$$

The coherence bandwidth depends on $|c_1|$ and $|c_2|$, and when $|c_1| = |c_2|$ it becomes

$$W_c = \frac{1}{4|\tau_1 - \tau_2|}. \tag{4.60}$$

The measured coherence bandwidth in mobile radio channels is a few hundred kHz.

4.4.2 Frequency-selective Fading

If there is a difference in time delay for a multipath channel, the channel has a transfer function that causes a (linear) distortion of the received signal. For example, let us consider the two-path model. Taking the Fourier transform of Eq. (4.48), we have

$$H(\omega) = c_1 e^{-j\omega\tau_1} + c_2 e^{-j\omega\tau_2}. \tag{4.61}$$

As shown in Fig. 4.7, the signal level fades at specific frequencies depending on the parameters c_1, c_2, and $\tau_1 - \tau_2$. Here we assume that the coefficients c_1 and c_2 are given by

$$c_1, c_2 = T_c(t) + jT_s(t), \tag{4.62}$$

where $T_c(t)$ and $T_s(t)$ are defined by Eqs. (4.11) and (4.12), respectively. Then the channel becomes time-varying and frequency-selective. The varying speed is determined by the maximum Doppler frequency.

In the special case of $\tau_1 = \tau_2$, the channel becomes a Rayleigh fading channel with a flat transfer function.

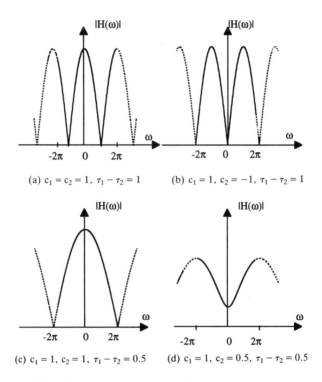

(a) $c_1 = c_2 = 1$, $\tau_1 - \tau_2 = 1$

(b) $c_1 = 1$, $c_2 = -1$, $\tau_1 - \tau_2 = 1$

(c) $c_1 = 1$, $c_2 = 1$, $\tau_1 - \tau_2 = 0.5$

(d) $c_1 = 1$, $c_2 = 0.5$, $\tau_1 - \tau_2 = 0.5$

Figure 4.7. Transfer function of two-path frequency-selective fading channels.

4.5   THE NEAR–FAR PROBLEM

The received signal level varies depending on the path length. The dynamic range of the received signal level can become as large as 70 dB for path length of 100 m to 10 km; this value is found using Eq. (4.3) with $\alpha = 3.5$. The near-far problem stems from the wide–dynamic range of the received level in mobile radio communications. Consider systems where two mobile stations are communicating with a base station or with other mobile stations as illustrated in Figs. 4.8(a) and (b). At the base station the signal levels received from the mobile stations A and B are quite different due to the difference in the path lengths. Let us assume that the mobile stations are using adjacent channels (Fig. 4.9). The out-of-band radiation of the signal from the mobile station B interferes with the signal from station A in the adjacent channel. The effect, called adjacent channel interference, becomes serious when the difference in the received signal levels is high. For this reason the out-of-band radiation must be kept small.

In a conventional mobile radio channel with channel separation of 25 kHz, it is recommended that the relative interference power in the adjacent

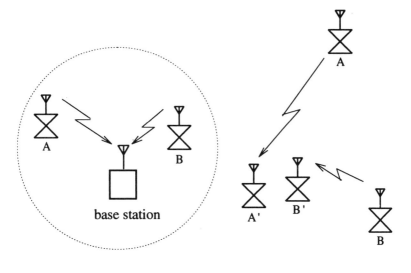

(a) System with base station   (b) System without base station

Figure 4.8. Two situation in the near–far problem.

channel be less than $-70\,\mathrm{dB}$ [5]. The tolerable relative adjacent channel interference level can be different depending on the system. A system with automatic transmit power control, where the received signal level is controlled within a given range, can tolerate a high relative adjacent channel interference level.

The near–far problem becomes more important for code-division multiple access systems where spread spectrum signals are multiplexed on the same frequency using low cross correlation codes (Section 7.3).

4.6 COCHANNEL INTERFERENCE

The key concept of cellular systems is spatial reuse of channels: the same channel is assigned to different cells, where the probability of cochannel interference between those cells is less than a given value. The probability of cochannel interference is defined as the probability that the desired signal level (envelope)r_d drops below a value proportional to the interfering (or undesired) signal level (envelope) r_u; that is,

$$P_c = \mathrm{Prob}(r_d \le \alpha r_u), \tag{4.63}$$

where α is the protection ratio.

Let us assume that the desired and the interfering signals are independent

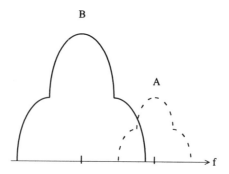

Figure 4.9. Adjacent channel interference.

of each other. We denote the probability density functions for the desired and the interfering signals as $p_1(r_1)$ and $p_2(r_2)$, respectively. Then P_c is given as

$$P_c = \int_0^\infty \text{Prob}(r_1 = x)\,\text{Prob}(r_2 \geq x/\alpha)\,dx$$

$$= \int_0^\infty p_1(r_1) \int_{r_1/\alpha}^\infty p_2(r_2)\,dr_2\,dr_1. \qquad (4.64)$$

In the following, the probability of cochannel interference is calculated for Rayleigh fading, shadowing, and the combined Rayleigh fading and shadowing signals.

4.6.1 Rayleigh Fading

The probability distribution function of the amplitude of a signal subjected to the Rayleigh fading is given by Eq. (4.20). In this case, the probability of cochannel interference can be calculated as

$$P_{cR} = \int_0^\infty \frac{r_1}{b_1} e^{-r_1^2/2b_1} \int_{r_1/\alpha}^\infty \frac{r_2}{b_2} e^{-r_2^2/2b_2}\,dr_2\,dr_1$$

$$= \frac{1}{1 + \alpha^{-2} b_1/b_2}, \qquad (4.65)$$

where b_1 and b_2 are the average power of the desired and undesired signals, respectively, and α is the protection ratio.

4.6.2 Shadowing

The probability distribution function of shadowing is given by Eq. (4.7). We can calculate the probability of the cochannel interference P_{cs} (see Appendix 4.1):

$$P_{cs} = \mathrm{Prob}(x_1 \leq x_2 + \beta)$$

$$= \frac{1}{\sqrt{\pi}} \int_b^\infty e^{-u^2} du$$

$$= \tfrac{1}{2}\mathrm{erfc}(b), \tag{4.66}$$

where x_1 and x_2 are the desired and undesired signal envelopes in decibels, respectively, and β is the protection ratio in decibels. In Eq. (4.66) b is given as

$$b = (x_{1m} - x_{2m} - \beta)/2\sigma$$

where $x_{1m} = \langle x_1 \rangle$, $x_{2m} = \langle x_2 \rangle$ and σ is the standard deviation in decibels.

4.6.3 Combined Fading and Shadowing

The probability density function of the signal envelope r under combined fading and shadowing is given using Eqs. (4.7) and (4.20) as

$$p_i(r) = \int_{-\infty}^\infty \frac{r}{b} e^{-r^2/2b} \frac{1}{\sqrt{2\pi}\sigma} e^{-(x-x_{im})^2/2\sigma^2} dx \qquad (i = 1, 2), \tag{4.67}$$

where b is a function of x, as $b = 10^{x/10}$.

The interference probability under this condition is given as

$$P_{cf\&s} = \int_0^\infty \mathrm{Prob}(r_1 = y)\,\mathrm{Prob}(r_2 \geq y/\gamma)\, dy$$

$$= \int_0^\infty p_1(r_1) \int_{r_1/\gamma}^\infty p_2(r_2)\, dr_2\, dr_1, \tag{4.68}$$

where r_1 and r_2 are the desired and undesired signal envelopes and γ is the protection ratio. Equation (4.68) can be reduced to (see Appendix 4.2)

$$P_{cf\&s} = \frac{1}{\sqrt{\pi}} \int_{-\infty}^\infty [1 + 10^{(x_{1m}-x_{2m}-R+2\sigma u)/10}]^{-1} e^{-u^2} du, \tag{4.69}$$

where u is an internal variable and $R = 20 \log \gamma$.

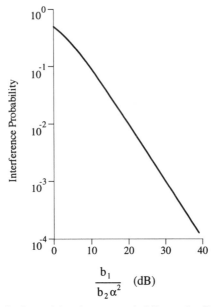

Figure 4.10. Cochannel interference probability under Rayleigh fading.

4.6.4 Discussion

The cochannel interference probabilities are shown in Figs. 4.10–4.12. Typically, for a fast fading environment, the cochannel interference under shadowing (Eq. 4.66) is used as a measure of system quality, for example, bit error rate. For a slow fading environment, the cochannel interference under both fading and shadowing is taken into consideration.

The above results for the cochannel interference probability can be applied to the adjacent channel interference by using appropriate values for the protection ratio.

APPENDIX 4.1 INTERFERENCE PROBABILITY UNDER SHADOWING

Denote the desired and undesired signal levels in decibels as x_1 and x_2. Then the probability density functions of x_1 and x_2 are given from Eq. (4.7) as

$$p_1(x_1) = \frac{1}{\sqrt{2\pi}\sigma}e^{-(x_1-x_{1m})^2/2\sigma^2} \qquad (-\infty < x_1 < \infty), \qquad \text{(A4.1a)}$$

$$p_2(x_2) = \frac{1}{\sqrt{2\pi}\sigma}e^{-(x_2-x_{2m})^2/2\sigma^2} \qquad (-\infty < x_2 < \infty). \qquad \text{(A4.1b)}$$

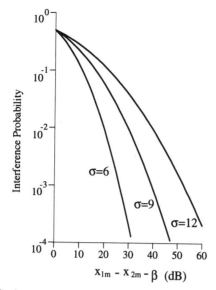

Figure 4.11. Cochannel interference probability under log-normal shadowing.

We define a new variable $z = x_1 - x_2$. Then we have

$$P_{cs} = \text{Prob}(z \le \beta) \qquad (A4.2)$$

The probability density function of z is given by

$$p(z) = \int_{-\infty}^{\infty} p_1(z + x_2) p_2(x_2)\, dx_2 \qquad (-\infty < z < \infty). \qquad (A4.3)$$

Using Eqs. (A4.1a) and (A4.1b), we have

$$p(z) = \frac{1}{2\pi\sigma^2} \int_{-\infty}^{\infty} \exp\left(-\frac{(z + x_2 - x_{1m})^2 - (x_2 - x_{2m})^2}{2\sigma^2}\right) dx_2$$

$$= \frac{1}{2\pi\sigma^2} \int_{-\infty}^{\infty} \exp\left(-\frac{[x_2 + \frac{1}{2}(z - x_{1m} - x_{2m})]^2}{\sigma^2}\right)$$

$$\exp\left(-\frac{(z - x_{1m} + x_{2m})^2}{4\sigma^2}\right) dx_2. \qquad (A4.4)$$

Performing the integration over x_2, and using $\int_{-\infty}^{\infty} e^{-x^2}\, dx = \sqrt{\pi}$, we have

$$p(z) = \frac{1}{2\sqrt{\pi}\sigma} \exp\left(\frac{(z - x_{1m} + x_{2m})^2}{4\sigma^2}\right). \qquad (A4.5)$$

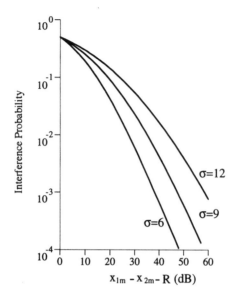

Figure 4.12. Cochannel interference probability under fading and shadowing.

Then we get

$$P_{cs} = \frac{1}{2\sqrt{\pi}\sigma} \int_{-\infty}^{\beta} \exp\left(-\frac{(z - x_{1m} + x_{2m})^2}{4\sigma^2}\right) dz$$

$$= \frac{1}{\sqrt{\pi}} \int_{(x_{1m}-x_{2m}-\beta)/2\sigma}^{\infty} e^{-u^2} du. \tag{A4.6}$$

APPENDIX 4.2 INTERFERENCE PROBABILITY UNDER COMBINED FADING AND SHADOWING

With Eqs. (4.67) and (4.68), we have

$$P_{cf\&s} = \int_0^\infty \left[\int \int_{-\infty}^\infty \frac{r_1}{b_1} \exp\left(-\frac{r_1^2}{2b_1}\right) \frac{1}{\sqrt{2\pi}\sigma} \exp\left(-\frac{(x_1 - x_{1m})^2}{2\sigma^2}\right) dx_1 \right.$$

$$\left. \int_{r_1/\gamma}^\infty \int_{-\infty}^\infty \frac{r_2}{b_2} \exp\left(-\frac{r_2^2}{2b_2}\right) \frac{1}{\sqrt{2\pi}\sigma} \exp\left(-\frac{(x_2 - x_{2m})^2}{2\sigma^2}\right) dx_2 \, dr_2 \right] dr_1.$$
$$\tag{A4.7}$$

The integration over r_2 is readily calculated and we have

$$P_{cf\&s} = \int_{-\infty}^{\infty} \left[\int_{-\infty}^{\infty} \frac{r_1}{b_1} \exp\left(-\frac{b_1 \gamma^{-2} + b_2}{b_1 b_2} \frac{r_1^2}{2} \right) \frac{1}{\sqrt{2\pi}\sigma} \exp\left(-\frac{(x_1 - x_{1m})^2}{2\sigma^2} \right) dx_1 \right.$$

$$\left. \int_{-\infty}^{\infty} \frac{1}{\sqrt{2\pi}\sigma} \exp\left(-\frac{(x_2 - x_{2m})^2}{2\sigma^2} \right) dx_2 \right] dr_1. \tag{A4.8}$$

Performing the integration over r_1, we get

$$P_{cf\&s} = \frac{1}{2\pi\sigma^2} \int_{-\infty}^{\infty} \int_{-\infty}^{\infty} \frac{1}{1 + \gamma^{-2} b_1/b_2} \exp\left(-\frac{(x_1 - x_{1m})^2 + (x_2 - x_{2m})^2}{2\sigma^2} \right) dx_1 \, dx_2. \tag{A4.9}$$

Using new variables $y_1 = x_1 - x_{1m}$ and $y_2 = x - x_{2m}$ and writing $b_1 = 10^{x_1/10}$ and $b_2 = 10^{x_2/10}$, Eq. (A4.9) can be rewritten as

$$P_{cf\&s} = \frac{1}{2\pi\sigma^2} \int_{-\infty}^{\infty} \int_{-\infty}^{\infty} [1 + 10^{(y_1 - y_2 + x_{1m} - x_{2m} - R)/10}]^{-1}$$

$$\exp\left(-\frac{(y_1 + y_2)^2 + (y_1 - y_2)^2}{4\sigma^2} \right) dy_1 \, dy_2. \tag{A4.10}$$

Using new variables $y_1 + y_2 = \sqrt{2}t_1$ and $y_1 - y_2 = \sqrt{2}t_2$ and integrating over t_1, we have a single integral as

$$P_{cf\&s} = \frac{1}{\sqrt{2\pi}\sigma} \int_{-\infty}^{\infty} [1 + 10^{(x_{1m} - x_{2m} - R + \sqrt{2}t_2)/10}]^{-1} \exp\left(-\frac{t_2^2}{2\sigma^2} \right) dt_2. \tag{A4.11}$$

If $u = t_2/\sqrt{2}\sigma$, we have Eq. (4.69).

REFERENCES

1. Y. Okumura et al., "Field strength and its variability in UHF and VHF land-mobile radio service," *Review of Electrical Communication Laboratory*, **16** (1968).
2. M. Hata, "Empirical formula for propagation loss in land mobile radio services," *IEEE Trans. Vehicular Technology*, **VT-29**, 317–325 (August 1980).
3. W. C. Jakes, ed., *Microwave Mobile Communications*, Wiley, New York, 1974.
4. W. C. Y. Lee, *Mobile Communications Engineering*, McGraw-Hill, New York, 1982.

5. CCIR Recommendation 478-2.

6. R. L. Mitchel, "Performance of the log-normal distribution," *Journal of the Optical Society of America*, **58**, 1267–1272 (September 1968).

ELEMENTS OF DIGITAL MODULATION

Modulation is a process in which a baseband signal is converted into an RF signal. Demodulation is the inverse process, in which the baseband signal is recovered from the RF signal. An RF signal has two degrees of freedom, namely, amplitude and phase, or in-phase and quadrature components. Modulation is carried out by varying these components according to the baseband signal. Digital modulation is different from analog in the sense that the digital signal takes finite discrete levels. Due to this feature, some particular techniques can be applied to the modulation and demodulation in digital communications.

The requirements for digital modulation are narrow bandwidth of the modulated signal, low bit error rate, and easy implementation of the modulation/demodulation circuits.

5.1 DIGITALLY MODULATED SIGNALS

This section describes briefly some general topics of digital modulation and demodulation. The reader may refer to many books [1–7] for further description.

A modulated signal can be expressed generally as

$$
\begin{aligned}
f(t) &= A(t) \cos[\omega_c t + \varphi(t)] \\
&= \mathrm{Re}[A(t)\, e^{j[\omega_c t + \varphi(t)]}] \\
&= \mathrm{Re}[f_b(t)\, e^{j\omega_c t}],
\end{aligned}
\tag{5.1}
$$

where $A(t)$ is the amplitude of the envelope, $\varphi(t)$ is the phase, ω_c is the (angular) frequency of the carrier signal, and $f_b(t) = A(t)e^{j\varphi(t)}$ is the complex baseband signal. We can rewrite Eq. (5.1) as

$$
f(t) = x(t) \cos(\omega_c t) - y(t) \sin(\omega_c t),
\tag{5.2}
$$

where

$$x(t) = A(t) \cos \varphi(t) \text{ and } y(t) = A(t) \sin \varphi(t)$$

are the in-phase and quadrature components, respectively. The carrier signal has two degrees of freedom, namely, $A(t)$ and $\varphi(t)$ or $x(t)$ and $y(t)$. The time differential of $\varphi(t)$ is the instantaneous (angular) frequency.

From the above expression one cannot distinguish a digitally modulated signal from an analog modulated signal. The difference between the two appears only in detailed expressions for $A(t)$ and $\varphi(t)$ or $x(t)$ and $y(t)$. Modulation techniques shared by digital and analog systems—AM, PM, and FM (Section 2.1)—are called ASK (amplitude shift keying), PSK (phase shift keying), and FSK (frequency shift keying), respectively. The QAM (quadrature amplitude modulation), in which the modulating signals $x(t)$ and $y(t)$ are independent of each other, is used widely for digital modulation.

5.2 LINEAR MODULATION VERSUS CONSTANT ENVELOPE MODULATION

ASK and QAM fall into the category of linear modulation. Linear modulation by definition can be generated by multiplying the baseband modulating signal(s) by a carrier signal. Thus the baseband spectrum of a linear modulated signal is symmetrically placed at the upper and lower sides of the carrier frequency. On the other hand, PSK with constant amplitude and FSK can be called nonlinear or constant-envelope modulation techniques. In this type of system the modulating process includes a nonlinear operation, viz., a trigonometric function of the modulating signal. The spectrum of the nonlinear modulation signal is quite different from the baseband signal spectrum.

The phase of a digitally modulated PSK signal is selected from a set of phases corresponding to the input digital signal. It can be either a linear or a constant-envelope (nonlinear) modulation. Most actual PSK signals are linearly modulated.

Linear modulation requires a linear channel. For example, a linear amplifier must be used for a transmit amplifier, otherwise the spectrum of the modulated signal spreads and the amplitude and/or phase component is distorted from nonlinearity of the circuit. In contrast to this, the constant-envelope signal never requires a linear channel; for example, a saturated power amplifier can be used. Although the waveform of the modulated signal becomes distorted at the output of the power amplifier, it is recovered by passing through a band-pass filter. This filter removes the higher-order frequency components generated in the saturated amplifier. The above discussion is assessed in Appendix 5.1.

A modulated signal can be represented on a two-dimensional plane where the in-phase and quadrature components correspond to the horizontal and vertical axes. An expression on this plane is equivalent to the expression on the complex amplitude plane (Section 2.1.11).

5.3 DIGITAL MODULATIONS

5.3.1 Phase Shift Keying (PSK)

A PSK signal takes one of fixed phases corresponding to the digitally modulating signal. Phase constellations of binary PSK (BPSK), quatenary PSK (QPSK), and 8-level PSK (8PSK) are shown in Fig. 5.1. A symbol of BPSK, QPSK, and 8PSK represents 1, 2, and 3 bits of digital signal, respectively.

The fact that the signal phase of a PSK signal takes specific values makes it possible to generate a carrier signal by frequency multiplication, as is shown later. From the same fact, differential or delay detection (where the relative phase change between successive symbol times is detected) can be applied to a PSK signal.

BPSK. The BPSK signal is expressed as

$$s(t) = \sum_{n=-\infty}^{\infty} a_n h(t - nT) \cos \omega_c t, \qquad a_n = \pm A, \qquad (5.3)$$

where a_n takes A or $-A$ (volt) corresponding to the digital signal, $h(t)$ is the impulse response of the baseband filter, and ω_c is the carrier frequency. The block diagram of a BPSK modulator is shown in Fig. 5.2.

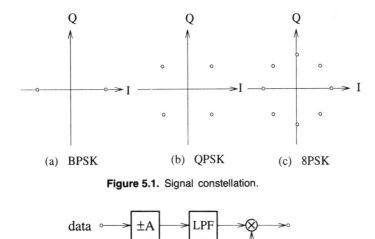

(a) BPSK (b) QPSK (c) 8PSK

Figure 5.1. Signal constellation.

data →| ±A | →| LPF | →⊗→

$\cos \omega_c t$

Figure 5.2. Block diagram of BPSK modulator.

QPSK. The QPSK signal is described as

$$s(t) = \sum_{n=-\infty}^{\infty} a_n h(t - nT_s) \cos \omega_c t + \sum_{n=-\infty}^{\infty} b_n h(t - nT_s) \sin \omega_c t$$

$$= \text{Re}\left\{ \sum_{n=-\infty}^{\infty} (a_n - jb_n)h(t - nT_s)\, e^{j\omega_c t} \right\}, \qquad a_n, b_n = \pm A, \qquad (5.4)$$

where T_s is the symbol duration and a_n, b_n are the input data. Figure 5.3 shows the block diagram of a QPSK modulator.

π/2 Shifted BPSK. The signal phase constellation is shown in Fig. 5.4. The phase is selected such that the signal never has the same phase for two

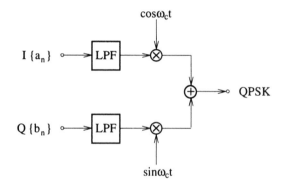

Figure 5.3. Block diagram of QPSK modulator.

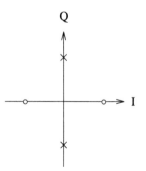

Figure 5.4. Signal constellation of π/2 shifted BPSK.

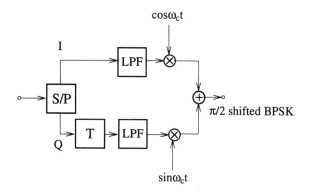

Figure 5.5. Block diagram of $\pi/2$ shifted BPSK modulator.

successive symbols. The block diagram of the modulation is shown in Fig. 5.5. The $\pi/2$ shifted BPSK signal is expressed as

$$s(t) = \sum_{n=-\infty}^{\infty} a_{2n} h(t - 2nT) \cos \omega_c t$$

$$+ \sum_{n=-\infty}^{\infty} a_{2n+1} h[t - (2n+1)T] \sin \omega_c t, \qquad a_n = \pm A, \qquad (5.5a)$$

where T is the bit duration. It must be noted that the bandwidth of the low-pass filter whose impulse response is $h(t)$ is the same as that of the BPSK. Thus the power spectra are the same for $\pi/2$ shifted BPSK and BPSK [see Eq. (2.73)].

Offset QPSK. If we decrease the bandwidth of the low-pass filter by one-half in Fig. 5.5, we have an offset QPSK. The offset QPSK signal can be expressed as

$$s(t) = \sum_{n=-\infty}^{\infty} a_n h(t - nT_s) \cos \omega_c t + \sum_{n=-\infty}^{\infty} b_n h[t - (n + \tfrac{1}{2})T_s] \sin \omega_c t. \quad (5.5b)$$

If we let $T_s = 2T$ in the above equation, this expression is equivalent to Eq. (5.5a). Thus the difference between $\pi/2$ shifted BPSK and offset QPSK is only in the spectral bandwidth of the baseband signals. For example, assuming an NRZ signal, the impulse response becomes

$$h(t) = \begin{cases} 1, & 0 \le t \le T \\ 0, & \text{otherwise} \end{cases} \quad \text{for } \pi/2 \text{ shifted BPSK,}$$

and

$$h(t) = \begin{cases} 1, & 0 \le t \le 2T \\ 0, & \text{otherwise} \end{cases} \quad \text{for offset QPSK}$$

The offset QPSK signal phase never takes fixed points determined by the current symbol data. It scatters to different points owing to the interference between successive symbols. In this sense, the wording of "phase shift keying" may be inappropriate. Nevertheless, the demodulated signal with coherent detection shows no intersymbol interference since the in-phase and quadrature signals are orthogonal with each other. Figure 5.6 compares the waveforms of $\pi/2$ shifted BPSK, QPSK, and offset QPSK; NRZ signaling is assumed. The offset QPSK signal can be obtained from the $\pi/2$ shifted BPSK signal if the pulse duration is doubled. Similarly, the $\pi/2$ shifted BPSK

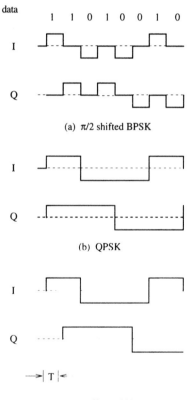

Figure 5.6. Comparison of waveforms for $\pi/2$ shifted BPSK, QPSK and offset QPSK.

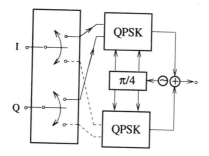

Figure 5.7. Block diagram of $\pi/4$ shifted QPSK modulator.

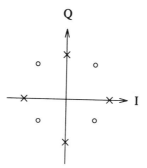

Figure 5.8. Signal constellation of $\pi/4$ shifted QPSK.

signal can be obtained from the offset QPSK if RZ signaling with 50% duty ratio is applied instead of the NRZ signaling.

$\pi/4$ Shifted QPSK. The $\pi/4$ shifted QPSK signal can be generated by combining the output of two QPSK modulators whose carrier phases are shifted by $\pi/4$ radian (Fig. 5.7). The signal phase constellation is shown in Fig. 5.8. The signal points are selected in turn from the two QPSK signal sets, which are marked with circles and crosses. The spectra of QPSK, offset QPSK, and $\pi/4$ shifted QPSK are the same.

The $\pi/4$ QPSK is described in [9–11]; in [10] it is called four-level DPEK-symmetric (phase exchange keying). The terminology "$\pi/4$ shifted QPSK" is used in [11], where the signal is generated by phase shifting the QPSK signal by $\pi/4$ for each symbol. For the $\pi/4$ shifted QPSK the fact that the signal phase change takes place at every symbol time guarantees extraction of a clock timing signal from it.

Differential encoding of the transmit digital signal is widely used for PSK. The purpose of differential encoding is to avoid phase ambiguity in the carrier recovery or to introduce differential detection, as described later.

M-*ary PSK.* This signal is expressed as

$$s(t) = \sum_{n=\infty}^{\infty} Ah(t - nT)\cos(\omega_c t + 2\pi a_n/M), \qquad a_n = 0, 1, \dots, M-1. \quad (5.6)$$

5.3.2 Frequency Shift Keying (FSK)

In this system, the instantaneous frequency is proportionally varied corresponding to the input digital signal. It is also called digital FM. Some people differentiate FSK from digital FM by using term "FSK" when NRZ signaling is adopted for the input digital signal and term "digital FM" if the input digital signal is band-limited. An FSK signal whose phase is continuous is called a continuous phase FSK (CPFSK). A discontinuous phase FSK signal is generated by switching the output from the two oscillators having different frequencies. The phase discontinuity in this system causes spectrum spreading. For this reason, discontinuous phase FSK is seldom used for mobile communication. CPFSK should be understood throughout this book unless stated otherwise.

An FSK signal has a constant envelope. Its spectrum is wider than the spectrum of a linear modulated signal. The shape of the spectrum is not the same as that of the baseband signal and it changes according to the modulation index as well as to the baseband signal. For example, Fig. 5.9 shows spectra of digital FM with the modulation index as a parameter. When the modulation index is small enough, the spectrum becomes similar to that of double sideband AM with the carrier signal component. When the modulation index is high, the spectrum has two peaks that correspond to the maximum frequency deviation. It is worthy of note that the spectra of an FSK signal can be controlled by band-limiting the input baseband signal: out-of-band radiation is effectively suppressed. This fact becomes important for digital mobile communications, where low out-of-band radiation is required.

The empirical formula for the bandwidth of an analog FM signal is well known as Carson's bandwidth:

$$W = 2(\Delta F_m + f_b) \qquad (5.7a)$$

$$= 2(m + 1)f_b, \qquad (5.7b)$$

where ΔF_m is the maximum frequency deviation, f_b is the highest frequency of the baseband signal, and m ($\equiv \Delta F_m/f_b$) is the modulation index. The Carson bandwidth shows that a bandwidth at least twice the baseband bandwidth is required for an FM signal.

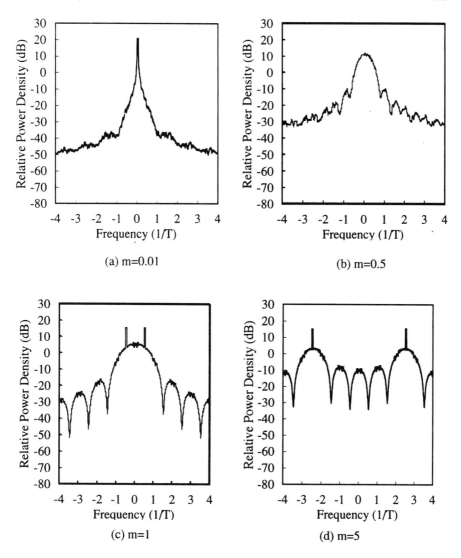

Figure 5.9. FSK signal spectra for different modulation indices.

For the FSK signal the modulation index is usually defined as

$$m = \frac{\Delta F_m}{1/(2T)} = 2\Delta F_m T, \tag{5.8}$$

where T is the symbol duration. Sometimes different definitions and notation are used for the modulation index such as h.

The phase of an FSK signal is given as

$$\varphi(t) = \int_{-\infty}^{t} \omega(\tau) \, d\tau$$

$$= \int_{-\infty}^{t} \sum_{n=-\infty}^{\infty} k_F a_n h(\tau - nT) \, d\tau$$

$$= \sum_{n=-\infty}^{\infty} k_F a_n q(t - nT), \tag{5.9}$$

where $q(t) = \int_{-\infty}^{t} h(\tau) \, d\tau$, k_F is a proportional constant, a_n takes one of the discrete values corresponding to the input digital signal (e.g., $a_n = \pm 1$ for the 2-level FSK), and $h(t)$ is the impulse response of the baseband filter. The actual maximum frequency deviation, $\Delta F_m = \max[\omega(t)/2\pi]$ depends on the impulse response, as well as on data patterns. Instead of the actual ΔF_m we use

$$\Delta F_m \equiv \frac{k_F a_{max}}{2\pi}, \tag{5.10}$$

where a_{max} is the highest of the discrete values. $h(t)$ is normalized so that

$$\int_{-\infty}^{\infty} h(t) \, dt = T. \tag{5.11}$$

The maximum phase shift caused by a symbol pulse becomes $2\pi \Delta F_m T = m\pi$. Equations (5.10) and (5.11) are valid for NRZ signaling, i.e., $h(t) = 1$ for $|t| \leq T/2$ and $h(t) = 0$ for $|t| > T/2$. $\int_{-\infty}^{\infty} h(t) \, dt$ is the value of the transfer function $H(\omega)$ at the dc frequency, i.e., $H(0) = \int_{-\infty}^{\infty} h(t) \, dt$. In case of $H(0) = 0$, for example in the bipolar or the modified duobinary [Eq. (3.47)] FM system, we cannot use the definition of the modulation index: a specific definition must be made.

When the impulse response $h(t)$ satisfies Nyquist's third criterion (Eq. 3.20), from Eqs. (5.8)–(5.11) we have

$$\varphi(iT) = \sum_{n=-\infty}^{i} k_F a_n \int_{nT-T/2}^{nT+T/2} h(t - nT) \, dt$$

$$= k_F T \sum_{n=-\infty}^{i} a_n$$

$$= \pi m \sum_{n=-\infty}^{i} a_n / a_{max}. \tag{5.12}$$

Nyquist's third criterion (Section 3.1.3) ensures that the phase shift of the FSK during a symbol time [i.e., $\Delta\varphi(iT) = \varphi(iT + T) - \varphi(iT)$] is determined by that symbol only: there is no intersymbol interference in $\Delta\varphi(iT)$.

With a special value of m, the phase $\varphi(iT)$ takes some fixed points. Consider a 2-level FSK ($a_n/a_{max} = \pm 1$) with $m = 0.5$, in this case we get the signal phase constellation of $\pi/2$ shifted BPSK. For a 4-level FSK ($a_n/a_{max} = \pm 1/3, \pm 1$) with $m = 3/4$, we get the signal phase constellation of $\pi/4$ shifted QPSK.

5.3.3 Constant Envelope PSK

A constant envelope PSK signal can be obtained by removing the time integration in Eq. (5.9): $q(t)$ instead of $h(t)$ represents the impulse response of the baseband filter.

5.3.4 Quadrature Amplitude Modulation (QAM)

In this system, the in-phase and quadrature components $x(t)$ and $y(t)$, respectively, are amplitude-modulated independently from each other. QPSK can be also understood as QAM. A 4-level signal for each $x(t)$ and $y(t)$ results in 4×4 signal levels. The signal constellation for 16QAM is shown in Fig. 5.10. When we apply the duobinary partial response (Section 3.2.6) to $x(t)$ and $y(t)$, we get the 3×3 signal constellation as is shown in Fig. 5.11. A higher-level QAM such as 256QAM was developed for digital transmission in public telephone channels or microwave channels.

5.4 POWER SPECTRAL DENSITY OF DIGITALLY MODULATED SIGNALS

The power spectral density (PSD) of a signal $z(t)$ is defined as

$$S_z(\omega) = \left\langle \lim_{T_m \to \infty} \frac{1}{2T_m} \left| \int_{-T_m}^{T_m} z_T(t) e^{-j\omega t} dt \right|^2 \right\rangle \tag{5.13a}$$

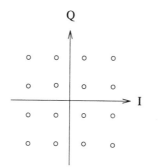

Figure 5.10. Signal constellation of 16QAM.

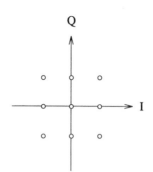

Figure 5.11. Signal constellation of a duobinary 4QAM.

or

$$= \int_{-\infty}^{\infty} R_z(\tau) e^{-j\omega\tau} d\tau, \tag{5.13b}$$

where $\langle \cdot \rangle$ means an ensemble average, $z_T(t)$ is defined over the time range $|t| < T_m$, and $R_z(\tau)$ is the autocorrelation function of $z_T(t)$,

$$R_z(\tau) = \left\langle \lim_{T_m \to \infty} \frac{1}{2T_m} \int_{-T_m}^{T_m} z_T(t) z_T(t + \tau) \, dt \right\rangle \tag{5.14}$$

For a modulated signal, we have $z_T(t) = \mathrm{Re}[f_b(t) e^{j\omega_c t}]$. Therefore, the autocorrelation function $R_z(\tau)$ can be written as

$$R_z(\tau) = \tfrac{1}{2}\mathrm{Re}[R_b(\tau) e^{j\omega_c \tau}], \tag{5.15}$$

where $R_b(\tau)$ is the autocorrelation of the complex baseband signal $f_b(t)$ and is defined as

$$R_b(\tau) = \left\langle \lim_{T_m \to \infty} \frac{1}{2T_m} \int_{-T_m}^{T_m} f_b(t + \tau) f_b^*(t) \, dt \right\rangle. \tag{5.16}$$

It can be shown that

$$R_b^*(\tau) = R_b(-\tau). \tag{5.17}$$

With the above relation Eq. (5.15) can be rewritten as

$$R_z(\tau) = \tfrac{1}{4}[R_b(\tau) e^{j\omega_c \tau} + R_b(-\tau) e^{-j\omega_c \tau}]. \tag{5.18}$$

Then the PSD of the modulated signal $z_T(t)$ becomes

$$S_z(\omega) = \tfrac{1}{4}\left[\int_{-\infty}^{\infty} R_b(\tau)\, e^{-j(\omega - \omega_c)\tau}\, d\tau + \int_{-\infty}^{\infty} R_b(\tau)\, e^{j(\omega + \omega_c)\tau}\, d\tau\right] \quad (5.19a)$$

$$= \tfrac{1}{4}[S_b(\omega - \omega_c) + S_b(-\omega - \omega_c)]. \quad (5.19b)$$

If $f_b(t)$ is real we have

$$S_z(\omega) = \tfrac{1}{4}[S_b(\omega - \omega_c) + S_b(\omega + \omega_c)], \quad (5.19c)$$

where $S_b(\omega)$ is the PSD of the complex baseband signal. Thus we can see that the PSD of the complex baseband signal is shifted by $\pm\omega_c$ due to the modulation.

5.4.1 Linear Modulation

For a linear modulation $z_T(t)$ can be expressed generally as

$$z_T(t) = x(t)\cos \omega_c t - y(t)\sin \omega_c t$$

$$= \text{Re}\{[x(t) + jy(t)]\, e^{j\omega_c t}\}, \quad -T_m \le t \le T_m, \quad (5.20)$$

where $x(t)$ and $y(t)$ are expressed as

$$x(t) = \sum_n a_n h(t - nT), \quad (5.21a)$$

$$y(t) = \sum_n b_n h(t - nT), \quad (5.21b)$$

and a_n and b_n take discrete levels corresponding to the input data signal and $h(t)$ is the impulse response of the baseband filter. If we assume random data, $x(t)$ and $y(t)$ become independent random variables. Owing to the independence of $x(t)$ and $y(t)$, $R_b(\tau)$ is given as

$$R_b(\tau) = R_x(\tau) + R_y(\tau), \quad (5.22)$$

where $R_x(\tau)$ and $R_y(\tau)$ are the autocorrelation functions of $x(t)$ and $y(t)$, respectively. In the present system $R_x(\tau) = R_y(\tau) = R_0(\tau)$. Using this relation and considering $x(t)$ and $y(t)$ are real, we have from Eqs. (5.19a) and (5.22)

$$S_z(\omega) = \tfrac{1}{2}\int_{-\infty}^{\infty} R_0(\tau)[e^{-j(\omega - \omega_c)\tau} + e^{-j(\omega + \omega_c)\tau}]\, d\tau$$

$$= \tfrac{1}{2}[S_0(\omega - \omega_c) + S_0(\omega + \omega_c)], \quad (5.23)$$

where $S_0(\omega) = \int_{-\infty}^{\infty} R_0(\tau) e^{-j\omega\tau} d\tau$ is the PSD of $x(t)$ or $y(t)$. Thus the PSD for a linear modulation is given by shifting the PSD of the baseband modulating signal to the carrier frequency. The PSD of the baseband modulating digital signal is given by Eq. (2.73) or (2.76).

5.4.2 Digital FM

For digital FM, unlike linear modulation, derivation of the PSD is not as easy. There is ample literature on the PSD for digital FM, in which different approaches are taken for the derivation. The simplest one is by Aulin and Sundberg [12] (Copyright © IEE 1983). Here their results are shown without the details of the derivation process.

Rewriting Eq. (5.9) with their notation we get

$$\varphi(t) = 2\pi h \sum_{i=-\infty}^{\infty} \alpha_i q(t - iT), \tag{5.24}$$

where h is the modulation index and symbols α_i take one of the following values:

$$\alpha_i = \pm 1, \pm 3, \ldots, \pm(M-1) \tag{5.25}$$

with a priori probabilities

$$p_i = \text{Prob}(\alpha_j = i), \ i = -M+1, -M+3, \ldots, M-1 \text{ for any integer } j \tag{5.26}$$

and

$$q(t) = \int_{-\infty}^{t} g(\tau) d\tau \tag{5.27}$$

where $g(t)$ is the impulse response of the baseband filter. The impulse response $g(t)$ is assumed to be causal and its duration is truncated within LT symbols; therefore

$$q(t) = \begin{cases} 0, & t \leq 0 \\ q(LT), & t \geq LT. \end{cases} \tag{5.28}$$

Here the modulation index h is selected so that the maximum absolute phase change caused by a pulse over the duration of LT, or equivalently the maximum phase change over any symbol period T, is $(M-1)h\pi$ with the normalization $q(LT) = 1/2$.

The complex time-limited zero-IF signal $S_{b_T}(t)$ is

$$s_{b_T}(t) = e^{j\varphi(t)}, \qquad -T_m \leq t \leq T_m. \tag{5.29}$$

The complex baseband autocorrelation function is then

$$R(\tau) = \left\langle \lim_{T_m \to \infty} \frac{1}{2T_m} \int_{-T_m}^{T_m} e^{j\varphi(t+\tau)} e^{-j\phi(t)} \, dt \right\rangle. \tag{5.30}$$

Omitting the derivation to be followed, we have the final result

$$S(\omega) = 2\mathrm{Re}\left\{ \int_0^{LT} R(\tau) e^{-j\omega\tau} \, d\tau + \frac{e^{-j\omega LT}}{1 - c_\alpha e^{-j\omega T}} \int_0^T R(\tau + LT) e^{-j\omega\tau} \, d\tau \right\}, \tag{5.31}$$

where

$$R(\tau) = R(\tau' + mT)$$

$$= \frac{1}{T} \int_0^T \prod_{n=1-L}^{m+1} \left\{ \sum_{\substack{k=-(M-1) \\ k:\mathrm{odd}}}^{M-1} p_k \exp\{j2\pi hk[q(t + \tau' - (n-m)T) - q(t - nT)]\} \right\} dt \tag{5.32}$$

and τ is over the period $0 - (L+1)T$. In this case the time difference τ has been written as

$$\tau = \tau' + mT, \qquad 0 \le \tau' \le T, \qquad m = 0, 1, \dots \tag{5.33}$$

and

$$c_\alpha = \sum_{\substack{k=-(M-1) \\ k:\mathrm{odd}}}^{M-1} p_k e^{jh\pi k}. \tag{5.34}$$

The discrete components of the PSD appear when $|c_\alpha| = 1$.

5.5   DEMODULATION

Demodulation or detection is a process where the transmit baseband signal is recovered from the modulated signal. The methods of demodulation are classified into coherent or synchronous detection and noncoherent detection, depending on whether a carrier signal is or is not used, respectively. Noncoherent detection includes envelope detection, differential detection, and frequency detection. Differential detection is also called differentially coherent detection or delay detection.

5.5.1   Coherent Detection

In coherent detection systems the baseband modulating signal is obtained by multiplying by a carrier signal, which is synchronized to the modulated

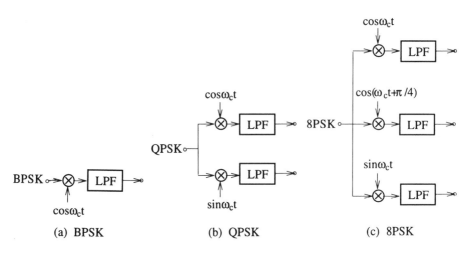

Figure 5.12. Coherent demodulator.

signal. This process is the inverse of linear modulation. Hence, coherent detection could be called linear demodulation.

Coherent demodulators for BPSK, QPSK, and 8PSK are shown in Figs. 5.12(a), (b) and (c), respectively; the decision circuit follows the demodulator.

Coherent detection has a superior error rate compared to noncoherent detection under static conditions. A problem with coherent detection for mobile radio channels is that the carrier recovery circuit becomes unstable due to low signal to noise power ratio and the random FM effect of fading. Due to the random FM effect, the bit error rate versus carrier to noise power ratio curve shows a floor. The floor error rate is called the irreducible error rate. The floor error rate is higher for coherent detection than for noncoherent detection.

5.5.1.1 Carrier Recovery. In contrast to the analog transmission system with coherent demodulation, typically, in digital transmission systems the carrier signal is not sent with the modulated signal but it is regenerated from the modulated signal. Regeneration of the carrier signal component for digital modulation is possible because the modulated signal phase takes specific discrete phase points at each symbol time.

In the regeneration process, first a discrete spectrum component is obtained that is synchronized to the carrier signal through appropriate operations on the received signal. Later, the discrete spectrum component signal is applied to a tank circuit or a phase-locked loop (PLL) circuit that is tuned to the discrete spectrum component in order to reduce jitter of the carrier signal. The Q-value of the tank circuit or, equivalently, the loop bandwidth of the PLL circuit must be compromised with the SNR and rise-up speed of the carrier signal regeneration. In order to boost the rise-up speed,

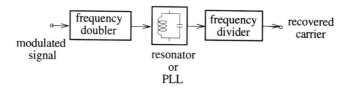

Figure 5.13. Block diagram of carrier regeneration circuit with frequency-multiplying method.

a preamble signal, which is rich in discrete spectrum components, is added before the information-carrying modulated signal. The exception to this is described in [13], where the received signal is stored and then demodulated.

The principles of carrier signal regeneration methods fall into three categories: frequency multiplication, inverse modulation, and remodulation. In the frequency multiplication method, modulated signal is subjected to a frequency multiplying circuit followed by a tank circuit or a PLL circuit and a frequency-dividing circuit. The block diagram of this method is shown in Fig. 5.13. From the signal phase constellation (Fig. 5.1), we see that the double, quadruple, and octuple frequency multiplication of the BPSK, QPSK, and 8PSK signals, respectively, results in a signal that takes one phase ($m\pi$; m integer) at each symbol time. Thus, the discrete spectrum component signal is obtained.

A method of coherent detection, where the frequency multiplication and division are equivalently performed at the baseband is known as the Costas loop demodulator (Fig. 5.14). Let the modulated signal be $s(t) = 2A \cos[\omega_c t + \varphi(t)]$; the outputs of the quadrature detector become $x(t) = A \cos[\varphi(t) + \theta_0]$ and $y(t) = A \sin[\varphi(t) + \theta_0]$, where θ_0 is the phase error between the regenerated carrier signal and the modulated signal. After multiplying $x(t)$ and $y(t)$, the input signal of the voltage controlled oscillator (VCO) becomes $(A^2/2) \sin[2\varphi(t) + 2\theta_0]$. This result shows a frequency doubling at baseband. For the BPSK signal, the phase of the frequency-doubled signal $2\varphi(t)$ takes a value of zero or 2π. Therefore the Costas

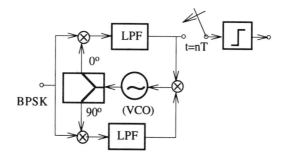

Figure 5.14. Costas loop demodulator for BPSK signal.

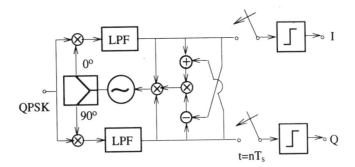

Figure 5.15. Costas loop demodulator for QPSK signal.

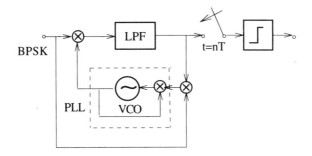

Figure 5.16. Coherent detection with inverse modulation.

loop circuit has a negative-feedback control to achieve $\theta_0 \to 0$, and the synchronization of the recovered carrier signal to the modulated signal is established. The Costas loop detection can be applied to the QPSK, by frequency quadrupling at baseband (Fig. 5.15).

A block diagram of coherent detection with inverse modulation is shown in Fig. 5.16. The input signal is inversely modulated with the baseband signal to remove the effect of modulation. The inverse modulation is equivalent to remodulation in BPSK system, since the signal phase takes values 0 or π. The recovered carrier signal is synchronized to the input signal through the use of a VCO.

The coherent detection with remodulation is shown in Fig. 5.17. The replica of the input signal is produced by modulation of the recovered carrier signal (VCO output) with the demodulated baseband signal. Thus the input modulated signal is tracked by the replica signal through the feedback control circuit including a VCO, a detector, and a modulator.

Phase ambiguity can still remain in the recovered carrier. For example, an ambiguity of $\pm \pi$ exists in a frequency dividing (1/2) circuit of the frequency multiplication method, as well as in the Costas loop circuit. In mobile radio communication, the carrier recovery circuit loses synchroniza-

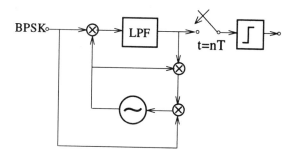

Figure 5.17. Coherent detection with remodulation.

tion because of degradation in signal to noise power ratio or random FM effects of fast fading. Although the synchronization is automatically rees-tablished, phase ambiguity of the carrier may exist for each recovery of synchronization. For DPSK, where the information is conveyed in the relative phase change between two successive symbols, phase ambiguity can be removed by differential encoding. With DPSK, error propagation exists: one error in the detection causes two successive symbol errors. The error propagation is avoided through use of coherent detection with an absolute phase carrier. For this purpose a pilot signal to determine the absolute phase must be transmitted either continuously or intermittently with the informa-tion signal.

5.5.1.2 Noise Power at the Output of the Coherent Detector. We

assume narrow-band band-pass noise, $n(t) = n_x(t) \cos \omega_c t - n_y(t) \sin \omega_c t$, with power spectral density of $N_0/2$,

$$S_n(\omega) = \begin{cases} N_0/2, & |\omega - \omega_c| \le \omega_a \\ 0 & \text{otherwise,} \end{cases} \tag{5.35}$$

where ω_a is an arbitrary constant which is high enough to cover the modulated signal bandwidth and is lower than the carrier frequency ω_c.

Multiplying the passband noise by $2 \cos \omega_c t$ and $2 \sin \omega_c t$, we have the in-phase and quadrature baseband noise components, $n_{xd}(t) = n_x(t)$ and $n_{yd}(t) = n_y(t)$. From Eqs. (2.133) and (5.35), power spectral density (PSD) for the demodulated noise is

$$S_{xd}(\omega) = S_{yd}(\omega) \equiv S_d(\omega) = \begin{cases} N_0, & -\omega_a < \omega < \omega_a \\ 0 & \text{otherwise.} \end{cases} \tag{5.36}$$

The noise power at the output of the receive baseband filter becomes

$$N = \frac{1}{2\pi} \int_{-\infty}^{\infty} S_d(\omega) |G(\omega)|^2 \, d\omega = \frac{N_0}{2\pi} \int_{-\infty}^{\infty} |G(\omega)|^2 \, d\omega \tag{5.37a}$$

or, from Parseval's theorem (Eq. 2.19),

$$N_0 \int_{-\infty}^{\infty} g^2(t)\, dt, \tag{5.37b}$$

where $G(\omega)$ is the transfer function of the filter and $g(t) \leftrightarrow G(\omega)$.

5.5.1.3 *Error Rate Analysis.* The error rate analysis of coherent detection is easy, since it is a linear process. However, this is true only when the recovered carrier is ideal. If we take the carrier recovery process into consideration, the analysis becomes difficult, since the system includes nonlinear operations. Burst errors occur when the carrier recovery circuit loses synchronization. The ideal carrier is assumed in the following analyses.

We assume symbol by symbol detection with a matched filter that meets Nyquist's first criterion (no intersymbol interference), and white Gaussian noise. The baseband signal obtained by demodulation is decided at each symbol time.

BPSK. Let us consider first the coherent detection of a BPSK signal (Fig. 5.18). The received signal is the sum of the BPSK signal and noise,

$$r(t) = s(t) + n(t)$$

$$= \sum_{n=-\infty}^{\infty} a_n h(t - nT) \cos \omega_c t + n_x(t) \cos \omega_c t - n_y(t) \sin \omega_c t \qquad (a_n = \pm A). \tag{5.38}$$

Multiplying $r(t)$ by the carrier signal $2 \cos \omega_c t$, we have the demodulated signal

$$r_d(t) = s_d(t) + n_d(t)$$

$$= \sum_{n=-\infty}^{\infty} a_n h(t - nT) + n_x(t).$$

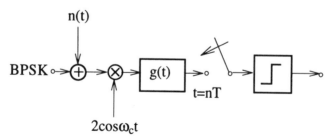

Figure 5.18. Coherent detection of a BPSK signal.

The demodulated signal is applied to a receive filter with an impulse response $g(t)$. Let us denote the output signal level at the sampling instant t_0 by $\pm d/2$, that is,

$$s_d(t_0) = \pm Ah(t) * g(t)|_{t_0} = \pm d/2.$$

A decision error occurs for the symbol m when $s_d(t_0)$ deviates by more than $d/2$ owing to the noise. The noise power $N = \sigma_n^2$ is given by Eq. (5.37a) or (5.37b). Hence, from Eq. (3.53) with $d_m = d/2$, the error rate is given as

$$P_e = \frac{1}{2}\operatorname{erfc}\left(\frac{d}{2\sqrt{2}\sigma_n}\right) = Q\left(\frac{d}{2\sigma_n}\right). \tag{5.39a}$$

Rewriting this equation using the signal to noise power ratio at the sampling instant $S/N = d^2/4\sigma_n^2$, we have

$$P_e = \frac{1}{2}\operatorname{erfc}\left(\sqrt{\frac{1}{2}\frac{S}{N}}\right) = Q\left(\sqrt{\frac{S}{N}}\right). \tag{5.39b}$$

When the receive filter is the matched filter, we have $g(t) = h(t_0 - t)$ (Eq. 3.70). Then from Eq. (3.76) we have

$$s_d(t_0) = \pm\frac{d}{2} = \pm A \int_{-\infty}^{\infty} h^2(t)\, dt \tag{5.40a}$$

and from Eq. (5.37b)

$$N = \sigma_n^2 = N_0 \int_{-\infty}^{\infty} h^2(t)\, dt. \tag{5.40b}$$

Hence we have

$$\frac{S}{N} = \frac{s_d^2(t_0)}{\sigma_n^2}$$

$$= \frac{\left[A \int_{-\infty}^{\infty} h^2(t)\, dt\right]^2}{N_0 \int_{-\infty}^{\infty} h^2(t)\, dt} = \frac{A^2}{N_0} \int_{-\infty}^{\infty} h^2(t)\, dt. \tag{5.41}$$

The average energy per symbol (or per bit) for the BPSK is given as

$$E_s = E_b = \frac{A^2}{2} \int_{-\infty}^{\infty} h^2(t)\, dt \qquad (5.42a)$$

or, using Parseval's theorem,

$$\frac{A^2}{2} \frac{1}{2\pi} \int_{-\infty}^{\infty} |H(\omega)|^2\, d\omega. \qquad (5.42b)$$

Inserting Eq. (5.42a) into Eq. (5.41), we have

$$\frac{S}{N} = 2\frac{E_b(=E_s)}{N_0}. \qquad (5.42c)$$

Hence we obtain

$$P_e = \tfrac{1}{2}\mathrm{erfc}(\sqrt{\lambda}) \qquad (5.43a)$$

$$= Q(\sqrt{2\lambda}), \qquad (5.43b)$$

where $\lambda = E_s/N_0 = E_b/N_0$.

Error rates for BPSK and other techniques are shown in Fig. 5.19.

QPSK. The QPSK signal is given by Eq. (5.4). The average energy per symbol can be calculated as

$$E_s = A^2 \int_{-\infty}^{\infty} h^2(t)\, dt. \qquad (5.44)$$

With a discussion similar to the analysis of the BPSK, we obtain the signal to noise power ratio at the output of the matched filter at the sampling instant for both the in-phase and quadrature signal as

$$\frac{S}{N} = \frac{A^2}{N_0} \int_{-\infty}^{\infty} h^2(t)\, dt, \qquad (5.45)$$

which is the same as that of the BPSK system. Rewriting the above equation with Eq. (5.44), we have

$$\frac{S}{N} = \frac{E_s}{N_0} (\equiv \lambda). \qquad (5.46)$$

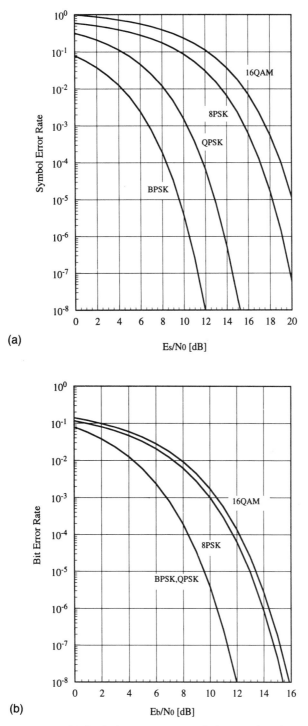

Figure 5.19. Error rate. (a) Symbol error rate vs. symbol energy to noise power density ratio. (b) Bit error rate vs. bit energy to noise power density ratio.

Using the relation $E_s = 2E_b$, we get

$$\frac{S}{N} = 2\frac{E_b}{N_0}. \tag{5.47}$$

A symbol error occurs when there is a decision error in the in-phase signal or quadrature signal. The error probability q for the in-phase or quadrature signal is $q = \frac{1}{2}\mathrm{erfc}(\sqrt{\lambda/2})$. Since the in-phase and quadrature noise are uncorrelated, the probability that a symbol is correctly decided becomes $(1-q)^2$. Thus we have the symbol error rate as

$$P_{es} = 1 - (1 - q)^2$$
$$= 2q - q^2 \tag{5.48a}$$
$$\approx 2q(q \ll 1), \tag{5.48b}$$

where

$$q = \frac{1}{2}\mathrm{erfc}(\sqrt{\lambda/2}) = Q(\sqrt{\lambda}) = \frac{1}{2}\mathrm{erfc}(\sqrt{E_b/N_0})$$

and $\lambda = E_s/N_0$ is the energy per symbol to noise power density ratio. The bit error rate depends on the assignment of the bit pairs of the four phase points. When we use a Gray code assignment (Fig. 5.20), a symbol error $m_1 \rightarrow m_2$ or $m_1 \rightarrow m_4$ causes one error in the two bits and a symbol error $m_1 \rightarrow m_3$ causes a two-bit error. Hence we have

$$P_{eb} = \frac{1}{2}q(1 - q) + \frac{1}{2}q(1 - q) + q^2$$
$$= q$$
$$= \frac{1}{2}\mathrm{erfc}(\sqrt{E_b/N_0}). \tag{5.49}$$

Thus, bit error rate as a function of E_b/N_0 is the same for BPSK and QPSK with the Gray code.

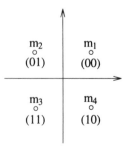

Figure 5.20. The Gray-coded QPSK.

Offset QPSK (OQPSK) and $\pi/4$ shifted QPSK have the same error rate as QPSK. This can be seen from the fact that the in-phase and quadrature signals are independent of each other, and decisions are made symbol by symbol using a matched filter. Since no intersymbol interference is considered, a demodulated pulse signal in the in-phase and quadrature components can be decided independently. Thus the timing offset between the in-phase and quadrature components in the OQPSK and the $\pi/4$ phase shift in the $\pi/4$ shifted QPSK have no effect on the decision-making process. Therefore, the signal to noise power ratio at the sampling instant at the output of the matched filter shows the same value for QPSK, OQPSK, and $\pi/4$ shifted QPSK.

Symbol and bit error rates for QPSK given with Eqs. (5.48b) and (5.49) are shown in Figs. 5.19(a) and (b), respectively.

M-*ary PSK* [1]. We consider a PSK signal whose phase takes M equidistant points (Fig. 5.21). A decision error occurs when the received signal phase deviates more than π/M due to the noise. The probability density function of the phase of a sinusoidal signal plus narrow-band band-pass Gaussian noise is given by Eq. (2.141). Then the (symbol) error rate can be calculated as

$$P_{es} = 1 - \int_{-\pi/M}^{\pi/M} p(\theta)\, d\theta$$

$$= 1 - \frac{1}{2\pi} \int_{-\pi/M}^{\pi/M} e^{-\lambda}\{1 + \sqrt{4\pi\lambda}\, \cos\theta e^{\lambda\cos^2\theta}[1 - Q(\sqrt{2\lambda}\, \cos\theta)]\}\, d\theta, \quad (5.50)$$

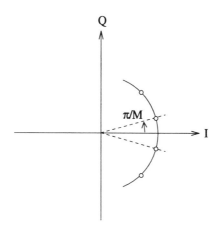

Figure 5.21. *M*-ary PSK signals.

where λ is the pulse energy (energy per symbol) to the noise power density ratio at the passband. For $\lambda \gg 1$ and $M \gg 2$, using $Q(x) \simeq 1/(x\sqrt{2\pi})e^{-x^2/2}$ for $x \gg 1$, Eq. (5.50) is approximated as

$$P_{es} \simeq 2Q\left(\sqrt{2\lambda \sin^2 \frac{\pi}{M}}\right) \tag{5.51a}$$

$$\simeq 2Q\left(\sqrt{\frac{2\pi^2\lambda}{M^2}}\right). \tag{5.51b}$$

For $\lambda \gg 1$, the probability that the phase deviates by more than $2\pi/M$ can be neglected. If we use Gray coding, we expect that the bit error rate, P_{eb} becomes $P_{es}/\log_2 M$. For $M = 2$ and $M = 4$, Eq. (5.50) should become Eq. (5.43a or b) and Eq. (5.48a), respectively.

Symbol and bit error rates given by Eq. (5.51a) are shown for 8PSK in Figs. 5.19(a) and (b), respectively.

16QAM [1]. Let us consider 16QAM. The signal configuration and decision regions are shown in Figs. 5.22(a) and (b), respectively. From Fig. 5.22(b) the probability that the symbol m_1 is correctly decided becomes

$$P(C|m_1) = \text{Prob}\left(n_x > -\frac{d}{2}\right)\text{Prob}\left(n_y > -\frac{d}{2}\right)$$

$$= \left[1 - Q\left(\frac{d/2}{\sigma_n}\right)\right]^2, \tag{5.52}$$

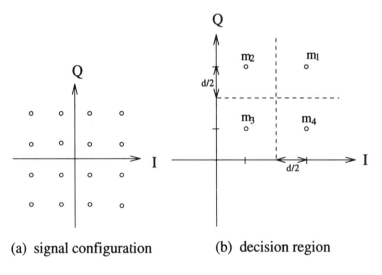

(a) signal configuration   (b) decision region

Figure 5.22. 16QAM.

where σ_n^2 is the noise power at the output of the baseband filter. Assuming $p = 1 - Q[(d/2)/\sigma_n]$, and using previous arguments, we get

$$P(C|m_1) = p^2,$$
$$P(C|m_2) = P(C|m_4) = p(2p - 1),$$
$$P(C|m_3) = (2p - 1)^2.$$

Assuming equal occurrence of each symbol and using the symmetry of the signal configuration, the probability of correct decision becomes

$$P(C) = \tfrac{1}{16} \sum_{i=1}^{16} P(C|m_i)$$

$$= \left(\frac{3p - 1}{2}\right)^2. \tag{5.53}$$

Hence the symbol error probability is given as

$$P_{es} = 1 - P(C)$$
$$= \tfrac{9}{4}(p + \tfrac{1}{3})(1 - p). \tag{5.54a}$$

For $p \approx 1$, we have

$$P_{es} \approx 3(1 - p)$$

$$= 3Q\left(\frac{d}{2\sigma_n}\right) \tag{5.54b}$$

At the sampling instant the signal power of the symbol m_1 is

$$S_1 = \left(\frac{3d}{2}\right)^2 + \left(\frac{3d}{2}\right)^2 = \tfrac{9}{2}d^2.$$

Similarly, we have

$$S_2 = S_4 = \tfrac{5}{2}d^2, \qquad S_3 = \tfrac{1}{2}d^2.$$

Hence the average power $\overline{S}$ becomes

$$\overline{S} = \tfrac{1}{4}(S_1 + S_2 + S_3 + S_4) = \tfrac{5}{2}d^2. \tag{5.55}$$

Thus

$$P_{es} \approx 3Q\left(\sqrt{\frac{S}{5N}}\right), \tag{5.56}$$

where $N = 2\sigma_n^2$.

In the following, the average $\bar{S}/N$ is expressed by the input signal energy per symbol and the noise power spectral density of $N_0/2$.

The 16QAM signal can be expressed as

$$s(t) = \sum_{n=-\infty}^{\infty} a_n h(t - nT) \cos \omega_c t + \sum_{n=-\infty}^{\infty} b_n h(t - nT) \sin \omega_c t, \qquad (5.57)$$

where a_n and b_n take values of $\pm A, \pm 3A$. Multiplying Eq. (5.57) by $2 \cos \omega_c t$ and $2 \sin \omega_c t$, we have the demodulated in-phase and quadrature signals,

$$s_{dx}(t) = \sum_{n=-\infty}^{\infty} a_n h(t - nT),$$

$$s_{dy}(t) = \sum_{n=-\infty}^{\infty} b_n h(t - nT).$$

Applying these signals to the matched filter, we have a signal level at the sampling instant of

$$A \int_{-\infty}^{\infty} h^2(t) = \frac{d}{2}. \qquad (5.58)$$

The average symbol energy E_s is given as

$$E_s = \tfrac{1}{4}[A^2 + (-A)^2 + (3A)^2 + (-3A)^2] \times \int_{-\infty}^{\infty} h^2(t)\, dt$$

$$= 5A^2 \int_{-\infty}^{\infty} h^2(t)\, dt. \qquad (5.59)$$

From Eqs. (5.55) and (5.40b) we have

$$\frac{\bar{S}}{N} = \frac{\frac{5}{2}\left[2A \int_{-\infty}^{\infty} h^2(t)\, dt \right]^2}{2N_0 \int_{-\infty}^{\infty} h^2(t)\, dt} = \frac{5A^2}{N_0} \int_{-\infty}^{\infty} h^2(t)\, dt.$$

Inserting Eq. (5.59) into the above equation, we get

$$\frac{\bar{S}}{N} = \frac{E_s}{N_0} = \frac{4E_b}{N_0}, \qquad (5.60)$$

where we have used $E_b = E_s/4$ (Eq. 2.65).

Hence, from Eqs. (5.56) and (5.60) we have

$$P_{es} \approx 3Q\left(\sqrt{\frac{E_s}{5N_0}}\right)$$

$$= 3Q\left(\sqrt{\frac{4E_b}{5N_0}}\right). \tag{5.61}$$

Assuming Gray coding, the bit error rates, P_{eb}, become $P_{eb} \approx P_{es}/4$. Symbol and bit error rate performances for 16QAM given by Eq. (5.61) are shown in Figs. 5.19(a) and (b), respectively.

5.5.2 Envelope Detection

In a narrow sense, noncoherent detection means envelope detection. The detector consists of the matched filter followed by the envelope detector, a sampler, and a decision device (Fig. 5.23). This detection is optimal when the phase difference between the input signal carrier and the recovered carrier is unknown. Let us assume that envelopes of modulated signals are orthogonal to each other for different symbol data. Thus, both signal and noise appear at the output of the matched filter, while only noise appears at the output of the other filter.

Using the probability density functions of envelopes of the sinusoidal signal plus noise (Eq. 2.140) and the noise (Eq. 2.125), the error rate can be analyzed.

For an ASK signal, where a modulated signal is expressed as

$$s(t) = \sum_{n=-\infty}^{\infty} a_n h(t - nT) \cos(\omega_c t + \theta_0) \qquad (a_n = A \text{ or } 0),$$

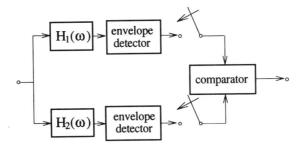

Figure 5.23. Noncoherent demodulation.

the error rate is given in [1] as

$$P_e \approx \tfrac{1}{2}e^{-\lambda/2}, \qquad \lambda \gg 1, \tag{5.62}$$

where $\lambda \equiv E_b/N_0$ is the energy per bit to noise power density ratio.

For an FSK signal the error rate is given in [1] as

$$P_e = \tfrac{1}{2}e^{-\lambda/2}. \tag{5.63}$$

5.5.3 Differential Detection

The phase difference between two symbols can be detected. Detection of the phase difference is carried out by multiplying the preceding signal, which is delayed by a symbol time, with the present signal (Fig. 5.24). This is alternatively termed differentially coherent detection, delay detection, or phase comparison detection. Here M-ary PSK is considered.

The operation of the differential detection is analyzed in the following. The M-ary PSK signal is given by Eq. (5.6). Denote the output signal of the (band-pass) matched filter as

$$s_m(t) = \sum_{n=-\infty}^{\infty} Ag(t - nT)\cos\left(\omega_c t + \frac{2\pi}{M}a_n\right) \quad (a_n = 0, 1, 2, \ldots, M-1), \tag{5.64}$$

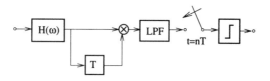

(a) for BPSK

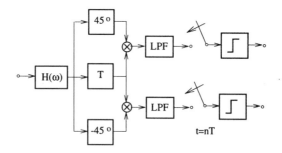

(b) for QPSK

Figure 5.24. Differential detection system.

where

$$g(t) = h(t) * h(t_0 - t).$$

The output signal of the delay line becomes

$$s_d(t) = \sum_{n=-\infty}^{\infty} A g(t - nT) \cos\left(\omega_c(t - T) + \frac{2\pi a_{n-1}}{M}\right). \tag{5.65}$$

Then for BPSK ($M = 2$) the output signal of the low-pass filter is

$$s_o(t) = \sum_{m=-\infty}^{\infty} \sum_{n=-\infty}^{\infty} \tfrac{1}{2} A^2 g(t - mT) g(t - nT) \cos[\omega_c T + (a_m - a_{n-1})\pi],$$
$$a_n = 0 \text{ or } 1, \tag{5.66}$$

where it is assumed that the low-pass filter attenuates the higher harmonics and its bandwidth is wide enough to pass the baseband signal.

If $g(nT) = 0$ for $n \neq 0$, that is, the system satisfies Nyquist's first criterion, we have at the sampling instant

$$s_o(nT) = \tfrac{1}{2} A^2 g^2(0) \cos[\omega_c T + (a_n - a_{n-1})\pi]. \tag{5.67}$$

When $\omega_c T = 2n\pi$ ($n = 1, 2, \ldots$) we have

$$s_o(nT) = \tfrac{1}{2} A^2 g^2(0) \cos[a_n - a_{n-1})\pi], \tag{5.68}$$

then

$$s_o(nT) \begin{cases} > 0, & a_n = a_{n-1} \\ < 0, & a_n \neq a_{n-1} \end{cases} \tag{5.69}$$

We can detect whether a data change has or has not occurred between the symbols by determining the polarity of $s_o(nT)$. Since we detect the data change, the data must be differentially encoded (Section 3.2.7) at the transmitter.

For QPSK the demodulated signals can be expressed as

$$s_{ox}(nT) = A_0 \cos\left((a_n - a_{n-1})\frac{\pi}{2} + \frac{\pi}{4}\right) \tag{5.70a}$$

$$s_{oy}(nT) = A_0 \sin\left((a_n - a_{n-1})\frac{\pi}{2} + \frac{\pi}{4}\right), \qquad a_n = 0, 1, 2, 3 \tag{5.70b}$$

The phase difference $(a_n - a_{n-1})\pi/2 \pmod{2\pi}$ becomes $0, \pm\pi/2, \pi$, and can be determined by a threshold detection on $s_{ox}(nT)$ and $s_{oy}(nT)$.

Error Rate. Consider a signal and noise at $t = t_1$ and t_2 as

$$s(t_1) = A \cos[\omega_c t_1 + \varphi(t_1)] + n_x(t_1) \cos[\omega_c t_1 + \varphi(t_1)]$$
$$- n_y(t_1) \sin[\omega_c t_1 + \varphi(t_1)], \qquad (5.71a)$$

$$s(t_2) = A \cos[\omega_c t_2 + \varphi(t_2)] + n_x(t_2) \cos[\omega_c t_2 + \varphi(t_2)]$$
$$- n_y(t_2) \sin[\omega_c t_2 + \varphi(t_2)]. \qquad (5.71b)$$

The phase difference between $s(t_1)$ and $s(t_2)$ is

$$\angle s(t_2) - \angle s(t_1) = \omega_c(t_2 - t_1) + \varphi(t_2) - \varphi(t_1)$$
$$+ \tan^{-1} \frac{n_y(t_1)}{A + n_x(t_1)} - \tan^{-1} \frac{n_y(t_2)}{A + n_x(t_2)}, \qquad (5.72)$$

where the term $\varphi(t_2) - \varphi(t_1)$ expresses the phase shift with modulation and the term

$$\psi \equiv \tan^{-1} \frac{n_y(t_1)}{A + n_x(t_1)} - \tan^{-1} \frac{n_y(t_2)}{A + n_x(t_2)}$$

corresponds to the deviation due to noise. The probability density function of the phase θ for the signal and noise is given by Eq. (2.141). If the noise values at $t = t_1$ and t_2 are uncorrelated, the probability density of the phase difference ψ is given as

$$p(\psi) = \int_{-\pi}^{\pi} p(\theta_1) p(\theta_1 + \psi) \, d\theta_1 \qquad (-\pi \le \psi \le \pi).$$

It is given by Pawula, Rice and Roberts [14] as follows:

$$\text{Prob}\{\psi_1 \le \psi \le \psi_2\} = F(\psi_2) - F(\psi_1), \qquad (5.73)$$

where

$$F(\psi) = -\frac{\sin \psi}{4\pi} \int_{-\pi/2}^{\pi/2} \frac{e^{-\lambda(1 - \cos \psi \cos t)}}{1 - \cos \psi \cos t} \, dt \qquad (5.74)$$

and $\lambda = A^2/2\sigma_n^2$ is the average signal to noise power ratio at the sampling instants. When a matched filter receiver is considered, λ becomes energy per symbol to noise density ratio.

A symbol error occurs when $|\psi| > \pi/M$. Thus, the symbol error rate is given as

$$P_e = \int_{\pi/M}^{\pi} p(\psi)\,d\psi + \int_{-\pi}^{-\pi/M} p(\psi)\,d\psi$$

$$= 2\int_{\pi/M}^{\pi} p(\psi)\,d\psi$$

$$= 2\,\mathrm{Prob}\{\pi/M \le \psi \le \pi\}.$$

Using Eqs. (5.73) and (5.74), we have

$$P_e = \frac{\sin(\pi/M)}{2\pi}\int_{-\pi/2}^{\pi/2} \frac{e^{-\lambda[1-\cos(\pi/M)\cos t]}}{1-\cos(\pi/M)\cos t}\,dt. \tag{5.75}$$

For BPSK ($M = 2$), we have

$$P_e = \tfrac{1}{2}e^{-\lambda}. \tag{5.76}$$

An approximate formula for Eq. (5.75) is given [15] as

$$P_e \approx \sqrt{\frac{1+\cos(\pi/M)}{2\cos(\pi/M)}}\,\mathrm{erfc}\sqrt{\{\lambda[1-\cos(\pi/M)]\}}. \tag{5.77}$$

For Gray-coded QPSK, the bit error rate P_{eb} is given as

$$P_{eb} = 2\times\frac{1}{2}\mathrm{Prob}\!\left(\frac{\pi}{4}<\psi<\frac{3}{4}\pi\right) + 2\mathrm{Prob}\!\left(\frac{3}{4}\pi<\psi<\pi\right)$$

$$= \mathrm{Prob}\!\left(\frac{\pi}{4}<\psi<\pi\right) + \mathrm{Prob}\!\left(\frac{3}{4}\pi<\psi<\pi\right)$$

$$= \frac{1}{4\sqrt{2\pi}}\int_{-\pi}^{\pi} \frac{\exp\!\left(-\lambda(1-\frac{1}{\sqrt{2}}\cos t)\right)}{1-(1/\sqrt{2})\cos t}\,dt. \tag{5.78}$$

Error rates for BPSK, QPSK, and 8PSK with differential detection are shown in Fig. 5.25. We must remember that the above results are given with no intersymbol interference and no correlation between noise at the sampling instant. These conditions are satisfied for a linear PSK system, where the square root of the Nyquist-I transfer characteristics is given for the transmit and receive bandpass filters and where the noise power spectrum is flat.

Satisfying the condition $\omega_c T = 2n\pi$ is rather easy for the case $\omega_c \gg 1/T$. In this case, if $\omega_c T \ne 2n\pi$, we can adjust the time delay $T \to T + \Delta T$ to

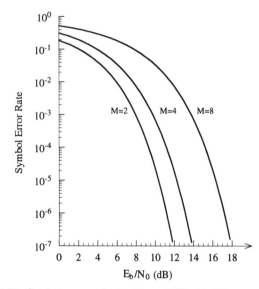

Figure 5.25. Symbol error rate for *M*-ary PSK with differential detection.

achieve $\omega_c(T + \Delta T) = \omega_c T(1 + \Delta T/T) = 2n\pi$. Since $\omega_c \gg 1$, $\Delta T/T$ can be set small. Small adjustments of time delay never cause significant problems in detecting the phase difference of two symbols.

On the other hand, the phase error, which is generated when $\omega_c T \neq 2n\pi$, causes a significant degradation of error rate performance. The phase error is produced by deviation of the time delay or the carrier frequency. The symbol error rate when the phase error $\Delta\theta$ is taken into consideration can be given as

$$
P_e\{\Delta\theta\} = \mathrm{Prob}\left(\frac{\pi}{M} - \Delta\theta < \psi < \pi\right) + \mathrm{Prob}\left(-\pi < \psi < -\frac{\pi}{M} - \Delta\theta\right)
$$

$$
= \frac{\sin(\pi/M - \Delta\theta)}{4\pi}\int_{-\pi/2}^{\pi/2}\frac{e^{-\lambda[1-\cos(\pi/M-\Delta\theta)\cos t]}}{1 - \cos(\pi/M - \Delta\theta)\cos t}\,dt
$$

$$
+ \frac{\sin(\pi/M + \Delta\theta)}{4\pi}\int_{-\pi/2}^{\pi/2}\frac{e^{-\lambda[1-\cos(\pi/M+\Delta\theta)\cos t]}}{1 - \cos(\pi/M + \Delta\theta)\cos t}\,dt. \qquad (5.79)
$$

The symbol error rate performances for BPSK and QPSK are shown in Fig. 5.26 with the phase error as a parameter.

5.5.4 Frequency Discriminator Detection

A frequency discriminator detection system is shown in Fig. 5.27. The input signal and noise are fed into a band-pass filter, followed by a limiter, a

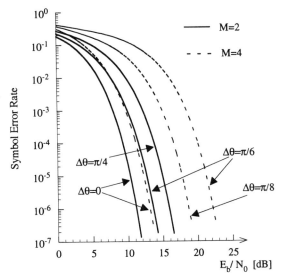

Figure 5.26. Effects of phase error on symbol error rate for BPSK and QPSK with differential detection.

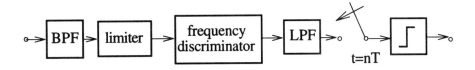

Figure 5.27. Frequency discriminator detection system.

frequency discriminator, a low-pass filter, and a sample-and-decision circuit. This demodulator, except for the sample-and-decision circuit, is widely used for analog FM systems in mobile radio communication. Implementation of the limiter circuit is rather easy compared with an automatic gain control (AGC) circuit, especially for application to mobile radio communication, where there is fast fading and the received signal level has a large dynamic range.

The analysis of the error rate performance of the frequency discriminator detection system is generally difficult, since the demodulator is a highly nonlinear circuit. This is especially true when the effects of the band-pass filter and the post-detection filters on intersymbol interference must be considered at the same time. The digital FM signal can be expressed as

$$s(t) = A_0 \cos[\omega_c t + \varphi(t)], \tag{5.80}$$

where

$$\varphi(t) = k_F \int_{-\infty}^{t} \sum_{n=-\infty}^{\infty} a_n h(t - nT) \, dt \qquad (5.81a)$$

$$= k_F \sum_{n=-\infty}^{\infty} a_n g(t - nT) \qquad (5.81b)$$

and a_n takes discrete values, $h(t)$ is the impulse response of the premodulation filter, and $g(t) = \int_{-\infty}^{t} h(t) \, dt$.

The signal at the output of the band-pass filter is written as

$$s_{BPF}(t) = \text{Re}\{A_0[\cos\varphi(t) + j\sin\varphi(t)] * h_P(t) \, e^{j\omega_c t}\}, \qquad (5.82)$$

where $h_P(t)$ is the low-pass equivalent impulse response of the IF band-pass filter. We assume that $h_P(t)$ is real, that is, $H_P(-\omega) = H_P^*(\omega)$, where $H_P(\omega) \leftrightarrow h_P(t)$. Then $s_{BPF}(t)$ is expressed as

$$s_{BPF}(t) = A(t)\cos[\omega_c t + \phi(t)], \qquad (5.83)$$

where

$$\frac{A^2(t)}{A_0^2} = \{\cos\varphi(t) * h_P(t)\}^2 + \{\sin\varphi(t) * h_P(t)\}^2 \qquad (5.84)$$

and

$$\phi(t) = \tan^{-1}\frac{\sin\varphi(t) * h_P(t)}{\cos\varphi(t) * h_P(t)}. \qquad (5.85)$$

The noise signal at the output of the band-pass filter is given as

$$n_{BPF}(t) = n_x(t)\cos[\omega_c t + \phi(t)] - n_y(t)\sin[\omega_c t + \phi(t)], \qquad (5.86)$$

where $n_x(t)$ and $n_y(t)$ are Gaussian random variables with zero mean and variance $\sigma_n^2 = (N_0/2\pi)\int_{-\infty}^{\infty}|H_P(\omega)|^2 \, d\omega$, and $N_0/2$ is the input noise power spectral density.

The output signal of the band-pass limiter is expressed as

$$s_{LIM}(t) = \cos[\omega_c t + \phi(t) + \eta(t)], \qquad (5.87)$$

where

$$\eta(t) = \tan^{-1}\frac{n_y(t)}{A(t) + n_x(t)}. \qquad (5.88)$$

Hence, the output signal of the frequency discriminator is given as

$$d(t) = \dot{\phi}(t) + \dot{\eta}(t), \tag{5.89}$$

where $(\cdot)$ denotes time differentiation. $\phi(t)$ corresponds to the information-bearing signal. The noise term $\dot{\eta}(t)$ can be given as

$$\dot{\eta}(t) = \frac{[A(t) + n_x(t)]\dot{n}_y(t) - n_y(t)[\dot{A}(t) + \dot{n}_x(t)]}{[A(t) + n_x(t)]^2 + n_y^2(t)}. \tag{5.90}$$

$\dot{\eta}(t)$ is no longer Gaussian.

The output signal of the post-demodulation filter with impulse response $h_d(t)$ becomes

$$\begin{aligned} d_{LPF}(t) &= d(t) * h_d(t) \\ &= s_d(t) + n_d(t), \end{aligned} \tag{5.91}$$

where $s_d(t) \equiv \dot{\phi}(t) * h_d(t)$ and $n_d(t) \equiv \dot{\eta}(t) * h_d(t)$.

A decision error occurs when $d_{LPF}(nT)$ deviates over the threshold value(s). In order to get the error rate theoretically, we must know the statistical property of the intersymbol interference in signal $s_d(t)$ and the probability density function of the noise $n_d(t)$. Since $s_d(t)$ and $n_d(t)$ are produced through nonlinear processes, it is impossible to find an expression for the error rate. The intersymbol interference depends on the premodulation filter, modulation index, the band-pass filter, and the post-demodulation filter, as well as the data sequence.

The behavior of the noise term $\dot{\eta}(t)$ was investigated by S.O. Rice [16] using the concept of "clicks." $\dot{\eta}(t)$ is expressed with the continuous part and discontinuous or the click part as

$$\dot{\eta}(t) = \dot{\eta}_c(t) + \dot{\eta}_d(t). \tag{5.92}$$

The click part $\dot{\eta}_d(t)$ is expressed as

$$\dot{\eta}_d(t) = \sum_i 2\pi \, \delta(t - t_i) - \sum_j 2\pi \, \delta(t - t_j), \tag{5.93}$$

where the first and the second terms express positive and negative clicks, respectively. The click is an impulse that occurs randomly as shown schematically in Fig. 5.28. The probabilities that clicks occur in a time T_0 is assumed to obey a Poisson distribution,

$$p_N = \frac{(\lambda T_0)^N e^{-\lambda T_0}}{N!}, \tag{5.94}$$

where N is the number of clicks and λ is the rate of occurrence of a click in an infinitely small time interval or, equivalently, the average number of

Figure 5.28. Schematic drawing of output noise of a frequency discriminator.

clicks in a unit time. The probability of occurrence of a click in a unit time
is given [16] for a positive click as

$$N_+(t) = \frac{r}{2}\left[\left(1 + \frac{f_i^2(t)}{r^2}\right)^{1/2} \operatorname{erfc}\left(\rho(t) + \rho(t)\frac{f_i^2(t)}{r^2}\right)^{1/2}\right.$$
$$\left. - \frac{|f_i(t)|}{r}e^{-\rho(t)}\operatorname{erfc}\left(\frac{|f_i(t)|}{r}\sqrt{\rho(t)}\right)\right] \tag{5.95}$$

and for a negative click as

$$N_-(t) = N_+(t) + |f_i(t)|e^{-\rho(t)}, \tag{5.96}$$

where $f_i(t) = \dot{\phi}(t)/2\pi$ is the instantaneous frequency of the signal and $f_i(t) \geq 0$
is assumed. $\rho(t)$ is the signal to noise power ratio at the output of the
band-pass filter and r is a parameter that is defined as

$$r = \frac{1}{2\pi}\frac{\langle \dot{n}_x^2(t) \rangle}{\langle n_x^2(t) \rangle} = \frac{1}{2\pi}\frac{\langle \dot{n}_y^2(t) \rangle}{\langle n_y^2(t) \rangle}. \tag{5.97}$$

$\rho(t)$ is written as

$$\rho(t) = \frac{A^2(t)}{2\sigma_n^2} \tag{5.98}$$

and σ_n^2 can be rewritten as

$$\sigma_n^2 = \frac{N_0}{2}\frac{1}{2\pi}\int_{-\infty}^{\infty}|H_P(\omega)|^2 \, d\omega$$

$$= \frac{N_0}{2}\int_{-\infty}^{\infty}|H_P(2\pi f)|^2 \, df \qquad (f = \omega/2\pi)$$

$$= N_0 B_{IF}, \tag{5.99}$$

where $B_{IF} = \int_0^\infty |H_P(2\pi f)|^2 \, df$ is the equivalent noise bandwidth of the band-pass filter. It can be seen from Eq. (5.96) that $N_-(t) \geq N_+(t) \, [f_i(t) \geq 0]$. For $f_i(t) < 0$, it is understood that $N_+(t)$ and $N_-(t)$ correspond to the negative and positive clicks, respectively.

5.5.4.1 Integrate-and-dump Post-demodulation Filter System. The error rate theories developed so far are different for the systems under consideration. Some of them, [4,17–19] treat a system with a rectangular pulse shape (NRZ signaling) for the transmit baseband signal and the integrate-and-dump filter as the post-detection filter. In this system, no intersymbol interference occurs, as long as the effect of the band-pass filter is ignored.

By integrating or equivalently filtering the output signal $d(t)$ of the frequency discriminator over a time period $(n-1)T \leq t \leq nT$, we have

$$d_{LPF}(nT) = \phi(nT) - \phi(nT - T) + \eta(nT) - \eta(nT - T). \quad (5.100)$$

The signal term $\phi(nT) - \phi(nT - T)$ is different from $\varphi(nT) - \varphi(nT - T)$ due to intersymbol interference, which depends on the data sequence. Let us denote

$$\Delta\phi(nT) \equiv \phi(nT) - \phi(nT - T)$$
$$= \Delta\varphi(nT) + \delta\varphi(nT| \cdots a_{n-1}a_n a_{n+1} \cdots), \quad (5.101)$$

where $\delta\varphi(nT| \cdots a_{n-1}a_n a_{n+1} \cdots)$ denotes the intersymbol interference given the data sequence. From Eq. (5.12) we have $\Delta\varphi(nT) = \pi m a_n / a_{max}$.

The noise term $\Delta\eta(nT) \equiv \eta(nT) - \eta(nT - T)$ can be expressed from Eqs. (5.92) and (5.93) as

$$\Delta\eta(nT) = \eta_c(nT) - \eta_c(nT - T) + 2\pi[N^+(nT) - N^-(nT)], \quad (5.102)$$

where $N^+(nT)$ and $N^-(nT)$ are the numbers of positive and negative clicks occurred during a time period of $(n-1)T \leq t \leq nT$. The probability of $N^+(nT)$ or $N^-(nT)$ can be given from Eq. (5.94) with substitution of $N = N^+(nT)$ or $N = N^-(nT)$, and

$$\lambda T_0 = \int_{(n-1)T}^{nT} N_+(t) \, dt \qquad \text{positive click} \qquad (5.103a)$$

or

$$= \int_{(n-1)T}^{nT} N_-(t) \, dt \qquad \text{negative click.} \qquad (5.103b)$$

When $\eta_c(nT)$, $\eta_c(nT - T)$ and $N^+(nT)$, $N^-(nT)$ are independent from each other, the probability density function $\Delta\eta(nT)$ is given by the convolution of the probabilities for those variables.

Tjhung and Wittke [18] calculated the error rate for 2-level FM by numerical calculation of the intersymbol interference $\delta\varphi(nT| \cdots a_{n-1}a_n a_{n+1} \ldots)$ with a random data sequence of 30 bits. Pawula [19,22] obtained a closed form formula of bit error rate for a narrow-band system by assuming a data sequence of 3 bits. His result is cited with his notation as [19] (Copyright © IEEE 1981)

$$P_e = P_{\text{continuous}} + P_{\text{click}}, \qquad (5.104)$$

where

$$P_{\text{click}} = \frac{h}{4}e^{-R_d} + \int_0^\pi \frac{d}{dx}\left(\tan^{-1}\frac{-m\cos x}{1 - n\cos(2x + \delta)}\right)$$

$$\times \exp\left(-R_a\frac{[1 - n\cos(2x + \delta)]^2 + m^2\cos^2 x}{(1 - n\cos\delta)^2 + m^2}\right)\frac{1}{4\pi}dx, \quad (5.105)$$

$$P_{\text{continuous}} = \tfrac{1}{4}[P\{\psi > \Delta\phi|111\} + P\{\psi > \Delta\phi|010\} + 2P\{\psi > \Delta\phi|011\}] \quad (5.106)$$

and 111, 010, 011 ared bit patterns.

$$P\{\psi > \Delta\phi\} = \int_{\Delta\phi}^\pi p(\psi)\,d\psi \qquad (5.107)$$

where

$$p(\psi) = \frac{e^{-U}}{2\pi}\left[\cosh V + \tfrac{1}{2}\int_0^\pi d\alpha(U\sin\alpha + W\cos\psi)\cosh(V\cos\alpha)\,e^W\sin\alpha\cos\psi\right].$$

$$(5.108)$$

Parameters $\Delta\phi$, U, and V are given for the bit patterns as

"111" BIT PATTERN

$$\Delta\phi = \pi h, \qquad U = R_d, \qquad V = 0. \qquad (5.109)$$

"010" BIT PATTERN

$$\Delta\phi = 2\tan^{-1}\frac{m}{1 - n\cos\delta}, \qquad U = R_a, \qquad V = 0. \qquad (5.110)$$

"011" BIT PATTERN

$$\Delta\phi = \frac{\pi h}{2} + \tan^{-1}\frac{m}{1 - n\cos\delta}, \qquad U = \frac{R_a + R_d}{2}, \qquad V = \frac{R_a - R_d}{2} \quad (5.111)$$

and $h = 2f_d T$, $f_1 = 1/2T$,

$$W = (U^2 - V^2)^{1/2},$$

$$m = \frac{2h^2|H(f_1)|}{1 - h^2}\cot\frac{\pi h}{2}; \qquad n = \frac{2h^2|H(2f_1)|}{4 - h^2},$$

$$\delta = \angle H(2f_1) - 2\angle H(f_1), \qquad (5.112)$$

$$R_a = \frac{E_b}{N_0}\left(\frac{\sin(\pi h/2)^2}{\pi h/2}\right)\frac{(1 - n\cos\delta)^2 + m^2}{T\displaystyle\int_{-\infty}^{\infty}|H(f)|^2\,df}, \qquad (5.113)$$

$$R_d = \frac{E_b}{N_0}\frac{|H(f_d)|^2}{T\displaystyle\int_{-\infty}^{\infty}|H(f)|^2\,df}, \qquad (5.114)$$

where $H(f)$ is the transfer function of the band-pass filter, N_0 is the *one-sided* spectral density of the input noise, f_d is the frequency deviation due to modulation, T is the bit duration, and h is the modulation index (the reader should not confuse the notation used here and in other parts of this book).

5.5.4.2 *General Post-demodulation Filter System.*

A pulse shape with a narrower spectrum than that of the rectangular pulse shape is desired for the modulating signal. This is because we can get a narrower spectrum for the modulated signal as described in Section 6.2. In this case, the integrate-and-dump filter is not appropriate for the post-demodulation filter. Schilling and colleagues [20] treated the system with a Gaussian post-demodulation filter. Following this, Akaiwa and Okamoto [21] analyzed the Nyquist- and partial response-filtered multilevel FM systems as follows.

Let us assume that the error rate is given as

$$P_e = P_g + P_c \qquad (5.115a)$$

$$= \frac{1}{N}\sum_{n=1}^{N}[P_g(nT) + P_c(nT)], \qquad (5.115b)$$

where P_g is the error rate with Gaussian noise and P_c is the error rate with click noise. (This phenomenological assumption is supported by computer simulation experiments.) $P_g(nT)$ and $P_c(nT)$ denote error rates at the symbol time $t = nT$, due to the Gaussian and click noises, respectively. N is the length of the data sequence under consideration.

Error Rates Due to Gaussian Noise. To account for the effect of the modulated signal on demodulated noise we express the noise signal at the band-pass filter in a way different from Eq. (5.86):

$$n_{BPF}(t) = n_x(t) \cos(\omega_c t) - n_y(t) \sin(\omega_c t). \tag{5.116}$$

The band-pass limited signal is then

$$s_{LIM}(t) = \cos[\omega_c t + \phi(t) + \psi(t)], \tag{5.117}$$

where

$$\psi(t) = \tan^{-1} \frac{n_y(t) \cos \phi(t) - n_x(t) \sin \phi(t)}{A(t) + n_x(t) \cos \phi(t) + n_y(t) \sin \phi(t)}. \tag{5.118}$$

When the signal to noise power ratio is high, using $|A(t)| \gg |n_x(t)|, |n_y(t)|$, $\psi(t)$ is linearly approximated as

$$\psi(t) \approx \frac{n_y(t) \cos \phi(t) - n_x(t) \sin \phi(t)}{A(t)}. \tag{5.119}$$

Hence $\psi(t)$ and also the discriminator output $d\psi(t)/dt$ becomes a Gaussian variable. The equivalent circuit for the approximated noise is represented in Fig. 5.29. The expected noise power at the output of the post-demodulation filter becomes (Appendix 5.2)

$$\langle N_g(t) \rangle = N_0 \int_{-\infty}^{\infty} |N(t, \tau)|^2 \, d\tau, \tag{5.120}$$

where

$$|N(t, \tau)|^2 = \{h_d(t) * [g(t - \tau)a(t) \cos \phi(t)]\}^2 + \{h_d(t) * [g(t - \tau)a(t) \sin \phi(t)]\}^2. \tag{5.121}$$

$g(t)$ is the low-pass equivalent impulse response of the band-pass filter, $h_d(t)$ is the impulse response of the differentiating circuit including the post-demodulation filter, and $a(t) \equiv 1/A(t)$.

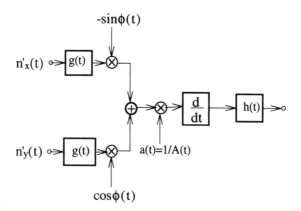

Figure 5.29. Equivalent circuit for noise at high signal to noise power ratio.

Now the error rate due to the Gaussian noise at time slot $t = nT$ is given as

$$P_g(nT_s) = Q(d_n^+/\sigma_n) + Q(d_n^-/\sigma_n), \qquad (5.122)$$

where

$$d_n^\pm = 2\pi[f_m \pm \Delta f_m(nT)], \qquad (5.123)$$

$$\sigma_n = [\langle N_g(t)\rangle]^{1/2}, \qquad (5.124)$$

$$Q(y) = \frac{1}{\sqrt{2\pi}} \int_y^\infty \exp(-x^2/2)\, dx, \qquad (5.125)$$

and $2\pi f_m$ is half the signal distance at the output of the post-detection filter. The term $2\pi \Delta f_m(nT)$ represents intersymbol interference at the sampling instant $t = nT$. These values can be obtained by numerical calculation assuming a data sequence. σ_n^2 is the expected noise power at the sampling instant. When the data take the highest (lowest) value, the first (second) term in the right-hand side of Eq. (5.122) may be ignored.

Error Rates Due to Click Noise. A positive click (impulsive) noise that occurs at $t = t_1$ generates a waveform $2\pi h(t - t_1)$ at the output of the post-detection low-pass filter; $h(t)$ is the impulse response of the filter. An error occurs if the following relations hold (Fig. 5.30):

$$2\pi h(nT - t_1)\ (\geq 0) \geq 2\pi f_m - 2\pi \Delta f_m(nT), \qquad (5.126a)$$

$$2\pi h(nT - t_1)\ (<0) \leq -[2\pi f_m + 2\pi \Delta f_m(nT)]. \qquad (5.126b)$$

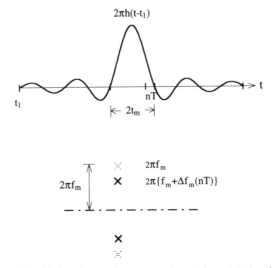

Figure 5.30. Click noise at the output of post-demodulation filter.

For simplicity of discussion, we only consider the time period $(2t_m)$ where $h(t) \geq 0$; then Eqs. (5.126a) and (5.126b) correspond to the positive and negative clicks, respectively. Furthermore, we neglect the possibility that two or more clicks cause an error, since the probability of that is low. (This cannot be valid for a wide-band digital FM system with signal to noise power ratio under the threshold value.) Hence the error rate due to the click noise is given as

$$P_c(nT) = \int_{D_{n1}^+}^{D_{n2}^+} N^+(t - nT)\, dt + \int_{D_{n1}^-}^{D_{n2}^-} N^-(t - nT)\, dt \qquad (5.127)$$

where $D_{n1}^{\pm}$, $D_{n2}^{\pm}$ are parameters that satisfy

$$h(D_{n1}^+) = h(D_{n2}^+) = f_m - \Delta f_m(nT) \geq 0, \qquad (5.128a)$$

$$h(D_{n1}^-) = h(D_{n2}^-) = f_m + \Delta f_m(nT) \geq 0. \qquad (5.128b)$$

For an $h(t)$ that is symmetrical with respect to time, we have $D_{n1}^+ = D_{n2}^+$ and $D_{n1}^- = D_{n2}^-$. No positive (negative) clicks cause errors at the symbol which takes the highest (lowest) level.

Error Rates in the Case of No Intersymbol Interference. If the bandwidth B_{IF} of the band-pass filter is wide, we can ignore the intersymbol interference due to the filter and the effect that modulation has on the Gaussian noise. This condition is supported by experiment if the bandwidth is wider than the Carson bandwidth:

$$B_{IF} \geq 2(\Delta F_m + f_b), \qquad (5.129)$$

where ΔF_m is the maximum frequency deviation and f_b is the highest frequency of the modulating baseband signal.

Furthermore, we assume that the baseband channel, including the premodulation filter and the post-detection filter, never produces intersymbol interference, except for the partial response system. The output signal of the post-demodulation filter is expressed as

$$s_0(t) = 2\pi f_m a_i h_T(t) * h_R(t), \qquad (i-1)T < t < iT, \qquad (5.130)$$

where the symbol data a_i take discrete values such that there is a signal distance of 2 between the neighboring values; $h_T(t)$ and $h_R(t)$ are the impulse response of the premodulation and post-demodulation filter, respectively. It is understood that

$$\int_{-\infty}^{\infty} h_T(t)\, dt = H_T(\omega = 0) = T, \qquad (5.131)$$

where $H_T(\omega) \leftrightarrow h_T(t)$. For the sake of simplicity, we let

$$h_T(t) * h_R(t)\big|_{t=iT} = 1. \qquad (5.132)$$

The signal distance $2d$ becomes

$$2d = 4\pi f_m. \qquad (5.133)$$

The transfer function for the baseband signal is

$$H(\omega) = H_T(\omega) H_R(\omega) = k_0 H_I(\omega), \qquad (5.134)$$

where $H_R(\omega) \leftrightarrow h_R(t)$, k_0 is a constant, and $H_I(\omega)$ is the transfer function which satisfies Nyquist's first criterion. For $H_I(\omega)$ with $H_I(\omega = 0) = 1$ we get $h_I(t = 0) = 1/T$ (Eq. 3.13). If we let

$$H_T(\omega) = T H_{IT}(\omega), \qquad (5.135a)$$

$$H_R(\omega) = H_{IR}(\omega), \qquad (5.135b)$$

where

$$H_{IT}(\omega = 0) = H_{IR}(\omega = 0) = 1, \qquad (5.136)$$

$$H_{IT}(\omega) H_{IR}(\omega) = H_I(\omega), \qquad (5.137)$$

then the condition given by Eq. (5.132) is satisfied.

When the bandwidth of the band-pass filter is so wide that the interference due to it is ignored, the amplitude of the output signal of the band-pass filter $A(t)$ becomes a constant value A. For a high signal to noise power ratio, the output signal of the frequency discriminator becomes

$$n_d(t) \approx \frac{d}{dt}\psi(t), \tag{5.138}$$

where

$$\psi(t) = \frac{n_y(t)\cos\varphi(t) - n_x(t)\sin\varphi(t)}{A}. \tag{5.139}$$

The autocorrelation function of $\psi(t)$ becomes

$$R_\psi(\tau) \equiv \langle\psi(t)\psi(t+\tau)\rangle = \frac{1}{A^2}R_{n_x}(\tau)\mathrm{Re}\{R_\varphi(\tau)\}, \tag{5.140}$$

where

$$R_{n_x}(\tau) = \langle n_x(t)n_x(t+\tau)\rangle = \langle n_y(t)n_y(t+\tau)\rangle$$

and

$$R_\varphi(\tau) = \langle e^{j\varphi(t)}\,e^{-j\varphi(t+\tau)}\rangle.$$

Taking the Fourier transform of Eq. (5.140), we have the power spectral density of $\psi(t)$: the convolution integral of the power spectral density of $n_x(t)$ and the FM signal $e^{j\varphi(t)}$.

When we assume a wide-band band-pass filter, the effect of FM on the noise power spectral density can be neglected. Hence the power spectrum density of $n_d(t)$ is given

$$S_{n_d}(\omega) = \frac{N_0\omega^2}{A^2}, \qquad -\pi B_{IF} < \omega < \pi B_{IF}, \tag{5.141}$$

where $N_0/2$ is the input noise power density and B_{IF} is the band-pass filter bandwidth. The average noise power at the output of the post-demodulation filter is

$$\langle N_g\rangle = \frac{1}{2\pi}\int_{-\infty}^{\infty}\frac{N_0\omega^2|H_R(\omega)|^2}{A^2}\,d\omega. \tag{5.142}$$

We define a normalized noise bandwidth by

$$W_{eq} = \frac{\displaystyle\int_{-\infty}^{\infty} \omega^2 |H_R(\omega)|^2 \, d\omega}{\displaystyle\int_{-\omega_s/2}^{\omega_s/2} \omega^2 \, d\omega \left(= \frac{2}{3}\left(\frac{\omega_s}{2}\right)^3 \right)},$$

where $\omega_s = 2\pi/T$.

Rewriting Eq. (5.142),

$$\langle N_g \rangle = \frac{1}{24} \frac{N_0 f_s}{A^2/2} (2\pi f_s)^2 W_{eq}$$

$$= \frac{1}{24} \frac{1}{E_s/N_0} (2\pi f_s)^2 W_{eq}. \tag{5.143}$$

Rewriting $d_n^{\pm} = d_0 = 2\pi f_m$ and $\sigma_n = \sigma_0 = [\langle N_g \rangle]^{1/2}$ in Eqs. (5.123) and (5.124) and taking into consideration error probability Prob(a_m) for the highest or lowest levels, we have

$$P_g = 2[1 - \tfrac{1}{2}\text{Prob}(a_m)]Q\left(\frac{d_0}{\sigma_0}\right), \tag{5.144}$$

where

$$\frac{d_0}{\sigma_0} = \frac{2f_m}{f_s}\sqrt{\frac{6}{W_{eq}}\frac{E_s}{N_0}} = \sqrt{\frac{6}{W_{eq}}\frac{E_s}{N_0}}\, m, \tag{5.145}$$

where m is the modulation index. Considering the Nyquist-I, duobinary, and class II partial response (PR) digital FM systems with their baseband filter characteristics [Eqs. (3.12), (3.38), and (3.48)] equally split between the premodulation and post-demodulation filter, and normalizing them to get $|H_R(\omega = 0)| = 1$, we have the normalized noise bandwidth

$$W_{eq} = \begin{cases} 1 + 3\left(1 - \dfrac{8}{\pi^2}\right)\alpha^2 & \text{(Nyquist-I)} & \text{(5.146a)} \\[2ex] 6\left(\dfrac{1}{\pi} - \dfrac{8}{\pi^3}\right) & \text{(duobinary)} & \text{(5.146b)} \\[2ex] \dfrac{1}{2} - \dfrac{3}{\pi^2} & \text{(class II PR),} & \text{(5.146c)} \end{cases}$$

where α is the roll-off factor.

For the integrate-and-dump filter, the integral $\int_{-\infty}^{\infty} \omega^2 |H_R(\omega)|^2 \, d\omega$ diverges and we cannot define the normalized bandwidth. For this system the noise power must be calculated taking into consideration the band-pass filter,

$$\langle N_g \rangle = \frac{1}{2\pi} \int_{-\infty}^{\infty} \frac{N_0 \omega^2 |G(\omega)H_R(\omega)|^2 \, d\omega}{A^2} , \tag{5.147}$$

where $G(\omega)$ is the low-pass equivalent transfer function of the band-pass filter. From this fact we can see that the noise power is higher for the integrate-and-dump filter system than for other systems when the band-pass filter bandwidth is much wider than the post-detection filter bandwidth. This is usually true for multilevel FM or partial response systems.

Error Rates Due to Click Noise. Assuming $\Delta f_m(nT) = 0$, $D_{n1}^+ = D_{n1}^- = D_{n1}$ and $D_{n2}^+ = D_{n2}^- = D_{n2}$ in Eqs. (5.128a) and (5.128b), we have the error rate due to click noise as

$$P_c = \frac{1}{N} \sum_{n=1}^{N} \left[\int_{D_{n1}}^{D_{n2}} [N_+(t - nT) + N_-(t - nT)] \, dt \right]. \tag{5.148}$$

The error rates are still dependent on the modulating signal $f_i(t)$ in Eq. (5.95). If we assume a dc modulating signal, $f_i(t) = a_n f_m$, then $\rho(t)$ becomes a constant value and $N_+(a_n f_m)$ and $N_-(a_n f_m)$ denotes $N_+(t)$ and $N_-(t)$ in Eqs. (5.95) and (5.96). In this case we have

$$P_c = \sum_{a_n} \text{Prob}(a_n) \, (D_{n2} - D_{n1}) \, [N_+(a_n f_m) + N_-(a_n f_m)] \tag{5.149}$$

5.5.4.3 The Optimum Modulation Index.

In an FM system with discriminator detection, the wide-band gain is well known: the demodulated signal to noise power ratio increases with the increase of the maximum frequency deviation or the modulation index [see for example Eq. (5.145)]. Figure 5.31 shows error rates for a 2-level FM as a function of the modulation index. The error rates are calculated using Eqs. (5.144), (5.146a), and (5.149) without intersymbol interference due to a band-pass filter. The band-pass filter bandwidth B_{IF} is chosen as the Carson bandwidth; that is, $B_{IF} = (m + 1 + \alpha)f_b$, where m is the modulation index, α is the roll-off factor, and f_b is the bit rate frequency. Therefore, the carrier to noise power ratio at the output of the band-pass filter becomes $C/N = E_b f_b / N_0 B_{IF} = (E_b/N_0) \times 1/(m + 1 + \alpha)$. When we increase the modulation index from a low initial value, the error rate first decreases due to the decrease of the Gaussian noise. After reaching the minimum value it increases due to the increase of the click noise (threshold effect). The optimum modulation index for the 2-level FM at a bit error rate of $10^{-2} \sim 10^{-3}$ is around 0.5. If we consider a lower error

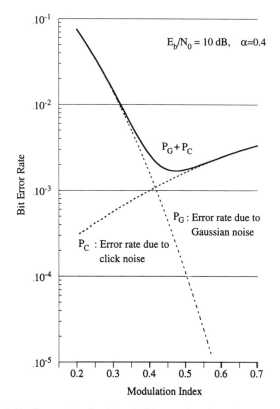

Figure 5.31. Error rates for 2-level FM as a function of modulation index.

rate and hence a higher E_b/N_0, the optimum modulation index takes a higher value.

In the above argument, we made use of the flat transfer characteristics of the band-pass filter. For a high modulation index, the 2-level FM spectrum shows two peaks at the frequencies of $f_c \pm \Delta f_m$, where Δf_m is the maximum deviation (see Fig. 5.9). In this case we can use a band-pass filter whose transfer characteristic is matched to the spectrum and then the threshold effect is softened.

5.5.5 Error Rates in Fading Channels

In a multipath fading channel, e.g., mobile radio channel, the signal is subjected to random fluctuations in power levels, phase and the Doppler frequency shifts. Furthermore, intersymbol interference appears when the signal bandwidth becomes comparable to the coherence bandwidth of the channel. These phenomena degrade performance.

5.5.5.1 Error Rates Due to Level Fluctuation. The average error rate is given as

$$\langle P_e \rangle = \int_0^\infty P_e(\gamma) p(\gamma) \, d\gamma, \tag{5.150}$$

where γ is the signal to noise power ratio S/N or symbol energy to noise power density ratio E_s/N_0; $P_e(\gamma)$ is error rate for a given γ; and $p(\gamma)$ is the probability density of γ. If we denote the constant signal envelope by u, we have

$$\gamma \equiv \frac{E_s}{N_0} = \frac{u^2 T}{2N_0}, \tag{5.151}$$

where T is the symbol duration.

Consider the Rayleigh fading discussed in Section 4.3, from Eq. (4.20) with $b = u_0^2/2$; we have

$$p(u) = \frac{2u}{u_0^2} e^{-u^2/u_0^2}, \qquad 0 \le u \le \infty, \tag{5.152}$$

where $u_0^2 = \langle u^2 \rangle$.

From Eqs. (5.151) and (5.152) we have

$$p(\gamma) = \frac{1}{\gamma_0} e^{-\gamma/\gamma_0}, \qquad 0 \le \gamma \le \infty. \tag{5.153}$$

For a Rayleigh distribution signal envelope u, the transformation of the variable $\gamma = ku^2$ does not depend on the coefficient k. Therefore the distribution density function $p(\gamma)$ can also be applied to $E_b/N_0 = (E_s/N_0) \cdot (T_b/T)$, where T_b is the bit duration.

For BPSK and Gray-coded QPSK with coherent demodulation, the bit error rate $P_e(\gamma)$ is given as $P_e(\gamma) = \frac{1}{2}\mathrm{erfc}(\sqrt{E_b/N_0})$ [Eq. (5.43a) and (5.49)]. Inserting $P_e(\gamma)$ into Eq. (5.150) and using Eq. (5.153), we have the average error rates

$$\langle P_e \rangle = \int_0^\infty \frac{1}{2}\mathrm{erfc}(\sqrt{\gamma}) \frac{1}{\gamma_0} e^{-\gamma/\gamma_0} \, d\gamma$$

$$= \frac{1}{2}\left(1 - \frac{1}{\sqrt{1 + 1/\gamma_0}}\right) \tag{5.154a}$$

$$\approx \frac{1}{4\gamma_0} \qquad (\gamma_0 \gg 1). \tag{5.154b}$$

For ASK and FSK with noncoherent detection, and BPSK with differential detection, we can express $P_e(\gamma)$ as [Eqs. (5.62), (5.63), and (5.76)]

$$P_e(\gamma) = \tfrac{1}{2}e^{-\alpha\gamma},\tag{5.155}$$

where

$$\alpha = \begin{cases} \tfrac{1}{2} & \text{ASK and FSK with noncoherent detection} \\ 1 & \text{BPSK with differential detection.} \end{cases}\tag{5.156}$$

The average error rate is given by

$$\langle P_e \rangle = \int_0^\infty \frac{1}{2}e^{-\alpha\gamma}\frac{1}{\gamma_0}e^{-\gamma/\gamma_0}\,d\gamma$$

$$= \frac{1}{2}\frac{1}{1+\alpha\gamma_0}\tag{5.157a}$$

$$\approx \frac{1}{2\alpha\gamma_0} \qquad (\alpha\gamma_0 \gg 1)\tag{5.157b}$$

5.5.5.2 Error Rates Due to the Random FM Effect.
Received signals experience random fluctuations in phase or frequency (random FM effect), as well as in envelope in fading channels. The error due to the random FM effects cannot be reduced by increasing the signal power; it is called the "irreducible error." The analysis of the irreducible error rates for coherent detection is difficult, since it includes a carrier recovery system, which is a highly nonlinear circuit. Here frequency discriminator detection and differential detection are considered.

For discriminator detection, an error occurs when the instantaneous frequency deviation due to the random FM exceeds a threshold value at the sampling instant. The probability density function of the instantaneous frequency $\dot\theta$ is given by Eq. (4.46). For a vertically polarized signal, we have

$$p(\dot\theta) = \frac{(\pi f_m)^2}{[\dot\theta^2 + 2(\pi f_m)^2]^{3/2}},\tag{5.158}$$

where f_m is the maximum Doppler frequency. The probability that $\dot\theta$ exceeds a threshold $\Delta\omega_d$ is then

$$\langle P \rangle = \text{Prob}(\dot{\theta} > \Delta\omega_d)$$

$$= \int_{\Delta\omega_d}^{\infty} \frac{(\pi f_m)^2}{[\dot{\theta}^2 + 2(\pi f_m)^2]^{3/2}} \, d\dot{\theta}$$

$$= \frac{1}{2}\left[1 - \frac{1}{\sqrt{1 + 2^{-1}(f_m/\Delta f_d)^2}} \right] \tag{5.159a}$$

$$\approx \frac{1}{8}\left(\frac{f_m}{\Delta f_d} \right)^2 \qquad (f_m \ll \Delta f_d), \tag{5.159b}$$

where $\Delta f_d = \Delta\omega_d/2\pi$ is one-half of the frequency separation between signals. Considering the error in the highest (lowest) level, the error rate due to the random FM becomes

$$\langle P_e \rangle = 2\{1 - \tfrac{1}{2}P(b_M)\}\langle P \rangle, \tag{5.160}$$

where $P(b_M)$ denotes the probability that the signal takes the highest or lowest level.

From Eq. (5.159b) we can see that the frequency separation Δf_d must be large enough compared with the maximum Doppler frequency f_m. This means that low-speed data transmission requires a high modulation index.

For BPSK with band-pass matched filter and differential detection, the error rate for Rayleigh fading is given as (equations (4.2–47) in [23])

$$\langle P_e \rangle = \frac{1 + \lambda_0[1 - J_0(2\pi f_m T)]}{2(1 + \lambda_0)}, \tag{5.161}$$

where λ_0 is the average energy per bit to noise power density, $J_0(\cdot)$ is the zeroth-order Bessel function of the first kind, and T is the bit duration. For quasi-static fading ($f_m \to 0$), $\langle P_e \rangle$ becomes Eq. (5.157a) ($\alpha = 1$). For $\lambda_0 \to \infty$, $\langle P_e \rangle$ corresponds to the irreducible error rate. It becomes

$$\langle P_e \rangle = \frac{1 - J_0(2\pi f_m T)}{2} \tag{5.162a}$$

$$\approx \tfrac{1}{2}(\pi f_m T)^2, \qquad f_m T \ll 1, \tag{5.162b}$$

where the approximation $J_0(x) \approx 1 - (x/2)^2$ ($x \ll 1$) is used.

5.5.5.3 *Error Rates Due to Frequency-selective Fading.* As the signal bandwidth becomes as wide as the coherence bandwidth of the channel, the signal experiences frequency-selective fading; the transfer function of the channel is not flat over the signal bandwidth, causing intersymbol interference. The error rate due to the frequency-selective fading depends on the pulse waveform, the demodulation system, as well as the statistics of the fading. Therefore it is difficult to obtain general formulas for error rates due to frequency-selective fading. The frequency correlation function of the channel is defined as

$$R(f) = \langle H(f_0)H^*(f_0+f)\rangle, \tag{5.163}$$

where $H(f)$ is the channel transfer function and $\langle\cdot\rangle$ means ensemble average.

Bello and Nelin [24] studied the effect of the frequency-selective fading on the error rates of incoherent and differentially coherent (delay detection) matched filter demodulation of binary FSK and BPSK, including post-detection diversity. They assumed that the channel is time-invariant (quasi-stationary) and that its transfer function is a sample function from a stationary complex-valued zero mean Gaussian random process. Also, the intersymbol interference depends on the data patterns. In their analysis the intersymbol interference only from adjacent bits is considered.

Based on their analysis, Bailey and Lindenlaub [25] discussed the matched filtered differentially coherent BPSK receiver with post-detection diversity (Fig. 5.32) for a few pulse waveforms and frequency correlation functions.

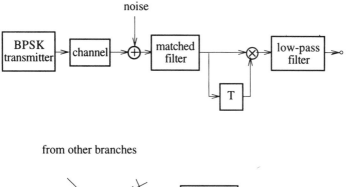

Figure 5.32. Matched filtered differentially coherent BPSK receiver with post-detection diversity.

The pulse waveforms were an NRZ pulse and raised cosine spectrum signal [Eq. (3.12) with $\alpha = 1$]. The considered frequency correlation functions are the Gaussian-shaped (G-F channel) as

$$R(f) = 2\sigma^2 e^{-4f^2/B_c^2} \tag{5.164}$$

and the sinc-function type (S-F channel) as

$$R(f) = 2R_0 T_m \text{sinc}(2fT_m), \tag{5.165}$$

where sinc $(x) = \sin(\pi x)/(\pi x)$.

They obtained a compact expression for irreducible error rates with the NRZ pulse waveform as

$$P_e \approx \frac{1}{4}\binom{2L-1}{L}[(2c_2)^L + 2(c_2 - c_1^2)^L]d^{2L}, \tag{5.166}$$

where L is the number of diversity branches and

$$d = \begin{cases} 1/(TB_c), & \text{G-F channel} \\ T_m/T, & \text{S-F channel,} \end{cases}$$

$$c_1 = \begin{cases} 1/(\pi\sqrt{\pi}), & \text{G-F channel} \\ 1/4, & \text{S-F channel,} \end{cases}$$

$$c_2 = \begin{cases} 1/\pi^2, & \text{G-F channel} \\ 1/6, & \text{S-F channel.} \end{cases}$$

The parameter d is the relative data rate: it is proportional to the ratio of the data rate $1/T$ to the channel coherence bandwidth. The coherent bandwidth can be defined by the frequency range where the correlation function decreases to $1/e$. Then the normalized coherence bandwidth W_{CN}, the coherence bandwidth normalized by data rate, becomes $W_{CN} = d^{-1}$ for the G-F channel and $W_{CN} = 0.7d^{-1}$ for the S-F channel. The irreducible error rates are shown as a function of W_{CN} in Fig. 5.33 for $L = 1$ (no diversity), $L = 2$, and $L = 4$.

The error rates due to the frequency-selective fading can be reduced by introducing diversity reception when data rate is relatively small.

5.5.5.4 Error Rates Due to Cochannel Interference. Evaluating the effect of cochannel interference on error rates becomes important for estimation of cochannel reuse distance and hence the spectrum efficiency in a cellular system. Nevertheless, it has not been well analyzed theoretically. Hirade and colleagues [26,27] discussed the error rates of digital FM with

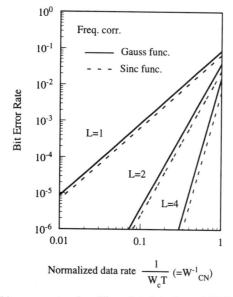

Figure 5.33. Irreducible error rates for differential detection of BPSK with diversity reception.

frequency discriminator and differential detection, taking the cochannel interference into consideration for fast Rayleigh fading channels. Their results for the irreducible error rates due to cochannel interference under the condition of quasi-static Rayleigh fading are given as

$$P_e = \frac{1}{2}\left[1 - \frac{\Lambda \sin(m\pi)}{\left\{ (\Lambda+1)^2 - \left[\Lambda\cos(m\pi) + \frac{1}{2}\left(\cos(m\pi) + \frac{\sin(m\pi)}{m\pi} \right) \right]^2 \right\}^{1/2}} \right],$$
$$\text{differential detection,} \qquad (5.167)$$

$$P_e = \frac{1}{2(\Lambda+1)}, \qquad \text{discriminator detection,} \qquad (5.168)$$

where Λ is the carrier to cochannel interference signal power ratio (C/I).

For $m = 0.5$ (minimum shift keying) and $\Lambda \gg 1$, Eq. (5.167) reduces to Eq. (5.168). If we compare this result with the error rate (Eq. 5.157a) due to noise from quasi-static Rayleigh fading, the same error rate is given for the same value of C/I and C/N. Usually values of the error rates for Gaussian noise with the same value of C/N and C/I is a safe (i.e., higher) estimation for relatively low error rates. This is because the probability that the interference signal level takes a high value is less than that of the Gaussian noise.

5.5.5.5 *Error Rates for QPSK in Fading Channels with Diversity Reception.* The theoretical analyses performed by Adachi and Parsons [28] and Adachi and Ohno [29] are valuable. They found exact expressions for the error rates for Rayleigh fading channels with diversity reception taking into account the effects of additive noise, random FM, cochannel interference, and frequency-selective fading, simultaneously. Their results for differentially encoded (π/4 shifted) QPSK with differential detection and post-detection combining diversity are cited in Figs. 5.34(a)–(d).

5.6 COMPUTER SIMULATION OF DIGITAL TRANSMISSION SYSTEMS

The main performance metrics of a digital transmission system are power spectrum of the transmitted signal, relative adjacent channel interference power, error rates, eye diagram, and cochannel interference. Those performance metrics must be evaluated for different modulation/demodulation schemes, combinations of transmit and receive filters, effects of carrier frequency offset, nonlinear distortion of a power amplifier, fading channel, diversity reception system, and so on. Some of these metrics can be estimated through theoretical analysis, otherwise they must be evaluated by experiment. Implementing an experimental system is time-consuming, especially for finding optimal parameters for the system. Furthermore, it is sometimes difficult for unskilled engineers.

Computer simulation experiments can be a powerful method for evaluating and designing digital modulation/demodulation systems, as well as other parts of a digital communication system. Operation of the experimental system is simulated by software programs on a computer. For example, Fig. 5.35 shows an experimental system to be simulated.

For the digital test data, a pseudo-noise data sequence that has a flat power spectrum is desirable, since it generates different data patterns. For this purpose the m (maximum length)-sequence is widely used. The m-sequence generators consisting of feedback shift registers are shown in Appendix 5.3 [30]. The length of the m-sequence is $2^N - 1$, where N is the length of the shift register. In computer simulations, the data length of 2^N is appropriate for the discrete Fourier transform (DFT). Therefore the test data is made by adding a "0" to the m-sequence: with this addition of a "0," the number of "1"s and "0"s becomes equal. The importance of the balance will be mentioned later.

In order for a digital computer to process signals, the signal must be sampled at discrete times. The sampling frequency must be high enough to satisfy the Nyquist sampling theorem (Section 2.1.8). From this viewpoint, simulation of a modulated signal at RF band is not appropriate. We then use the complex zero-IF expression of the signal (Section 2.1.11), as well as the RF circuit, for example the band-pass filter. The complex zero-IF

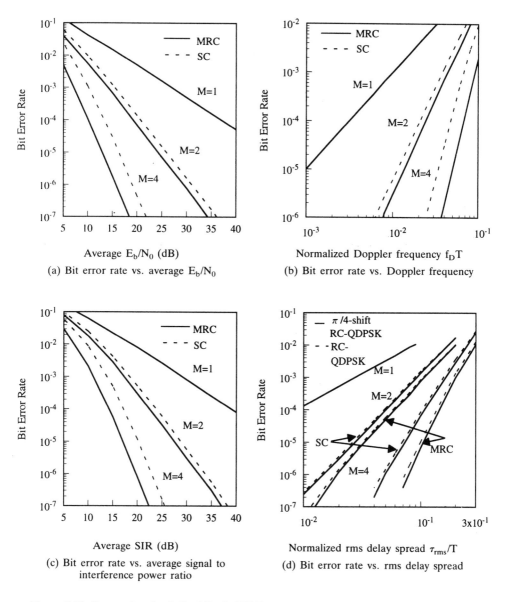

Figure 5.34. Error rates for ($\pi/4$ shifted) QPSK under flat fading condition with diversity reception. Parameter *M* is the number of the diversity branch [29]. (Copyright © IEEE 1991.)

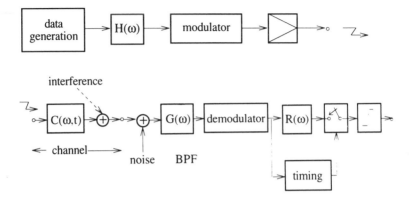

Figure 5.35. Example of a digital transmission system for a computer simulation experiment.

expression never loses generality if the band-pass signal is band-limited within a bandwidth smaller than the carrier frequency (a narrow band-pass signal). A nonlinearly modulated signal such as digital FM has, in principle, a spectrum which spreads into an infinite bandwidth, and hence the sampling theorem cannot be strictly satisfied. However, the effect of the aliasing errors can be made sufficiently small by choosing an appropriately high sampling frequency, since the spectrum density decreases with increase in frequency.

Filtering of a signal can be processed by convolution integral in the time domain or multiplication in the frequency domain. In many cases, frequency-domain processing is efficient due to the fast DFT technique. Attention must be paid to signal processing in the frequency domain. The DFT of a given signal block assumes a repetition of the signal block. The discontinuity of the signal from the end to the beginning of the signal block must be avoided: the effect of the discontinuity appears especially in unwanted spectrum spread and distortion of a signal at the beginning and the end of the signal block. To solve this problem, window function techniques are, in general, applied to the signal during processing. For a digitally modulated signal with linear modulation, the discontinuity between the signal at the end and the beginning of the signal block is not different from that between each data symbol, and hence it is not a problem. For a digital FM signal, however, it becomes a problem. This problem can be simply avoided by balancing the plus and minus excursion of the modulating signal. This is the reason why balancing the "1" and "0" for test data was recommended earlier.

The effects of nonlinear distortion in an amplifier can be treated with the complex zero-IF signal, as far as the fundamental frequency component is concerned. As discussed previously (Section 5.2), the odd-order distortion

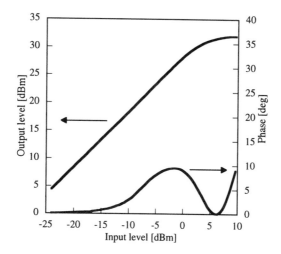

Figure 5.36. AM-AM and AM-PM conversion characteristics of a 900 MHz band power amplifier module (FMC-80802-20).

for linear modulation is important. The output versus input relation of the amplifier, that is, the output signal envelope and phase shift as a function of the input signal level, called AM-AM conversion and AM-PM conversion, respectively, must be described. For example, Fig. 5.36 shows measured AM-AM and AM-PM conversion characteristics of a 900 MHz band power amplifier module. To obtain the characteristics, we can use a network analyzer with variable output power. A computer-simulated power spectrum of a QPSK signal at the output of the power amplifier is shown in Fig. 5.37. The first, second, and third out-of-band radiation components correspond to the third-, fifth-, and seventh-order distortion, respectively. The average power efficiency can also be evaluated for a modulated signal, if the power efficiency is described as a function of the envelope of the input (unmodulated) signal.

The channel is simulated with a linear filter, which is either time-invariant or time-varying (fading). Cochannel or adjacent channel interference can be evaluated by adding the interfering signals. There are two ways to simulate the Rayleigh flat fading: one is to add many signals with different amplitude and frequencies, which are distributed over the range of Doppler frequency. The principle of this method is described in Section 4.3. A prime number frequencies is desirable in order to generate a randomized output signal. The other way is to use quadrature modulation as shown in Fig. 5.38. The principles of the fading generator are discussed in Section 2.2 [Eqs. (2.122) to (2.126)]. The power transfer function of the noise filter is given by the received power spectrum due to fast fading, for example Eq. (4.29). The frequency-selective fading is simulated using a time-varying filter whose amplitude and/or delay characteristics depend on frequency.

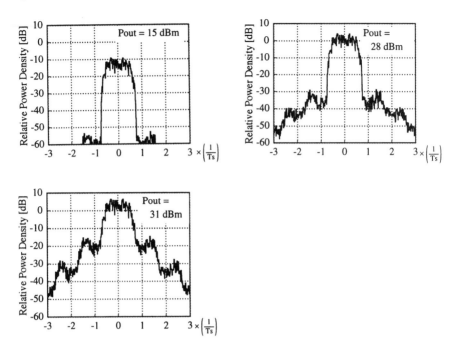

Figure 5.37. Computer-simulated spectra of $\pi/4$ shifted QPSK at the output of a quasi-linear power amplifier. Square-root of Nyquist-I filter with roll-off factor of 0.5 is used.

The noise is simulated by a random number generator, that has the required stochastic characteristic, for example, white Gaussian noise. The characteristic of the band-pass filter is described with an equivalent low-pass transfer function or impulse response. The operation of a demodulator can be described as discussed in Section 5.5. Sample and decision circuits including a clock recovery circuit can be described according to the actual system.

Many other parts of a digital transmission system, such as a diversity receiver, an automatic equalizer, and a carrier recovery circuit, can be incorporated by mathematically describing their operation on the signal. Measuring instruments such as an oscilloscope for monitoring eye-diagrams at various stages of the system, a spectrum analyzer, and an error counter are easily simulated. Implementation of a subroutine library for each part of the transmission system is highly recommended, since one can easily set up one's own system by picking up appropriate parts of the system. The importance of software simulators for digital communication systems will increase, since the complexity of systems will increase and computing power is improving.

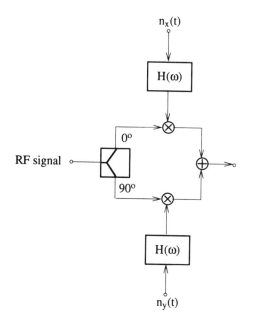

Figure 5.38. Rayleigh fading generator. $n_x(t)$ and $n_y(t)$ are independent white Gaussian noise.

APPENDIX 5.1   DISTORTION OF MODULATED SIGNAL APPLIED TO A NONLINEAR CIRCUIT

Let us express the input–output relation of the nonlinear circuit by series expansion as

$$z(t) = a_1 y(t) + a_2 y^2(t) + a_3 y^3(t) + \cdots, \tag{A5.1}$$

where $y(t)$ and $z(t)$ are the input and output signal waveforms, respectively. The distortion is represented by the higher-order components.

For linear modulation, we denote

$$y(t) = A(t) \cos(\omega_c t) \tag{A5.2}$$

and for a constant envelope modulation,

$$y(t) = A_0 \cos[\omega_c t + \varphi(t)]. \tag{A5.3}$$

We assume the modulated signal bandwidth is less than carrier frequency, ω_c. From this assumption, the spectra of higher-order distortion signals are assured not to overlap with each other. From Eq. (A5.1) and the fact that $(\cos \omega_c t)^{2n}$ has no component of $\cos \omega_c t$, we can see that the even-order distortion has no frequency component at the frequency of the input signal $y(t)$. Thus the even-order distorted signal can be eliminated by applying a band-pass filter that is tuned to the carrier frequency.

For a constant envelope modulation signal, the carrier frequency component $z_{\omega_c}(t)$ becomes

$$z_{\omega_c}(t) = A_0' \cos[\omega_c t + \varphi(t)], \tag{A5.4}$$

where

$$A_0' = a_1 A_0 + \tfrac{3}{4} a_3 A_0^3 + \tfrac{10}{16} a_5 A_0^5 + \cdots.$$

Thus we get a distortion-free signal at the output of the band-pass filter.

For a linear modulation signal,

$$z_{\omega_c}(t) = A'(t) \cos(\omega_c t), \tag{A5.5}$$

where

$$A'(t) = a_1 A(t) + \tfrac{3}{4} a_3 A^3(t) + \tfrac{10}{16} a_5 A^5(t) + \cdots. \tag{A.5.6}$$

The signal is distorted due to the odd-order components. The spectrum of the signal $A^n(t)$ is given by the n-fold convolution integral of the spectrum of $A(t)$. Thus its spectrum spreads n times wider than that of the input signal.

APPENDIX 5.2   DERIVATION OF THE EXPECTED GAUSSIAN NOISE POWER FOR A FREQUENCY DISCRIMINATOR

Denoting $a(t) = 1/A(t)$ and

$$b(t) = n_y(t) \cos \phi(t) - n_x(t) \sin \phi(t), \tag{A5.7}$$

we have the demodulated noise from Eq. (5.119) as

$$c(t) = a(t)b(t). \tag{A5.8}$$

Then we get the expected noise power at the output of the post-demodulation filter as

$$\langle N_g(t) \rangle = \left\langle \left\{ \int_{-\infty}^{\infty} h_d(t - s) c(s) \, ds \right\}^2 \right\rangle$$

$$= \int_{-\infty}^{\infty} \int_{-\infty}^{\infty} h_d(t - s_1) h_d(t - s_2) \langle c(s_1) c(s_2) \rangle \, ds_1 \, ds_2$$

$$= \int_{-\infty}^{\infty} \int_{-\infty}^{\infty} h_d(t - s_1) h_d(t - s_2) a(s_1) \langle b(s_1) b(s_2) \rangle \, ds_1 \, ds_2, \tag{A5.9}$$

where $h_d(t)$ is the total impulse responses of the differentiating circuit and the post-demodulation filter. Using Eqs. (A5.7), and the relations given by (2.119) and (2.121), we get

$$\langle b(s_1) b(s_2) \rangle = N_0 \int_{-\infty}^{\infty} g(s_1 - \tau) g(s_2 - \tau) [\cos \phi(s_1) \cos \phi(s_2)$$

$$+ \sin \phi(s_1) \sin \phi(s_2)] \, d\tau. \tag{A5.10}$$

With Eqs. (A5.9) and (A5.10), we have

$$\langle N_g(t) \rangle = N_0 \int_{-\infty}^{\infty} \left[\left(\int_{-\infty}^{\infty} h_d(t - s) g(s - \tau) a(s) \cos \phi(s) \, ds \right)^2 \right.$$

$$\left. + \left(\int_{-\infty}^{\infty} h_d(t - s) g(s - \tau) a(s) \sin \phi(s) \, ds \right)^2 \right] d\tau$$

$$= N_0 \int_{-\infty}^{\infty} \left[\{ h_d(t) * [g(t - \tau) a(t) \cos \phi(t)] \}^2 \right.$$

$$\left. + \{ h_d(t) * [g(t - \tau) a(t) \sin \phi(t)] \}^2 \right] d\tau. \tag{A5.11}$$

APPENDIX 5.3 *M*-SEQUENCE GENERATOR

Number of stage, n	Period $(2^n - 1)$	$a_0, a_1, \ldots, a_n$
2	3	111
3	7	1011
4	15	10011
5	31	100101
6	63	1000011
7	127	10001001
8	255	100011101
9	511	1000010001
10	1023	10000001001
11	2047	100000000101
12	4095	1000001010011
13	8191	10000000011011
14	16 383	100010001000011
15	32 767	1000000000000011

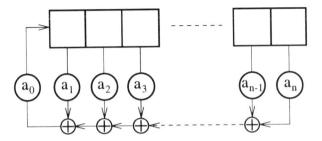

REFERENCES

1. B. P. Lathi, *Modern Digital and Analog Communication Systems*, Holt, Reinhart and Winston, New York, 1983.

2. J. G. Proakis, *Digital Communication*, 3rd ed., McGraw-Hill, New York, 1995.

3. M. Schwartz, *Information Transmission, Modulation and Noise*, 3rd ed., McGraw-Hill, New York, 1980.

4. R. W. Lucky, J. Saltz, and E. J. Weldon, Jr., *Principles of Data Communication*, McGraw-Hill, New York, 1968.

5. W. R. Bennet and J. R. Davey, *Data Transmission*, McGraw-Hill, New York, 1965.

6. M. Schwartz, W. R. Bennet, and S. Stein, *Communication Systems and Techniques*, McGraw-Hill, New York, 1966.

7. J. B. Anderson, T. Aulin, and C.-E. Sundberg, *Digital Phase Modulation*, Plenum Press, New York, 1986.

8. K. Fahrer, *Digital Communications, Satellite/Earth Engineering*, Prentice-Hall, Englewood Cliffs, N.J., 1983.

9. P. A. Baker, "Phase-modulation data sets for serial transmission at 2000 and 2400 bits per second," *AIEEE Trans., Part I (Commun. Electro.)*, 166–171 (July 1962).

10. F. G. Jenks, P. D. Morgan, and C. S. Warren, "Use of four-level phase modulation for digital mobile radio," *IEEE Trans. Electromag. Compat.*, **EMC-14**, 113–128 (November 1972).

11. K. Miyauchi, K. Izumi, S. Seki, and N. Ishida, "Characteristics of an experimental guided millimeter-wave transmission system," *IEEE Trans. Communications*, **COM-20**, 808–813 (August 1972).

12. T. Aulin and C. E. Sundberg, "An easy way to calculate power spectra for digital FM," *IEEE Proceedings*, **130**, part F, 519–525 (October 1983).

13. J. Namiki, "Block demodulation for short radio packet," *Trans. IECE*, **67B**, 54–61 (January 1984). [Translated to English in *Electronics and Communications in Japan*, **67-B**, 47–56 (1984).]

14. R. F. Pawula, S. O. Rice, and J. H. Roberts, "Distribution of the phase angle between two vectors perturbed by Gaussian noise," *IEEE Trans. Communications*, **COM-30**, 1828–1841 (August 1982).

15. R. F. Pawula, "Asymptotics and error rate bounds for *M*-ary DPSK," *IEEE Trans. Communications*, **COM-32**, 93–94 (January 1984).

16. S. O. Rice, "Noise in FM receivers," in *Time Series Analysis*, M. Rosenblatt, ed., Wiley, New York, 1963.

17. J. E. Mazo and J. Saltz, "Theory of error rates for digital FM," *Bell Syst. Tech. J.*, **45**, 1511–1535 (November 1966).

18. T. T. Tjhung and P. H. Wittke, "Carrier transmission of binary data in a restricted band," *IEEE Trans. Communication Technology*, **COM-18**, 295–304 (August 1970).

19. R. F. Pawula, "On the theory of error rates for narrow-band digital FM," *IEEE Trans. Communications*, **COM-29**, 1634–1643 (November 1981).

20. D. L. Schilling, E. Hoffman, and E. A. Nelson, "Error rates for digital signals demodulated by an FM discriminator," *IEEE Trans. Communication Technology*, **COM-15**, 507–517 (August 1967).

21. Y. Akaiwa and E. Okamoto, "An analysis of error rates for Nyquist — and partial response — baseband-filtered digital FM with discriminator detection," *Trans. IECE*, **J66-B**, 534–541 (April 1983).

22. R. F. Pawula, "Refinements to the theory of error rates for narrow-band digital FM," *IEEE Trans. Communications*, **COM-36**, 509–513 (April 1988).

23. W. C. Jakes, ed., *Microwave Mobile Communications*, Wiley, New York, 1974.

24. P. A. Bello and B. D. Nelin, "The effect of frequency selective fading on the binary error probabilities of incoherent and differentially coherent matched filter receivers," *IEEE Trans. Communications Systems*, **CS-11**, 170–186 (June 1963). [See also Corrections, *IEEE Trans. Commun. Tech.*, **COM-12**, 230 (December 1964).]

25. C. C. Bailey and J. C. Lindenlaub, "Further results concerning the effect of frequency-selective fading on differentially coherent matched filter receivers," *IEEE Trans. Communication Technology*, **COM-16**, 749–751 (October 1968).

26. K. Hirade, M. Ishizuka, and F. Adachi, "Error-rate performance of digital FM with discriminator-detection in the presence of co-channel interference under fast Rayleigh fading environment," *Trans. IECE*, **E61**, 704–709 (September 1978).

27. K. Hirade, M. Ishizuka, F. Adachi, and K. Ohtani, "Error-rate performance of digital FM with differential detection in land mobile radio channels," *IEEE Trans. Vehicular Technology*, **VT-28**, 204–212 (August 1979).

28. F. Adachi and J. D. Parsons, "Error rate performance of digital FM mobile radio with postdetection diversity," *IEEE Trans. Communications*, **37**, 200–210 (March 1989).

29. F. Adachi and K. Ohono, "BER performance of QDPSK with postdetection diversity reception in mobile radio channels," *IEEE Trans. Vehicular Technology*, **40**, 237–249 (February 1991).

30. H. Kaneko, *PCM Communication Technology*, Sanpou, 1976.

6

DIGITAL MODULATION/DEMODULATION FOR MOBILE RADIO COMMUNICATION

Digital modulation has been used for a long time in analog FM mobile radio communication systems. The role of digital transmission in the analog FM mobile radio system is to send low-rate data signals to control the system, e.g., paging a called mobile terminal and setting up and releasing of a dedicated channel. Spectrum efficiency of the digital modulation was not a key issue for such systems, since the data transmission rate is low: from several hundred to several thousand bits per second.

The history of research and development of spectrum-efficient digital modulation techniques for mobile radio communication is not so long. It started in 1978 with the invention of spectrum-efficient digital FM by de Jager and Dekker. After their invention many digital modulation methods were proposed in a very short period, described later in detail. The motivation for efficient digital modulation was digital voice communications.

This chapter describes some digital modulation methods in order of appearance—digital modulation for analog FM systems, spectrum-efficient digital FM, and linear modulation. Research and development in digital modulation techniques for mobile radio communications are still in progress at the time of writing.

Special requirements for digital mobile radio communication are stable performance of bit error rate in fast fading channels and cochannel interfering environments; low power consumption; as well as size and cost-effectiveness of equipment. Due to these special requirements, digital modulation/demodulation for mobile radio communication differs from that for other communication systems.

6.1 DIGITAL MODULATION FOR ANALOG FM MOBILE RADIO SYSTEMS

An important issue is how to incorporate the digital modulation/demodulation into an analog FM system. The easiest and actually adopted method is to transmit the data signal in the voice channel of the analog FM system. This is similar to data communication through wireline analog public telephone channels. Since the spectrum of the voice signal covers frequencies from about 0.3 to 3 kHz, the voice channel for analog FM is designed not to pass a dc signal: a dc circuit is rather difficult to implement.

For transmission of digital signal over the dc-blocked channel, we have two methods: one is to use a line code that has no dc component; the other is to use a modulated signal whose carrier frequency is in the middle of the voice band channel. The Manchester code described in Section 3.2.4 is useful for the former method. For the latter method, called the subcarrier system, minimum shift keying (MSK) with a carrier frequency around 1.5 kHz is sometimes used. The subcarrier system is shown in Fig. 6.1. The carrier signal is synchronized to the data clock signal for convenience in demodulation of the MSK signal, whose carrier and data frequencies are comparable.

Another subcarrier system uses a data modem (modulator/demodulator), which was developed for wireline public telephone channels [1]. The modem uses different modulation for different data transmission rates of from 300 bps to 28.8 kpbs. A high-speed modem employs sophisticated techniques including automatic channel equalization. For application of the data modem to the analog voice FM mobile radio channels, the highest data rate is limited to around 4800 bps, due to the fading phenomenon in mobile radio channels.

6.2 CONSTANT ENVELOPE MODULATION

This type of modulation is advantageous for application to mobile radio communications because a saturated power amplifier, which is principally power-efficient, can be employed as discussed in Section 5.2. Furthermore, a band-pass limiter circuit can be used at a receiver. The limiter circuit can easily be implemented to cope with fast fading and wide dynamic range effects in mobile radio channels. Therefore it is natural that many researchers have

Figure 6.1. Subcarrier MSK signal transmission for an analog FM system.

focused on constant envelope digital modulation. From the viewpoint of compact spectrum characteristics, a continuous phase frequency shift keying (CPFSK) with a low modulation index becomes important in mobile radio channels.

6.2.1 Minimum Shift Keying

Binary FM with rectangular (NRZ) pulse shaping and limiter-discriminator detection is the most primitive system. Rectangular pulse shaped binary CPFSK with a modulation index of 0.5 is known as MSK (minimum shift keying) or FFSK (fast frequency shift keying) [2,3]. An MSK signal can be expressed as

$$s(t) = A_0 \cos\left(\omega_c t + a_n \frac{\pi}{2T} t' + \frac{\pi}{2} \sum_{i=1}^{n-1} a_i\right), \qquad (n-1)T \le t \le nT, \quad (6.1)$$

where $t' = t - (n-1)T$, A_0 is the amplitude, ω_c is the carrier frequency, a_n takes the values ± 1 corresponding to the input digital signal, and T is the bit duration. The frequency becomes $f_M = f_c - 1/4T$ (mark) and $f_s = f_c + 1/4T$ (space), corresponding to the input data, where $f_c = \omega_c/2\pi$.

The name MSK comes from the fact that the modulation index is minimum for a pair of FSK signals that are orthogonal, corresponding to the 2-level signal. For orthogonal signals we can apply coherent detection without interference between the two signals. The coherent detector for the MSK signal is shown in Fig. 6.2. The low-pass (zonal) filter is intended to suppress the harmonic signal components. The noncoherent detector can be obtained if we place an envelope detector between the low-pass filter and the integrator.

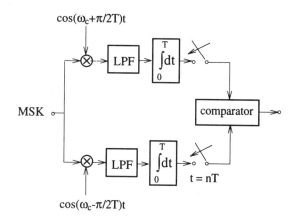

Figure 6.2. A coherent detector for MSK signals.

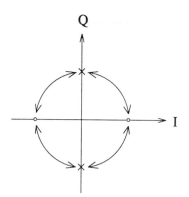

Figure 6.3. Phase constellation of MSK signals.

At the end of a symbol time the signal takes a phase from two groups in turn: one group includes $+\pi/2$ and $-\pi/2$ and the other includes 0 and π, as shown in Fig. 6.3. The phase constellation is the same as $\pi/2$ shifted BPSK. The signal phase moves on the circle with constant speed.

Equation (6.1) can be rewritten as

$$s(t) = x_n(t) \cos \omega_c t - y_n(t) \sin \omega_c t, \qquad (n-1)\,T \le t \le nT, \qquad (6.2)$$

where

$$x_n(t) = A_0 \cos\left(a_n \frac{\pi}{2T} t' + \frac{\pi}{2} b_{n-1}\right), \qquad (n-1)\,T \le t \le nT, \qquad (6.3)$$

$$y_n(t) = A_0 \sin\left(a_n \frac{\pi}{2T} t' + \frac{\pi}{2} b_{n-1}\right), \qquad (n-1)\,T \le t \le nT, \qquad (6.4)$$

and

$$b_{n-1} = \sum_{i=1}^{n-1} a_i. \qquad (6.5)$$

Using Eqs. (6.3)–(6.5) we define

$$x'_{2m}(t) \equiv x_{2m}(t) + x_{2m+1}(t)$$

$$= -a_{2m} \sin\left(\frac{\pi}{2} b_{2m-1}\right) \sin\left(\frac{\pi}{2T} t'_1\right), \qquad 2mT \le t \le 2(m+1)\,T, \quad (6.6)$$

where $t_1' = t - 2mT$, and

$$y_{2m}'(t) \equiv y_{2m-1}(t) + y_{2m}(t)$$

$$= a_{2m-1}\cos\left(\frac{\pi}{2}b_{2m-2}\right)\sin\left(\frac{\pi}{2T}t_2'\right), \qquad (2m-1)\,T \leq t \leq (2m+1)\,T. \tag{6.7}$$

where $t_2' = t - (2m-1)\,T$.

If we correspond 1 and 0 with 1 and -1, respectively, for a_n, we have the following expressions:

$$c_{2m} = c_{2m-1} \oplus \overline{d_{2m}}, \tag{6.8}$$

$$c_{2m+1} = c_{2m-2} \oplus d_{2m-1}, \tag{6.9}$$

where $\oplus$ means mod 2 addition, $d_n \in \{1,0\}$ denotes the input data, and c_n is the differentially encoded data, which is inverted every even time slot. Thus, as well as a conventional FM modulator, we have an MSK signal modulator with quadrature modulation as shown in Fig. 6.4. The impulse response is given by $h(t) = \sin[(\pi/2T)t]$ for $0 \leq t \leq 2T$ and $h(t) = 0$ otherwise.

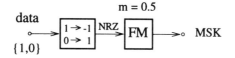

(a) FM modulator type

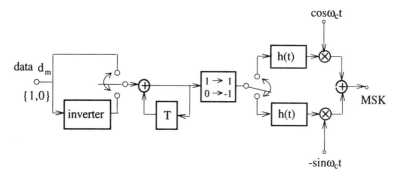

(b) quadrature modulator type

Figure 6.4. Modulators for MSK signals.

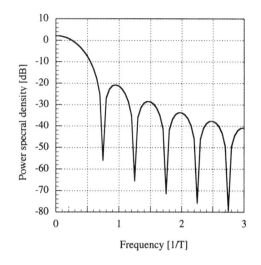

Figure 6.5. Power spectrum of MSK.

The power spectral density of MSK signals for random input data can be calculated by following the discussion for constant envelope modulation in Section 5.4.2. However, the easiest way is to follow the method for linear modulation discussed in Section 2.1.10 and Section 5.4.1, noting that the MSK signal can be generated with quadrature modulation. Assuming that the input data and therefore the in-phase and the quadrature component signals are random, we have the power spectral density as

$$S(\omega) = \frac{(4/\pi)^2 T}{[1 - (2\omega T/\pi)^2]^2} \cos^2(\omega T), \qquad -\infty \leq \omega \leq \infty. \qquad (6.10)$$

This is shown in Fig. 6.5.

Before we go further, let us compare the in-phase and quadrature waveforms of MSK, offset QPSK, and $\pi/2$ shifted BPSK in Fig. 6.6. An MSK signal can be understood as a special case of offset QPSK, whose pulse waveform $h(t)$ is $h(t) = \sin(\pi t/2T)$ for $0 \leq t \leq 2T$ and $h(t) = 0$ for $t < 0$ and $t > 2T$.

Expression of an MSK signal by Eqs. (6.2)–(6.9) and its modulator configuration in Fig. 6.4(b) suggests a coherent detector as shown in Fig. 6.7. This detector is different from the previous one (Fig. 6.2): the observation time is expanded to $2T$ instead of T. The eye-diagram of quadrature demodulated MSK is shown in Fig. 6.8. A recovered carrier phase slip by 180° never affects the received data because of the differential encoding embedded in the MSK modulation and the differential decoding at the receiver.

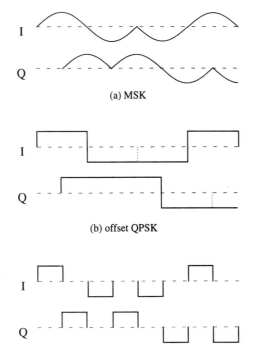

(a) MSK

(b) offset QPSK

(c) $\pi/2$ shifted BPSK

Figure 6.6. Comparison of the in-phase and quadrature waveforms between MSK, offset QPSK, and $\pi/2$ shifted BPSK.

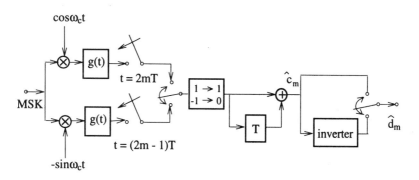

$g(t) = h(T-t)$: matched filter

Figure 6.7. Coherent detector for MSK signals.

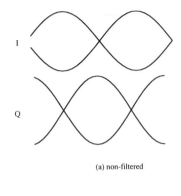

(a) non-filtered

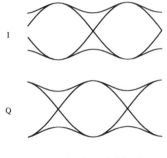

(b) matched filtered

Figure 6.8. Eye-diagram of a MSK signal with coherent detection.

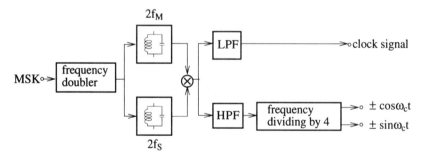

Figure 6.9. Carrier recovery circuit for MSK signals.

A carrier recovery circuit for an MSK signal is shown in Fig. 6.9. By applying an MSK signal to a frequency doubler, we have discontinuous frequency components at $2f_M$ and $2f_S$, where $f_M = f_c - 1/4T$ and $f_S = f_c + 1/4T$. The existence of the discontinuous frequency components in the frequency-doubled signal can be supported by confirming that $|C_\alpha| = 1$ with $M = 2$, $L = 1$ and $h(\equiv m) = 1$ (see Section 5.4.2), where the modulation

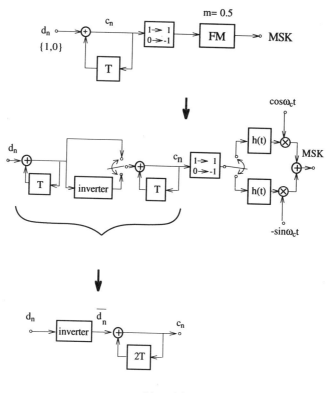

(a) modulator

Figure 6.10. Modulator and coherent detector for MSK signals with differential encoding.

index is doubled from frequency doubling. Each component is entered into a resonator to reduce noise components. The principle of operation of the carrier recovery is in the frequency quadruplication. A phase ambiguity of $\pm \pi/2$ or $\pm \pi$ occurs in the frequency quadri-divider. In order to cope with phase ambiguity of $+\pi/2$ and $-\pi/2$, differential encoding is introduced. The transmitter and receiver are shown in Fig. 6.10. The encoded signal c_n and the input data d_n are expressed as

$$d_n = c_{n-1} \oplus c_n \quad \text{or equivalently} \quad c_n = c_{n-1} \oplus d_n.$$

From the characteristics of the MSK signal, the phase shift during two symbol-times $(2T)$ becomes 180° or 0°, corresponding to $c_{n-1} \oplus c_n$ equaling 0 or 1, respectively. Here $c_{n-1} \oplus c_n = 0$ means $c_{n-1} = c_n$ (successive 1s or

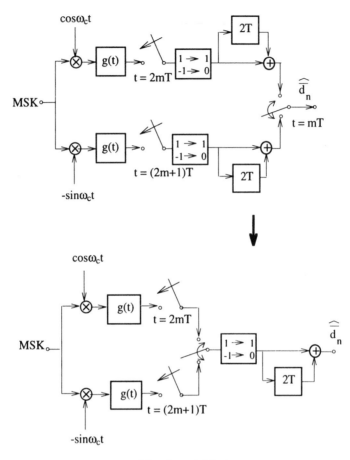

(b) coherent detector

Figure 6.10. (*continued*).

0s) and $c_{n-1} \oplus c_n = 1$ means $c_{n-1} \neq c_n$. In error-free transmission we have

$$\overline{d_{2n}} = \overline{c_{2n-1} \oplus c_{2n}} = I_{2n-2} \oplus I_{2n} = \hat{d}_{2n}, \tag{6.11a}$$

$$\overline{d_{2n+1}} = \overline{c_{2n} \oplus c_{2n+1}} = Q_{2n-1} \oplus Q_{2n+1} = \hat{d}_{2n+1}, \tag{6.11b}$$

where I_n and Q_n are the in-phase and quadrature data signals, respectively, decided at the receiver and $\hat{d}_n$ are the received data signals. We can see that the phase ambiguity of $\pm \pi/2$, as well as $\pm \pi$, has no effect on the received data, since inversion of I_n and Q_n (180° ambiguity) and/or interchange of I_n and Q_n ($\pm 90°$ ambiguity) never changes $\hat{d}_n$ in Eqs. (6.11a) and (6.11b).

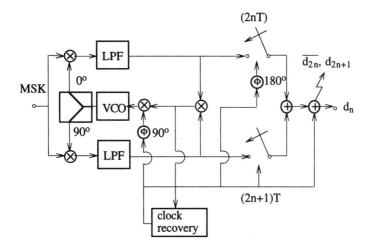

Figure 6.11. Coherent detector for MSK signals, which requires no differential encoding [22]. (Copyright © IEICE 1981.)

A coherent detector that requires no differential encoding was proposed in [23] (Fig. 6.11). The symbol $\oplus$ denotes mod 2 addition. The principle of operation of the coherent detector is the same as that shown in Fig. 6.7. The phase ambiguity of $\pi/2$ and $-\pi/2$ is solved by the clock signal-assisted carrier recovery system. The effect of carrier signal phase slip is compensated by the recovered clock signal.

Another coherent detector [5] for MSK signals is shown in Fig. 6.12. The circuit consists of only one path, similarly to a coherent detector for BPSK signals. The recovered local signal which is synchronized to the mark or space frequency is used for demodulation (deviated-frequency method). The relative phase of the MSK signal to the local signal phase is 0° or 180°.

The other method for carrier recovery of the MSK signal is described in [2], where the signal required for the correlation receiver is obtained together with the carrier signal.

The error rate of an MSK signal with coherent detection is different depending on the type of coherent detector. A coherent detector that observes a received signal for two symbol periods (Fig. 6.7) shows a superior performance by 3 dB in signal to noise power ratio compared to a coherent detector, that observes for one symbol period (Fig. 6.2). This is because the former uses an antipodal and the latter an orthogonal signaling system. The optimum error rate performance is obtained by matched filtering and coherent detection with 2-symbol-time observations, as discussed in Section 3.3. Eye-diagrams of nonfiltered and matched filtered MSK signals with coherent detection are shown in Fig. 6.8.

Two kinds of differential detection of MSK signals are known (Fig.

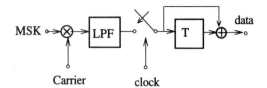

(a) coherent detector

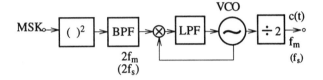

(b) carrier recovery circuit

Figure 6.12. Coherent detector for MSK signals with deviated-frequency-locking scheme.

6.13). The phase shift during one and two symbol times is $(\pi/2)a_n$ and $(\pi/2)(a_n + a_{n+1})(a_n = \pm 1)$, respectively. By adjusting the carrier phase difference between the two paths in the detector, the signal at the output of the zonal low-pass filter becomes, for a one-bit delay detector,

$$s_d(nT) = \sin\left(\frac{\pi}{2}a_n\right) = \begin{cases} 1, & a_n = 1 \\ -1, & a_n = -1, \end{cases} \qquad (6.12a)$$

and for a two-bit delay detector,

$$s_d(nT) = \cos\left[\frac{\pi}{2}(a_n + a_{n+1})\right] = \begin{cases} 1, & a_n + a_{n+1} = 0 \\ -1, & a_n + a_{n+1} = \pm 2. \end{cases} \qquad (6.12b)$$

Differential encoding is necessary for the two-bit delay detector, since the differential detection is equivalent to differential decoding on the data signal.

Contrary to coherent detection, matched filtered differential detection of an MSK signal shows intersymbol interference (Fig. 6.14). The matched filtered signal $s(t)$ becomes $s(t) = \pm h(t)*h(2T-t)$, where $h(t) = \sin(\pi t/2T)$ for $0 \le t \le 2T$ and $h(t) = 0$ otherwise. $s(2T)$, the in-phase signal, becomes T at the sampled time. At this instant $s(T)$, the quadrature component, takes a value of 0, $+2T/\pi$, and $-2T/\pi$. Thus, the phases of the matched filtered MSK signals scatter by 0 or $\pm\tan^{-1}(2/\pi)$. Due to the intersymbol interference,

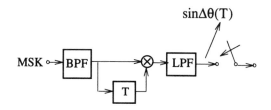

(a) one-bit delay detector

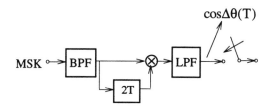

(b) two-bit delay detector

Figure 6.13. Differential detector for MSK signals.

theoretical analysis of the bit error rate for matched filtered MSK with differential detection is difficult.

Optimization of the receiver filter for differential detection of MSK is discussed by Kawas Kaleh [6]. A differential detection system for MSK with nonredundant error correction capability is proposed by Masamura and colleagues [7].

The other method for demodulation of MSK signals is limiter-discriminator detection. An integrate-and-dump filter is exclusively used as a post-detection filter, since it shows no intersymbol interference when receiving an NRZ signal, which is obtained at the demodulator output. The integrate-and-dump filter is not the matched filter for post-detection, since the noise spectrum is not flat at the output of the limiter–discriminator.

The matched filter receiver for either coherent or differential detection of MSK signals is not actually recommended for application to mobile radio communications. The reason is that the matched filter offers poor channel selectivity performance, which is imperative to compensate for the near–far problem in mobile radio communications. Better channel selectivity is obtained by using a band-pass filter, which shows a steeper transfer function, such as a Gaussian filter. Assuming a general band-pass filter, the error rates for MSK with differential detection, as well as limiter–discriminator detection are analyzed in [8], where calculated results are presented for a Gaussian band-pass filter.

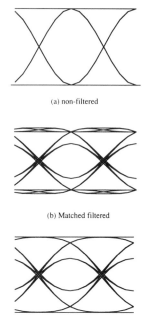

(a) non-filtered

(b) Matched filtered

(c) Matched filtered and amplitude-limited

Figure 6.14. Eye-diagram of an MSK with differential detection.

Even when a band-pass filter with good channel selectivity is adopted, MSK is not recommended for mobile radio communication, since its spectrum contains high out-of-band radiation: channel separation must be wide to get sufficient channel selectivity. In order to control the out-of-band radiation of MSK, pulse shapes have been proposed for the premodulation signal [9–11]. However, the results are not adequate for application to mobile radio communications, since the band-limited pulse waveforms are still not stringent enough. The effect of strict band-limited pulse waveforms on MSK-type signals is demonstrated in Fig. 6.15. The out-of-band radiation is decreased by using the Nyquist-III roll-off filter (Section 3.1.3), where the baseband signal spectrum is rigorously band-limited within $|\omega| \le (1 + \alpha)\,\pi/T$, where T is the bit duration and α is the roll-off factor. The characteristic of MSK, that the modulated signal takes fixed phases from 0, $\pm\pi/2$, and π, is preserved by using the Nyquist-III filter.

We have discussed MSK in detail, since it presents a good reference system for other constant envelope digital modulations.

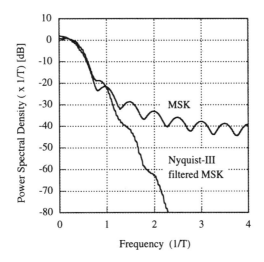

Figure 6.15. Effect of a band-limited premodulation signal on the spectrum for MSK-type signals.

6.2.2 Partial-Response Digital FM

Partial response digital FM was proposed many years ago, for example, duobinary FM by Lender [12]. However, rigorous band-limitation on the premodulation signal was not adopted. In 1978, de Jager and Dekker proposed a pioneering digital FM system called Tamed FM (TFM) [13], which meets the requirements for application to mobile radio communications: the out-of-band radiation is drastically decreased. Their success is attributed to a partial-response digital FM system, in which band-limitation of the premodulation baseband signal is strict and the modulation index is low. The proposal of TFM stimulated many researchers to develop other spectrum-efficient digital FM systems.

6.2.2.1 Duobinary FM. In this system, duobinary coding is employed for frequency modulation with limiter–discriminator detection. The baseband signal takes three levels due to the duobinary coding (Section 3.2.6). Figure 6.16 shows the spectra of duobinary FM for NRZ and Nyquist-I pulse shaping. The out-of-band radiation is decreased due to the band-limitation of the premodulation signal by the Nyquist-I filter. The error rate performance of noncoherent demodulation of duobinary coded MSK and TFM in fading mobile radio channels is described in [14].

6.2.2.2 Tamed Frequency Modulation. The block diagram of TFM is shown in Fig. 6.17. A class II partial response (Section 3.2.6) system is introduced to digital FM with modulation index of 0.5 and Nyquist-III pulse

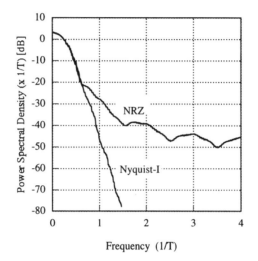

Figure 6.16. Power spectral densities of duobinary FM for NRZ and Nyquist-I pulse shaping (roll-off factor of zero).

shaping is employed. The transfer function becomes

$$H(\omega) = \cos^2\left(\frac{\omega T}{2}\right) N_{III}(\omega), \tag{6.13}$$

where T is the bit duration and $N_{III}(\omega)$ denotes the Nyquist-III transfer function. $N_{III}(\omega)$ is given as

$$N_{III}(\omega) = \frac{\omega T/2}{\sin(\omega T/2)} N_I(\omega),$$

where $N_I(\omega)$ denotes the Nyquist-I transfer function. The term $\cos^2(\omega T/2)$ represents class II partial response signaling and is given by the Fourier transform of its impulse response, that is, $\delta(t + T) + 2\delta(t) + \delta(t - T)$.

The phase shift during one bit duration is given as

$$\Delta\theta(nT) = \theta(nT + T) - \theta(nT) = \frac{\pi}{2}\left(\frac{1}{4}a_{n-1} + \frac{1}{2}a_n + \frac{1}{4}a_{n+1}\right), \tag{6.14}$$

where the data symbols a_n take values $+1$ or -1.

The power spectral density of TFM together with MSK is shown in Fig. 6.18. Tamed FSK is given by replacing the Nyquist-III filter with an NRZ pulse shaping filter. When we apply the TFM to the conventional FM channels, where the channel separation is 25 kHz and adjacent channel

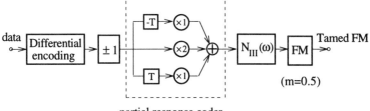

(a) modulator

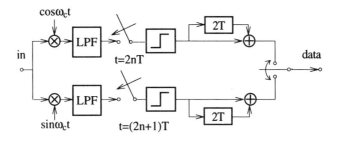

(b) coherent detector

Figure 6.17. Block diagram of a tamed FM system.

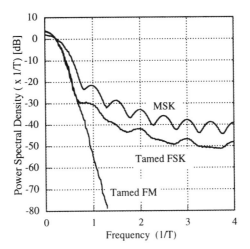

Figure 6.18. Power spectral densities of tamed FM, tamed FSK, and MSK signals.

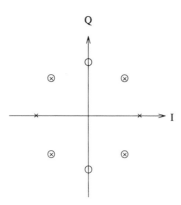

Figure 6.19. Phase constellation of tamed FM.

interference is around -60 dB, we can accommodate digital transmission data rates of 16 kbps. At the time when TFM was invented, a one-chip voice CODEC (coder/decoder) with ADM (adaptive delta modulation; Section 7.9.2) was available. The combination of 16 kbps ADM voice coding and TFM was a historical landmark for digital voice communication through mobile radio channels.

The TFM takes one of fixed phases, namely, $(\pi/8)(a_{n-1} + 2a_{n-1} + a_{n-1})$ $(a_n = \pm 1)$ at each symbol time (Fig. 6.19) since the modulation index is chosen to be 0.5 and Nyquist-III pulse shaping is employed. The signal phase is taken in turn from the two signal groups marked by the cross and circle in Fig. 6.19. By comparing the phase constellations for TFM and for MSK (Fig. 6.3), we can see that the TFM signal scatters due to the introduction of partial response signaling. Even with the scattering of the signal phases, an eye-opening is obtained for the in-phase and quadrature components of TFM, as shown in Fig. 6.20. Therefore the coherent demodulation system, which is similar to that of MSK (Fig. 6.10), can be applied to TFM. The difference between the two systems resides in the low-pass filter. In contrast to MSK, the waveform of in-phase or quadrature signal component of TFM is not time-limited within two symbol times. Also, it never meets Nyquist's first criterion. Thus the receiver with symbol-by-symbol decision does not yield optimum error rate performance.

El-Tanany and Mahmoud [15] investigated the optimum receiver filter for quadrature coherent demodulation of digital FM with modulation index of 0.5. Their results for TFM show that the performance degradation in terms of E_b/N_0, compared to MSK, is 0.75 dB and 1.2 dB at error rates of 10^{-3} and 10^{-6}, respectively. From the viewpoint of actual applications, the receiver filter must be designed taking into consideration the channel selectivity as well as the error rate performance.

Differential detection of TFM shows poor error rate performance [14] due

I

Q

Figure 6.20. Eye-diagram of tamed FM with coherent detection.

to intersymbol interference in the demodulated signal. Limiter–discriminator detection of TFM results in a demodulated signal with five levels due to the class-II partial response coding. Transmitted data can be recovered by decoding the five levels into two levels (see Section 3.2.6). In this case, the precoding $H_p(z) = 1/(1 + 2z^{-1} + z^{-2})$ (mod 2) is assumed. Intersymbol interference, which is in excess from partial response coding, appears because Nyquist-III filtering is used instead of Nyquist-I filtering.

The experimental and computer-simulated error rate performance of TFM is described in [16–18].

6.2.2.3 *Generalized Tamed Frequency Modulation.* In this system [19], generalized partial response (Section 3.2.6) signaling is used for the premodulation signal. The phase shift during a bit duration is given as

$$\Delta\theta(nT) = \theta(nT + T) - \theta(nT)$$

$$= \frac{\pi}{2}(Aa_{n-1} + Ba_n + Aa_{n+1}), \qquad a_n = \pm 1, \qquad (6.15)$$

under the constraint that $2A + B = 1$. TFM is obtained as the special case $A = 1/4$ and $B = 1/2$. The transfer function of the partial response filtering is given as

$$G(\omega) = B + 2A\cos(\omega T). \qquad (6.16)$$

The Nyquist-III filter is still employed.

Changing the parameters A and B and the roll-off factor for the Nyquist-III filter, the spectrum and the bit error rate performance can be varied. For example, the eye-diagram of the receiver baseband signal obtained with frequency discriminator detection is as shown in Fig. 6.21. We can see the almost ideal 3-level eye-opening at the middle of successive symbol times for $B = 0.62$ and roll-off factor of 0.36 in Fig. 6.21(c). This is due to the fact that Nyquist's second criterion is satisfied by choosing those parameters.

(a) B=0.5, r=0 (TFM)

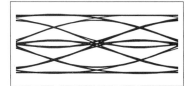

(b) B=0.54, r=0.2

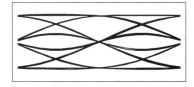

(c) B=0.62, r=0.36

Figure 6.21. Eye-diagram of generalized tamed FM with frequency discriminator detection.

Three-level detection instead of 5-level detection is adopted, since a 3-level decision yields better error rate performance than a 5-level decision. A rectifier and a 2-level threshold detector can be employed for the 3-level detection. This is because the 3-level eye-opening is produced from the interference between the successive symbols; that is, the effect is equivalent to the duobinary coding. The differential encoding for (generalized) TFM acts as precoding for the duobinary coding. Experimental results [19] with frequency discriminator detection of the generalized TFM show superior performance compared to coherent detection in fast fading conditions. An MLSE (maximum likelihood sequence estimation) was applied to the 3-level partial response system with frequency discriminator detection. The error rate performance of frequency-discriminator–MLSE detection of GTFM in a fast fading channel is reported in [20]. The modulation index does not need to be 0.5 as long as frequency discriminator detection is applied.

6.2.2.4 GMSK. GMSK (Gaussian filtered MSK) [21–23] was proposed by Hirade and Murota in 1979. The block diagram of the GMSK modulator is shown in Fig. 6.22. GMSK modulation is performed by inserting a Gaussian

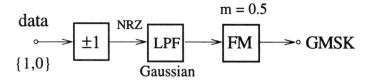

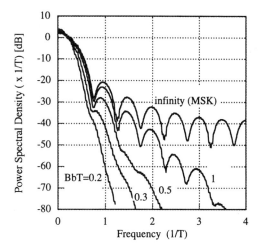

Figure 6.22. GMSK signal generator.

Figure 6.23. GMSK spectra.

low-pass filter as a premodulation filter to MSK modulation. When we assume an impulsive waveform, the transfer function of the premodulation is given as

$$H(\omega) = \frac{\sin(\omega T/2)}{\omega T/2} \exp\left[-(\ln\sqrt{2})\left(\frac{\omega}{2\pi B_b}\right)^2 \right] \qquad (6.17)$$

where B_b is the 3 dB bandwidth (Hz) of the Gaussian filter. The first term corresponds to an NRZ pulse waveform. The GMSK spectrum can be controlled by changing the bandwidth of the Gaussian filter (Fig. 6.23). The spectrum for $B_b T = 0.21$ is almost equivalent to the spectrum of TFM.

A GMSK system can be seen as partial-response digital FM system in which the degree of intersymbol interference changes continuously with the Gaussian filter bandwidth (B_b). The ability of GMSK to continuously control the band-limitation of the premodulation signal is an advantage over other partial-response FM systems. GMSK never takes fixed phase points, since the premodulation filter does not satisfy the Nyquist-III criterion.

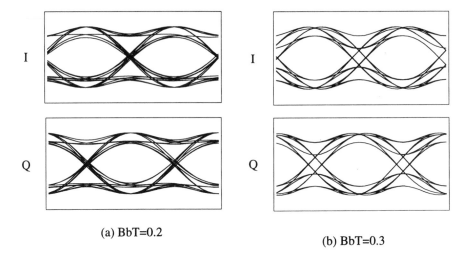

(a) BbT=0.2

(b) BbT=0.3

Figure 6.24. GMSK eye-diagram with coherent detection.

Coherent, differential, and frequency discriminator detection can be applied to GMSK, as well as to MSK. Eye-diagrams of GMSK with coherent detection are shown in Fig. 6.24. The eye-diagram is similar to that of TFM. As we increase B_bT, the eye-diagram becomes close to the MSK eye-diagram. Previously discussed demodulators for MSK can also be applied to GMSK.

Experimental results on bit error rates for coherent detection of GMSK with a Gaussian IF band-pass filter are shown in Fig. 6.25 [23]. The optimum normalized 3 dB bandwidth B_iT for the Gaussian band-pass filter is reported as $B_iT \approx 0.63$. The degradation of the bit error rate performance in terms of E_b/N_0 is 1.6 dB for $B_bT = 0.25$ compared with the ideal antipodal transmission system.

Differential detection of GMSK is discussed in [24–31]. One-bit and two-bit delay detection are considered. The received eye-patterns are shown in Fig. 6.26. The eye-opening becomes narrower due to intersymbol interference with a smaller B_bT value. Theoretical error rates for GMSK with one-bit and two-bit delay detection are described in [30]. Two-bit detection shows superior performance over one-bit detection. The optimum bandwidth ($B_iT = 0.9$–1.4 depending on B_bT, the detection scheme, and bit error rate) of the IF band-pass Gaussian filter is considerably wider for differential detection than for coherent detection. Differential detection of GMSK using decision feedback is described in [31]. Owing to the improved eye-opening with decision feedback, bit error rate performance is significantly improved.

Frequency discriminator detection of GMSK is discussed in [32–35]. With

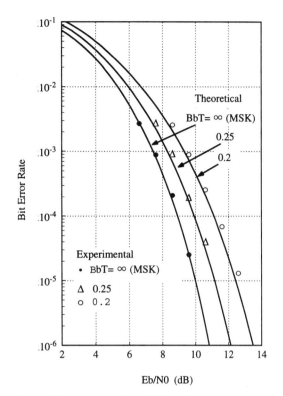

Figure 6.25. Experimental bit error rate for GMSK with coherent detection [23]. (Copyright © IEICE 1981.)

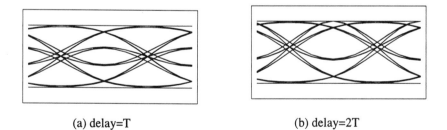

(a) delay=T   (b) delay=2T

Figure 6.26. GMSK eye-diagram with differential detection (B_bT = 0.25). Gaussian band-pass filter is used (B_iT = 1.25).

this detection, the modulation index can take an arbitrary value. The received eye-diagrams are shown in Fig. 6.27. In order to get wider eye-openings, adaptive multilevel threshold decision was introduced [32] as shown in Fig. 6.28. This scheme is equivalent to decision feedback detection. Bit error rate performance is shown in Fig. 6.29.

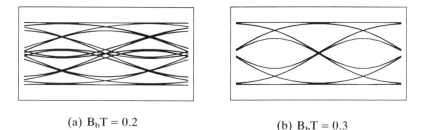

(a) $B_b T = 0.2$

(b) $B_b T = 0.3$

Figure 6.27. GMSK eye-diagram with frequency discriminator detection.

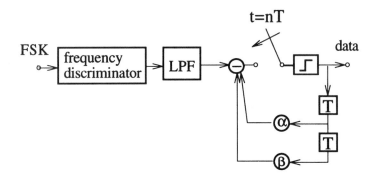

Figure 6.28. Adaptive multilevel threshold decision for GMSK with frequency discriminator detection.

A 3-level decision at the instant between successive symbol times was described in [33]. The error rate performance obtained with this method is almost the same as with the adaptive threshold multilevel decision. Error rate performance of differential and frequency discriminator detection of digital FM, including GMSK, is analyzed theoretically in [36].

Spectrum efficiency of a cellular system with GMSK modulation is analyzed in [37]. As a result of compromise between the spectrum bandwidth and cochannel interference, $B_b T = 0.25$ and forward error correction with rate of 4/5 yields the maximum spectrum efficiency.

6.2.2.5 CCPSK (Compact Spectrum Constant Envelope PSK). This scheme was proposed by Okai, Sugiyama, Asakawa, and Okumura in 1979 [38–39]. CCPSK is a constant envelope continuous phase PSK, where phase transitions in symbol times are governed by 3-bit input sequences. CCPSK signals take one of 12 signal points marked by circles and crosses in turn as shown in Fig. 6.30. The parameter φ takes an arbitrary value. If $\varphi = 45°$ or $0°$, we have the phase constellations for TFM and MSK, respectively.

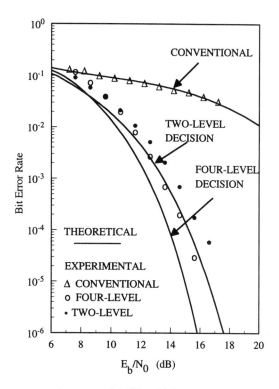

Figure 6.29. Bit error rate performance of GMSK with frequency discriminator detection using adaptive threshold detection [32]. (Copyright © IEEE 1984.)

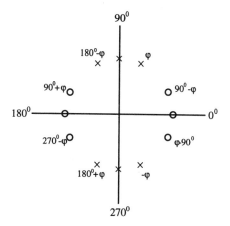

Figure 6.30. Phase constellation for CCPSK. Phase transition occurs from a ○-point marked to a ×-point and vice versa.

Similarly, we have a 2-level eye-opening for the in-phase and quadrature-phase components. CCPSK is a generalized 12PM3 modulation method [40], where 12 stands for 12 phase points, PM means phase modulation, and 3 corresponds to the fact that the phase transition is governed by 3-bit sequences. Some waveforms for phase transition are suggested to maintain phase continuity. The derivative of the phase waveform, that is, the frequency waveform, is not necessarily continuous and causes poor out-of-band radiation for CCPSK.

6.2.2.6 *Correlative Phase Shift Keying (CORPSK).* CORPSK is a special form of constant envelope PSK [18] that is defined by the following three points: (i) The information is encoded into phase shifts $\Delta\varphi_m$, so that the phase is continuous. Over one symbol interval the phase shift is

$$\Delta\varphi_m = \varphi[(m+1)\,T_s] - \varphi(mT_s) = c_m \frac{2\pi}{n}, \qquad (6.18)$$

where the integer c_m carries the information and the constant n is the number of possible phase points. (ii) Successive phase points are correlated due to correlative encoding of input data which takes L (≥ 2) levels; nonlinear as well as linear correlative encoding are considered. (iii) The phase waveform, as well as its time derivative (frequency waveform) is continuous.

Thus CORPSK includes many varieties of constant envelope PSK with correlative encoding. A new method of CORPSK is the four-level input, nonlinear correlative encoding system (CORPSK(4-5)). In this system phase points are the same as in QPSK. The phase transition of π or $-\pi$, both of which result in the same phase point, is nonlinearly controlled so that the phase path takes a smoother trajectory. CORPSK shows almost the same out-of-band radiation as TFM but has better bit error rate performance.

6.2.2.7 *Digitally Phase-Modulated (DPM) Signals.* Digital phase modulation with correlative coding was analyzed by Maseng [41]. In this system, modulated signals never take fixed points, unlike in phase shift keying systems. Hence, symbol-by-symbol phase detection on the in-phase and quadrature-phase plane is not applicable. Coherent demodulation with maximum likelihood sequence estimation is used for the receiver. A residual carrier component, which appears in the digital phase modulated signal, is an attractive feature of the DPM signal for coherent demodulation. Modems for DPM are easier to implement than for digital FM, since the phase of a DPM signal never depends on the preceding data, in contrast to digital FM. Compared to binary phase shift keying, a coding gain of several decibels is obtained with higher modulation index.

6.2.2.8 *Continuous Phase Modulation.* Continuous phase modulation (CPM) has a constant envelope continuous phase (digital) waveform.

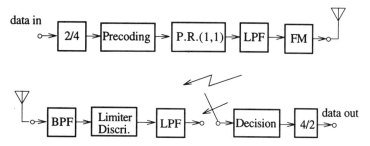

Figure 6.31. Duoquaternary FM with frequency discriminator detection.

Multilevel modulation, various modulation indices, and premodulation waveforms including partial response signaling are considered. The optimum (sequence) detector with coherent demodulation is assumed. The demodulated signal at a symbol period is correlated with those at previous periods due to an arbitrary modulation index. Even when the premodulation waveforms have no correlation between different symbol periods (full response system), the correlation operation is still performed. Therefore, the optimum detector observes the received signal for more than one symbol interval.

Spectrum and bit error rate performance are investigated extensively in [42–44]. Some CPM signals show superior performance compared to MSK and PSK signals. Multi-h modulation, where the modulation index is periodically varied, was proposed to increase distance between signal sequences [77].

6.2.2.9 Duoquatenary FM. If we apply the 4-level signal to a partial response filter whose impulse response is $h(t) = \delta(t) + \delta(t - T)$ ($T =$ symbol duration), we have the duoquatenary signal. Duoquatenary FM with limiter–frequency discriminator detection (see Fig. 6.31) was discussed in [50, 51]. A 7-level signal is obtained at the output of the post-demodulation filter. The spectrum and bit error rate performance are better than for 4-level FM.

6.2.3 Nyquist-Filtered Digital FM

6.2.3.1 Nyquist-Filtered Multilevel FM. In 1979, Akaiwa, Takase, Ikoma, and Saegusa [45, 46] proposed digital FM with frequency discriminator detection, where the premodulation and postmodulation low-pass filters as a whole meet Nyquist's first criterion (see Fig. 6.32). When the bandwidth of the band-pass IF filter is wide enough, intersymbol interference-free transmission is achieved. The premodulation filter, together a low modulation index and multilevel signaling, contributes to a narrow spectrum of the

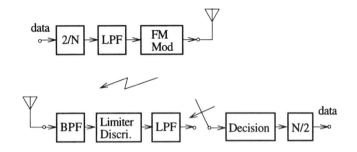

Figure 6.32. Block diagram of *N*-level digital FM with frequency discriminator detection.

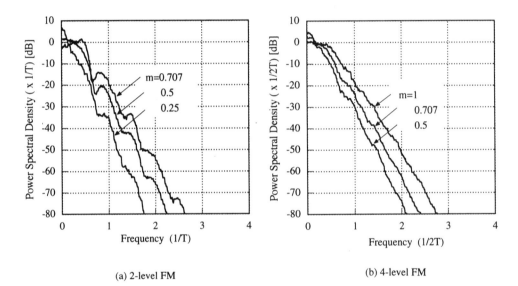

(a) 2-level FM

(b) 4-level FM

Figure 6.33. Spectra of square root of Nyquist-I filtered digital FM with modulation index as a parameter. Roll-off factor is 0.5. *T* is bit duration.

modulated signal. The post-demodulation filter helps achieve good bit error rate performance by band-limiting the demodulated noise.

The spectrum and bit error rate performance are shown in Figs. 6.33(a), (b), and 6.34(a), (b). The Nyquist-I filter characteristics are equally divided into the premodulation and post-demodulation filter. A sharp cut-off band-pass filter with the Carson bandwidth (Section 5.3.2) is assumed.

The effects of the IF band-pass filter bandwidth on received eye-diagrams and bit error rate are shown in Figs. 6.35 and 6.36, respectively. The crosses in Fig. 6.36 denote computer-simulated results and solid lines show theoretical results described in Section 5.5.4.

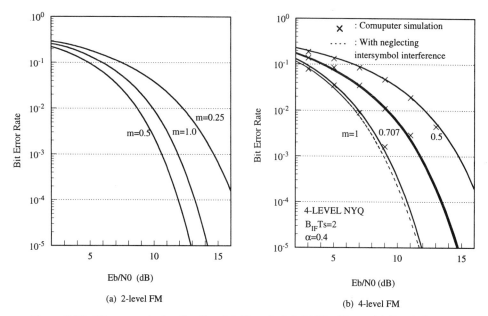

(a) 2-level FM

(b) 4-level FM

Figure 6.34. Bit error rate for the Nyquist filtered digital FM with modulation index as a parameter.

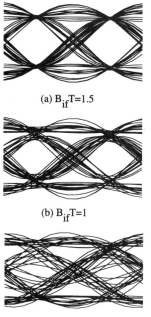

(a) $B_{if}T=1.5$

(b) $B_{if}T=1$

(c) $B_{if}T=0.9$

Figure 6.35. Effects of IF band-pass filter bandwidth on received eye-diagrams for Nyquist filtered 2-level FM. Modulation index = 0.5, roll-off factor is 0.4, and band-pass filter characteristics are rectangular shape.

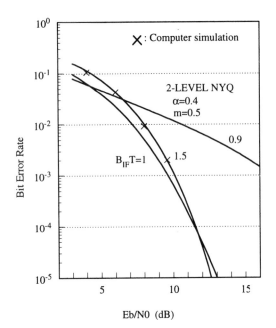

Figure 6.36. Effects of IF band-pass filter bandwidth on bit error rate for Nyquist-filtered 2-level FM.

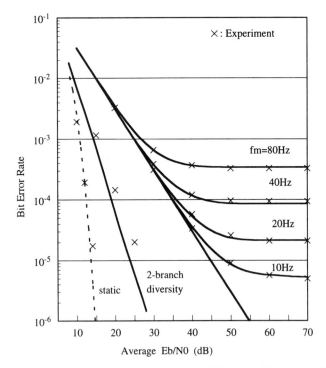

Figure 6.37. Bit error rates for Nyquist-filtered 4-level FM under Rayleigh fading. In plot, f_m denotes the maximum Doppler frequency.

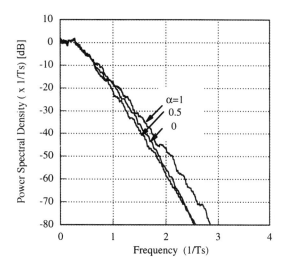

Figure 6.38. Spectra of Nyquist-III filtered 4-level FM with a modulation index of 0.75 with roll-off factor as a parameter.

The bit error rates for Nyquist filtered four-level FM under Rayleigh fading are shown in Fig. 6.37. Solid lines show theoretical results obtained by adding the error rates given by Eqs. (5.160) and (5.157b) without diversity reception.

If we adopt the Nyquist third characteristics (Section 3.1.3) to the premodulation filter and choose a proper value for the modulation index, we have constant envelope phase shift keyed signals: for 2-level and 4-level signaling systems, constant envelope $\pi/2$ shifted BPSK and $\pi/4$ shifted QPSK signals are obtained with modulation index of 0.5 and 0.75, respectively [47]. Since these signals take fixed signal phases, coherent and differential detection can be used to demodulate the signal. Spectrum and bit error rate performance of the 4-level system are shown in Figs. 6.38 and 6.39, respectively, and is almost equal to the performance of TFM.

6.2.3.2 *PLL-QPSK.* PLL-QPSK was proposed by Honma, Murata, and Tatsuzawa in 1979 [48]. The block diagram of PLL-QPSK is shown in Fig. 6.40. $\pi/4$ shifted QPSK with an NRZ pulse shape is fed to a PLL (phase-locked loop) circuit to produce a constant envelope narrow-band signal. A low-pass characteristic is given to the loop transfer function to suppress out-of-band radiation.

A limiter–frequency discriminator detector with an integrate-and-dump filter is employed. It detects the phase shifts of $\pm\pi/4$ and $\pm3\pi/4$ of the $\pi/4$ shifted QPSK signal. Therefore differential encoding at the transmitter is assumed. The differential encoding of the input data signal can be performed in FM modulation through its integrating process. Thus the PLL-QPSK signal can be produced by FM modulation as shown in Fig. 6.41(d). Consequently,

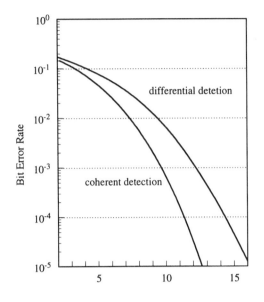

Figure 6.39. Bit error rates for Nyquist-III filtered 4-level FM with coherent and differential detection. Differential encoding is assumed. Modulation index is 0.75 and the roll-off factor is 0.5. Band-pass filter has a rectangular transfer function with normalized bandwidth of $0.8/T$ (T = bit duration).

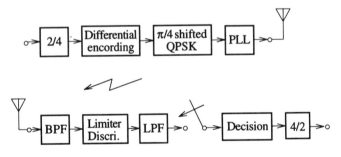

Figure 6.40. Block diagram of PLL-QPSK.

we can see from Fig. 6.41 that the PLL-QPSK is a special case of 4-level FM, where the modulation index is 0.75. The equivalent premodulation filter for the FM modulator and post-demodulation filter including the I&D filter, as a whole, satisfy Nyquist's first criterion.

6.2.4 Performance Comparison

A simultaneous comparison of spectrum and bit error rate performance between digital FMs is shown in Fig. 6.42 [52]. The spectrum performance

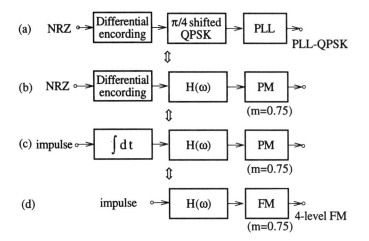

Figure 6.41. Equivalence between PLL-QPSK and 4-level FM.

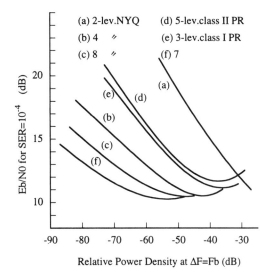

Figure 6.42. Performance comparison between digital FMs with limiter frequency discriminator detection. The curves are obtained by changing modulation index.

is represented by the attenuation of the power spectral density at a frequency that deviates by bit rate frequency from the center frequency. For the power spectral density attenuation of around $-60\,$dB, the duoquatenary system shows the best performance. The actual choice from those systems must take into consideration the circuit complexity and performance degradation due to imperfections in the circuit.

6.3 LINEAR MODULATION

The spectrum of linearly modulated signals is given by translating the modulating baseband signal spectrum into the carrier frequency (Section 5.4.1). Thus filtered linear modulation yields a narrow spectrum corresponding to the filtering at the baseband frequency. A penalty for the narrow spectrum is that the digital linear modulation loses its constant envelope property as with unfiltered signaling.

Figure 6.43 compares the spectra for filtered linear QPSK and premodulation filtered 4-level FM. The superiority of the linear modulation over the constant envelope (nonlinear) modulation is clear: two to three times narrower bandwidth at an out-of-band radiation of around -60 dB (for example) is obtained with the linear modulation.

However, this is true only in an ideal channel. If the transmitter shows nonlinear distortion, the linear modulation loses the spectral superiority, since the out-of-band radiation appears proportionally to the degree of odd-order nonlinear distortion (Appendix 5.1). For example, Fig. 6.44 shows spectra for $\pi/4$ shifted QPSK at the output of a quasi-linear transmit power amplifier. The out-of-band radiation increases with increase of the power efficiency of the amplifier: a higher power efficiency is achieved with a higher nonlinear distortion in the power amplifier. The power efficiency is a prime issue in mobile radio communication, especially for portable communication where batteries are the power sources.

Suppressing the out-of-band radiation after the power amplifier by the use of a band-pass filter is impractical in mobile communication since the center

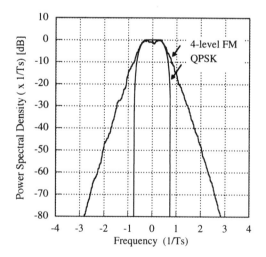

Figure 6.43. Spectra for QPSK and 4-level FM with modulation index of 1. Square root of Nyquist-I characteristic with a roll-off factor of 0.5 is assumed.

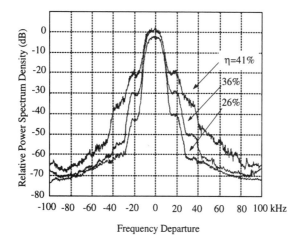

Figure 6.44. Spectra of $\pi/4$ shifted QPSK at the output of a 900 MHz, MOSFET quasi-linear power amplifier for different input power (η = power efficiency). Roll-off factor of 0.5 is used. Bit rate is 32 kbps.

frequency must be varied. Even when the center frequency is fixed, the realization of a low-loss band-pass filter is difficult, since the ratio of passband bandwidth to the center frequency is very small in typical mobile radio communications.

It is difficult to get a linear power amplifier that is power-efficient and at the same time shows low nonlinear distortion. This was one reason why linear modulation was discarded for application to mobile radio communications [53]. Another difficulty with the linear modulation for mobile radio communication was the implementation of the receiver circuit: in order to receive a linearly modulated signal, an AGC (automatic gain control) circuit was required that could cope with fast fading and wide dynamic range of the received signal in mobile radio channels. Such an AGC circuit is much harder to implement than a limiter circuit, which can be used for receiving a constant envelope signal. This is especially so for narrow-band signal transmission.

In 1985, Akaiwa and Nagata proposed a linear $\pi/4$ shifted QPSK modulation system (Fig. 6.45(a)) that resolved the above difficulties [54, 55]. A feedback control was introduced to compensate for the nonlinear distortion of a power-efficient amplifier and a limiter circuit was adopted in the receiver. The spectral performance is improved by 30 dB, as shown in Fig. 6.45(b), owing to the negative feedback control. Out-of-band radiation of -60 dB was achieved with a power efficiency of 35%. Bit error rate performance was comparable with performance for digital FM. Since then, many reports have been published on linear modulation systems for digital mobile communications [56–59]. In addition to a narrower signal spectrum, a linear modulation system employing linear circuits at the transmitter shows advantages such as

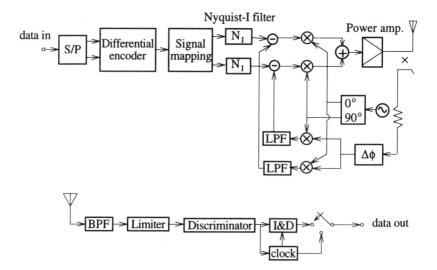

Figure 6.45. (a) Block diagram of linear $\pi/4$ shifted QPSK system with a linearized power amplifier and limiter–discriminator detection.

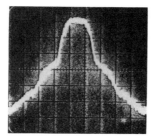 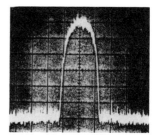

Figure 6.45. (b) Spectrum improvement with negative feedback control. Vertical axis 10 dB/division; horizontal axis 10 kHz/division.

ease of transmitter output power control and tailoring of the on–off burst signal to prevent spectrum spreading for TDMA signals.

In order for digital linear modulations to show a superior spectrum compared to narrow-band constant envelope modulations such as tamed FM, PLL-QPSK, and Nyquist-filtered multilevel FM, multilevel modulation must be introduced. The candidates are QPSK and its modifications OQPSK and $\pi/4$ shifted QPSK, 8PSK, 16QAM, and other higher-level modulations.

This section describes linear modulation/demodulation systems including nonlinearity compensating techniques.

(a) QPSK   (b) π/4 shifted QPSK   (c) Offset QPSK

Figure 6.46. QPSK signal trajectories.

6.3.1 π/4 Shifted QPSK

The same spectra are given by QPSK, OQPSK, and π/4 shifted QPSK. Their trajectories are shown in Fig. 6.46. OQPSK and π/4 shifted QPSK never pass through the origin, where the amplitude becomes zero. This property has advantages when the signals are applied to an amplitude limiter together with noise at a receiver: the phase perturbations due to noise becomes less. Furthermore, a lower dynamic range of the signal amplitude for OQPSK and π/4 shifted QPSK is beneficial from the viewpoint of transmitter and receiver circuit implementation.

Coherent detection can be applied to all the three modulation methods. Differential detection cannot be applied to OQPSK, since the signal phase never takes fixed points, as shown in Fig. 6.47, and consequently the demodulated signal eye never opens.

Based on the above consideration, π/4 shifted QPSK was selected and investigated for application to digital mobile radio communication in [55]. A limiter–frequency discriminator with integrate-and-dump detection was proposed (Fig. 6.45(a)). The principle of this system is to detect phase shifts of ±π/4 and ±3π/4 associated with the π/4 shifted QPSK signal, as explained mathematically in the following.

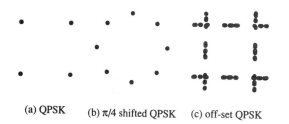

(a) QPSK   (b) π/4 shifted QPSK   (c) off-set QPSK

Figure 6.47. Signal phase points at sampling instants.

Figure 6.48. Eye-diagram of limiter frequency discriminator and integrate-and-dump detection of $\pi/4$ shifted QPSK. Nyquist-I filter with roll-off factor of 0.5 is assumed.

The output of the frequency discriminator is proportional to the instantaneous frequency $\omega_i(t)$. At the output of the integrate-and-dump filter, we have

$$s_d(nT + t) = \int_{nT}^{nT+t} \omega_i(\tau)\, d\tau, \qquad 0 \leq t \leq T, \tag{6.19}$$

where T is the symbol duration. At the sampling instant $(n+1)T$ we have

$$s_d(nT + T) = \Delta\theta(nT)$$
$$\equiv \theta(nT + T) - \theta(nT), \tag{6.20}$$

where the phase $\theta(t)$ is given by $\theta(t) = \int_{-\infty}^{t} \omega_i(t)\, dt$, and $\Delta\theta(nT)$ represents the phase shifts during one symbol duration.

Differential encoding is assumed in the transmitter. The received signal eye-diagram is shown in Fig. 6.48. Raised-cosine roll-off Nyquist-I filtering is employed at the transmitter. Intersymbol interference is caused by the sharp cut-off band-pass filter at the receiver.

The bit error rate performance is shown in Fig. 6.49. With the above detection system, errors occur more often for the phase shift of $\pm 3\pi/4$ than for $\pm\pi/4$, since the signal trajectory passes closer to the origin. An error occurs when the phase trajectory is deviated by noise such that the resultant signal's phase rotation may take a direction opposite to its original path. This results in a total phase shift of $\pm 5\pi/4$ instead of $\mp 3\pi/4$. To prevent this kind of error, the phase shift of $5\pi/4$ and $-5\pi/4$ is corrected as $-3\pi/4$ and $3\pi/4$, respectively (i.e., mod 2π decision). This is done by making 6-level instead of 4-level decisions at the output of the integrate-and-dump filter. A more general decision is made by applying the mod 2π operation also to the phase shift of $\pm\pi/4$ (i.e., 8-level decision). The bit error rate performance is improved with this technique, as shown in Fig. 6.49.

Differential detection is also employed to detect the phase shift of $\pm\pi/4$

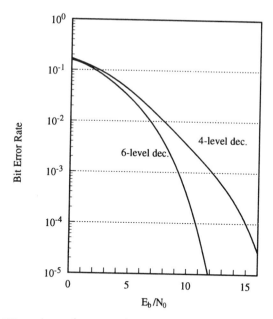

Figure 6.49. Bit error rate performance for $\pi/4$ shifted QPSK with limiter frequency discriminator and integrate-and-dump filter detection. Raised cosine roll-off Nyquist-I filter with roll-off factor of 0.5 is assumed at transmitter. Gray code mapping is assumed.

and $\pm 3\pi/4$. The output of the differential detector at the sampling instants become

$$s_x(nT + T) = \sin \Delta\theta(nT), \quad s_y(nT + T) = \cos \Delta\theta(nT).$$

The mod 2π operation on $\Delta\theta(nT)$ described above is automatically built into the sine and cosine functions.

Since the 4-level decision on the $\Delta\theta(nT)$ is equivalent to the 2-level decision on $\sin [\Delta\theta(nT)]$ and $\cos [\Delta\theta(nT)]$, the same bit error rate performance is obtained with frequency discriminator detection with mod 2π operation and integrate-and-dump filter, or differential detector. This is true as long as the effect of a post-detection filter for the differential detector is neglected. This assumption is usually valid since the post-detection filter is only used to remove the higher-order IF frequency components produced in the multiplication circuit. In this case the limiter does not affect the error rate performance.

Equal splitting of the raised-cosine roll-off Nyquist-I filter characteristics between the transmitter and receiver gives the optimum error rate performance for differential as well as coherent detection under additive white noise conditions. For differential detection the receiver filter must be placed before the differential detector. A band-pass filter whose low-pass equivalent

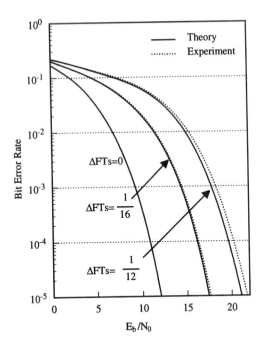

Figure 6.50. Bit error rate performance for $\pi/4$ shifted QPSK with differential detection and frequency offset as a parameter.

transfer function is the square root of the raised-cosine roll-off Nyquist-I filter meets the requirements [57].

The bit error rate performance of ($\pi/4$ shifted) QPSK using differential detection and a matched band-pass filter is investigated in the following. Experimental results are given by computer simulations.

Figure 6.50 shows bit error rates vs. bit energy to noise power ratio. The theoretical results are given by Eq. (5.78). In the same figure are shown the error rates with the center frequency offsets. Theoretical values are calculated with Eq. (5.79). The discrepancy between the theoretical and experimental results is due to the fact that the intersymbol interference caused by the frequency offset is not considered in the theory. The experimental bit error rate performance under the cochannel interference environment is shown in Fig. 6.51.

Under the Rayleigh fading condition, Fig. 6.52 and Fig. 6.53 show the bit error rate performance for average bit energy to noise power ratio and cochannel interference, respectively.

6.3.2 8-Level PSK

8-Level PSK has a signal spectrum that is two-thirds of a QPSK signal spectrum. It may be the highest-level modulation that falls into the category

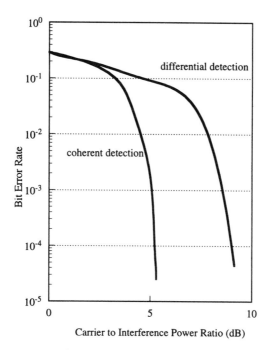

Figure 6.51. Bit error rate performance of $\pi/4$ shifted QPSK under cochannel interference environment.

of phase shift keying and is adopted for actual use in mobile radio communication: higher-level modulation is probably adopted from QAM such as 16QAM.

If we combine a rate 2/3 error correcting code and 8PSK, the bandwidth of the modulated signal is the same as for QPSK. The error rate performance is improved with the error correcting code.

Currently 8PSK has not been sufficiently investigated for application to mobile radio communications [60].

6.3.3 16QAM

16QAM is a bandwidth-efficient modulation method because of the encoding of 4 bits/symbol. 16QAM occupies half the bandwidth of QPSK. Because the signal constellation is spread in both amplitude and phase, noncoherent demodulation cannot be applied, which is a disadvantage in mobile radio communications. To make 16QAM applicable to mobile radio communications, coherent demodulation methods are proposed [61, 62] that are sufficiently effective under fast fading conditions. The principal idea is to send a pilot (carrier) inserted in the modulated signal for stable operation of coherent demodulation. The extracted pilot signal is used to compensate for

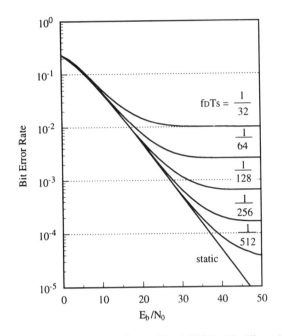

Figure 6.52. Bit error rate performance of $\pi/4$ shifted QPSK with differential detection under Rayleigh fading condition. f_0: maximum Doppler frequency, T_0: symbol duration.

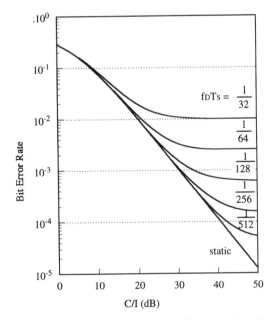

Figure 6.53. Error rate for $\pi/4$ shifted QPSK with differential detection in the presence of cochannel interference and Rayleigh fading condition. f_D: maximum Doppler frequency, T_s: symbol duration.

P: Pilot signal

Figure 6.54. Insertion of pilot signal in time domain.

the perturbed amplitude and phase of the received signal passing through the fading channel. Two approaches are known for insertion of the carrier signal: one is in the frequency domain [61] and the other is in the time domain [62]. (For a BPSK system, there is another class of the carrier signal insertion [63], namely, insertion in the orthogonal component of the BPSK signal.)

In the TTIB (transparent tone-in-band) method, the modulated signal is divided into two spectrum components in the frequency domain and they are shifted up and down to give a band gap into which to insert the pilot signal. At the receivers, the inverse process is carried out. The dual pilot tone can be inserted at the edge of the modulated signal spectrum [64]. This technique is potentially effective for compensating for frequency-selective fading. Using the pilot signals, estimation of the frequency-selective transfer function over the signal bandwidth is possible.

The method of inserting the pilot signal in the time domain was proposed by Sampei [62]. Into each of N-1 symbols is inserted a pilot symbol (Fig. 6.54). At the receiver the pilot signal is extracted and interpolated to produce a time-continuous carrier signal. This method requires less signal processing than others. The interval of insertion of the pilot signal should be short enough to cope with fading speeds. For a transmission speed of 64 kbps, coherent detection is stable with pilot signal insertion in every 16th symbol and a second-order interpolator. Field trial experiment results are described in [65].

6.3.4 Linear Power Amplifier

A power amplifier with high power efficiency and low nonlinear distortion is a key requirement for linear modulation to be feasible in mobile radio communications. Many efforts on nonlinearity compensation methods had been devoted to power-efficient amplifiers. For AM broadcasting and SSB radio communications, nonlinearity compensating techniques have long been investigated [70–73]. Nonlinearity compensated amplifiers can be classified into feedback, predistortion, and feed-forward amplifiers.

Feedback Amplifier. Feedback technique is well known for nonlinear compensation in power amplifiers, especially for audio amplifiers. Its

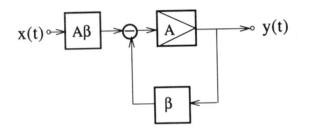

Figure 6.55. Feedback amplifier.

principle is as follows. Let $x(t)$ denote the input signal and $y(t)$ the output signal,

$$y(t) = Ax(t) + \Delta f_d[x(t)], \qquad (6.21)$$

where A is the (linear) amplification coefficient and the function $\Delta f_d[\cdot]$ denotes the nonlinear distortion. $\Delta f_d[\cdot]$ depends on the nonlinearity of the amplifier for a given $x(t)$. Let us consider a feedback amplifier shown in Fig. 6.55, where the coefficient $A\beta$ is introduced to keep the output power level unchanged with the feedback control. We have

$$y(t) = A[A\beta x(t) - \beta y(t)] + \Delta f_d[A\beta x(t) - \beta y(t)] \qquad (6.22)$$

or

$$y(t) = \frac{A^2\beta x(t) + \Delta f_d[\varepsilon(t)]}{1 + A\beta}, \qquad (6.23)$$

where $\varepsilon(t) = A\beta x(t) - \beta y(t)$.
   For $|A\beta| \gg 1$, $y(t)$ is approximated as

$$y(t) \approx Ax(t) + \frac{\Delta f_d[\varepsilon(t)]}{A\beta}. \qquad (6.24)$$

With low distortion, that is, $|\Delta f_d[\varepsilon(t)]| \ll |A^2\beta x(t)|$, using Eq. (6.23), $y(t)$ is approximated as

$$y(t) \approx \frac{A^2\beta x(t)}{1 + A\beta}$$

and $\varepsilon(t)$ is approximated as

$$\varepsilon(t) \approx x(t). \qquad (6.25)$$

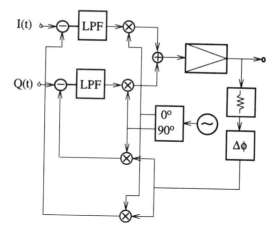

Figure 6.56. Cartesian loop modulation feedback amplifier.

This approximation means that the nonlinear distortion component is neglected in the feedback path. Noting Eq. (6.25) and comparing Eqs. (6.21) and (6.24), we can see that the nonlinear distortion is decreased by a factor of $A\beta$, the feedback gain factor.

Feedback control at RF frequencies becomes difficult because of the feedback circuit implementation. To overcome the difficulty, the RF signal is demodulated to produce baseband signals. The baseband signals are negatively fed back to the baseband signals or to modulating circuits. The demodulation of an RF signal in Cartesian coordinates (in-phase and quadrature components) and polar coordinates (amplitude and phase components) is shown in Figs. 6.56 and 6.57, respectively. The Cartesian loop (feedback control) method is suitable for the quadrature modulator since quadrature components are used. This method was applied to the 140 MHz, class AB power amplifier with $\pi/4$ shifted QPSK [55]. A power efficiency of 35% and out-of-band radiation density of -60 dB are obtained with a feedback gain of 30 dB. The results with this method applied to a 800 MHz power amplifier are reported in [66].

The polar loop method was applied to a $\pi/4$ shifted QPSK transmitter [67]. This method has advantages since it is immune to the signal phase shift caused while selecting different channel frequency in multichannel systems. Results when only the amplitude component is fed back are reported in [68]. An 850 MHz power amplifier with power efficiency of 50% and third-order distortion of -30 dB is obtained.

The advantage of the feedback method is that nonlinearity compensation performance can be maintained even when the amplifier parameters deviate due to temperature and age. The inherent disadvantage for the feedback control is the problem of possible instability. In this application care must

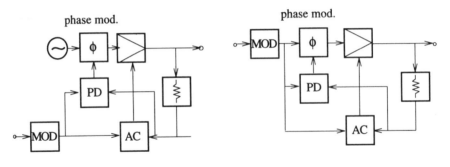

(a) Method 1   (b) Method 2

Figure 6.57. Polar coordinate demodulation feedback amplifier. PD = phase detector, AC = amplitude comparator.

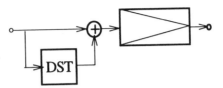

DST: distortion generator

Figure 6.58. Linearized power amplifier with predistorter.

be taken, especially when the channel frequencies are widely varied for channel selection. Currently, the question of which method, including the predistortor and feed-forward methods, is most suitable for a digital mobile radio amplifier cannot be answered. It is the opinion of the author that the RF feedback method with feedback gain around 15 dB is the most practical for mobile terminals, where ease of circuit implementation is important.

Predistorter. The predistorter adds a distorted signal to the input signal in advance to cancel the distortion generated in the amplifiers (Fig. 6.58). Assume the characteristic of an amplifier given by Eq. (6.21), then the predistorted input signal $x'(t)$ to the amplifier becomes

$$x'(t) = x(t) - \Delta f_d[x(t)]/A, \tag{6.26}$$

where $x(t)$ denotes the signal to be amplified. The output of the amplifier becomes

$$y(t) = Ax(t) - \Delta f_d[x(t)] + \Delta f_d\{x(t) - \Delta f_d[x(t)]/A\}. \tag{6.27}$$

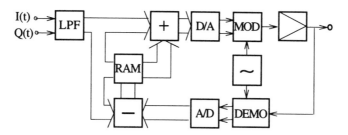

Figure 6.59. Adaptive predistorter amplifier.

When distortion is small $(|\Delta f_d[x(t)]| \ll A|x(t)|)$, $y(t)$ is approximated as

$$y(t) \approx Ax(t). \tag{6.28}$$

Thus the distortion is canceled.

The advantage of the predistorted amplifier is that, in principle, it is free from the instability problem, since it belongs to an open-loop control. However, the compensating performance degrades if the amplifier parameters deviate from the preset values. In order to resolve this disadvantage, an adaptive control method is proposed by Nagata [69] as shown in Fig. 6.59. A feedback path is introduced to detect the error and to automatically correct the predistorter. The predistorter is composed of RAM (random access memory) at baseband frequency. Output of the power amplifier is demodulated and compared with input signal to produce an error signal. The error signal is fed into the predistorter RAM. Thus the predistorter is adaptive to the change in characteristics of the power amplifier. It should be noted that the bandwidth of the closed loop is much narrower than the bandwidth of the modulating signal; this is not the case for the feedback compensation method. The correction algorithm is based on the iterative method. An algorithm was proposed in [74] to speed up the convergence time of this method.

Feed-Forward Method. The block diagram of a feed-forward amplifier is shown in Fig. 6.60. The distortion or error signal produced in the amplifier is detected by comparing the input and output signals. The detected error signals fed into the linear subamplifier to amplify it to the same level as that of the main power amplifier. The amplified error signal is subtracted from the output of the power amplifier. The advantages and disadvantages of the feed-forward amplifier are the same as those of the predistorted amplifier.

The adjustment of the delay time and gain in the second path is important for the success of the feed-forward amplifier. The linearity of the subamplifier must be high. This can decrease the overall power efficiency. An automatic error-correcting technique for the feed-forward amplifier was introduced as

main amp.

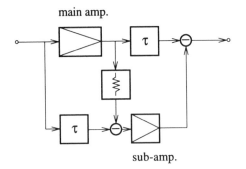

sub-amp.

Figure 6.60. Feed-forward amplifier.

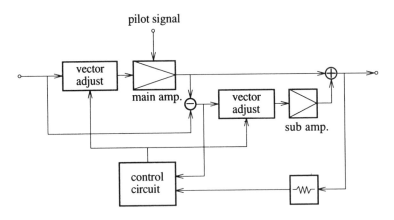

Figure 6.61. Adaptive feed-forward amplifier.

shown in Fig. 6.61. Figure 6.62 shows results for an 800 MHz power amplifier [75]. The nonlinear distortion is suppressed more than 30 dB with a 32 W (average) FET power amplifier. This result was obtained over a 20 MHz bandwidth. With this amplifier, experiments were performed on simultaneous amplification of 32 carriers, where each carrier has 1 W of power and is spaced at 100 kHz intervals. Leakage power into unused channels was less than −60 dB. These results show that this amplifier is suitable for a multicarrier amplifier which has application in cellular base stations.

In [71], linearly modulated signals are split into the amplitude component and the phase component with constant envelope and these are amplified independently. The constant envelope phase component is modulated by the amplitude components. This methods falls into the category of feed-forward amplifiers. The gain control is actually performed by controlling the supplied voltage of the power amplifier. Using a similar idea, another method controls the bias voltage corresponding to the input signal level.

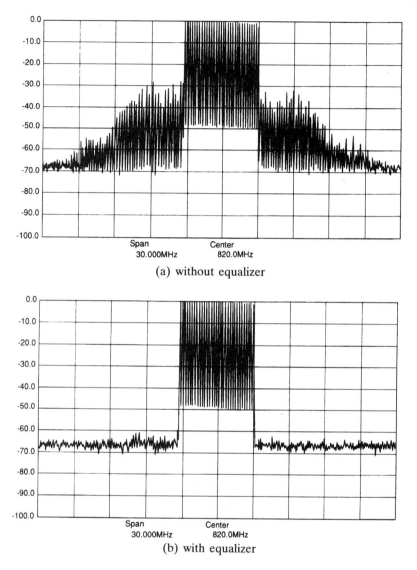

(a) without equalizer

(b) with equalizer

Figure 6.62. Spectrum improvement with feed-forward nonlinearity compensation [75]. (Copyright © NTT 1992.)

LINC Amplifier. The LINC (linear amplification with nonlinear components) system [76] is a combination of the principles of the predistorter and feed-forward techniques (Fig. 6.63). The input signal is generally given by

$$s_\alpha(t) = E(t) \cos[\omega_0 t + \theta(t)], \tag{6.29}$$

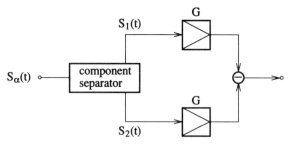

Figure 6.63. LINC amplifier.

where the envelope $E(t)$ and phase $\theta(t)$ correspond to the modulating baseband signal. Expressing $E(t) = E_m \sin \varphi(t)$, $s_\alpha(t)$ is written as

$$s_\alpha(t) = s_1(t) - s_2(t),$$

where

$$s_1(t) = \frac{E_m}{2} \sin[\omega_0 t + \theta(t) + \varphi(t)] \tag{6.30}$$

and

$$s_2(t) = \frac{E_m}{2} \sin[\omega_0 t + \theta(t) - \varphi(t)]. \tag{6.31}$$

$s_1(t)$ and $s_2(t)$ are constant envelope signals and hence each of them can be amplified through a saturated amplifier with a gain factor of G. The combined output becomes $Gs_1(t) - Gs_2(t) = GE_m \sin \varphi(t) \cos[\omega_0 t + \theta(t)] = Gs_\alpha(t)$. The key points to the success of this linear power amplifier are the component separation circuit and the balance of the two signal paths.

REFERENCES

1. CCITT V-series Recommendation.
2. R. Debuda, "Coherent demodulation of frequency shift keying with low deviation ratio," *IEEE Trans. Communications*, **COM-20**, 429–435 (June 1972).
3. S. Pasupathy, "Minimum shift keying: a spectrally efficient modulation," *IEEE Communication Magazine*, 14–22 (July 1979).
4. K. Murota and K. Hirade, "GMSK modulation for digital mobile radio telephony," *IEEE Trans. Communications*, **COM-29**, 1044–1050 (July 1981).
5. M. Ishizuka and K. Hirade, "Optimum Gaussian filter and deviated-frequency-locking scheme for coherent detection of MSK," *IEEE Trans. Communications*, **COM-28**, 850–857 (June 1980).

6. G. Kawas Kaleh, "A differentially coherent receiver for minimum shift keying," *IEEE Journal on Selected Areas in Communications*, **7**, 99–106 (January 1989).

7. T. Masamura, S. Samejima, Y. Morihiro, and F. Fuketa, "Differential detection of MSK with nonredundant error correction," *IEEE Trans. Communications*, **COM-27**, 912–918 (June 1979).

8. M. K. Simon and C. C. Wang, "Differential versus limiter–discriminator detection of narrow-band FM," *IEEE Trans. Communications*, **COM-31**, 1227–1234 (June 1983).

9. F. Amoroso, "Pulse and spectrum manipulation in the minimum (frequency) shift keying (MSK) format," *IEEE Trans. Communications*, **COM-24**, 381–384 (March 1976).

10. M. K. Simon, "A generalization of minimum-shift-keying (MSK)-type signaling based upon input data symbol pulse shaping," *IEEE Trans. Communications*, **COM-24**, 845–856 (August 1976).

11. N. Rabzel and S. Pasupathy, "Spectral shaping in minimum shift keying (MSK)-type signals," *IEEE Trans. Communications*, **COM-26**, 189–195 (August 1978).

12. A. Lender, "The duobinary techniques for high-speed data transmission," *IEEE Trans. Commun. & Electron.*, 214–218 (May 1963).

13. F. de Jager and C. B. Dekker, "Tamed frequency modulation, a novel method to achieve spectrum economy in digital transmission," *IEEE Trans. Communications*, **COM-26**, 534–542 (August 1978).

14. S. Elnoubi and S. C. Gupta, "Error rate performance of noncoherent detection of duobinary coded MSK and TFM in mobile radio communication systems," *IEEE Trans. Vehicular Technology*, **VT-30**, 62–72 (May 1981).

15. M. S. El-Tanany and S. A. Mahmoud, "Mean-square error optimization of quadrature receivers for CPM with modulation index 1/2," *IEEE Journal on Selected Areas in Communications*, **SAC-5**, 896–905 (June 1987).

16. C. B. Dekker, "The application of tamed frequency modulation to digital transmission via radio," *Proc. IEEE National Telecommunication Conference*, 55.3.1–55.3.7 (1979).

17. D. Muilwijk, "Tamed frequency modulation—a bandwidth-saving digital modulation method, suited for mobile radio," *Philips Telecommunication Review*, **37**, 35–49 (March 1979).

18. D. Muilwijk, "Correlative phase shift keying—a class of constant envelope modulation techniques," *IEEE Trans. Communications*, **COM-29**, 226–236 (March 1981).

19. K. Chung, "Generalized tamed frequency modulation and its application for mobile radio communications," *IEEE Trans. Vehicular Technology*, **VT-33**, 103–113 (August 1984).

20. K. Chung, "Discriminator – MLSE detection of a GTFM signal in the presence of fast Rayleigh fading," *IEEE Trans. Communications*, **COM-35**, 1374–1376 (December 1987).

21. K. Hirade and K. Murota, "A study of modulation for digital mobile telephony," *Proc. IEEE Vehicular Technology Conference*, 13–19 (March 1979).

22. K. Murota and K. Hirade, "GMSK modulation for digital mobile radio telephony," *IEEE Trans. Communications*, **COM-29**, 1044–1050 (July 1981).

23. K. Murota and K. Hirade, "Transmission performance of GMSK modulation," *Trans. IECE Japan*, **J64-B**, 1123–1130 (October 1981).

24. H. Suzuki, "Error-rate performance of GMSK modulation with differential detection," *IECE Technical Report*, **CS79-129**, 23–30 (1979).

25. S. Ogose and K. Murota, "Experimental studies on differentially-encoded GMSK with differential-detection," *IECE Technical Report*, **CS79-130**, 31–38 (1979).

26. S. Ogose and K. Murota, "Differentially encoded GMSK with 2-bit differential detection," *Trans. IECE*, **J64-B**, 248–254 (April 1981).

27. S. Ogose, "Error-rate performance of differentially encoded GMSK with 2-bit differential detection," *Trans IECE*, **J65-B**, 1253–1259 (October 1982).

28. H. Suzuki, "Optimum Gaussian filter differential detection of MSK," *IEEE Trans. Communications*, **COM-29**, 916–918 (June 1981).

29. S. Ogose, "Optimum Gaussian filter for MSK with 2-bit differential detection," *Trans. IECE*, **E-66**, 459–460 (July 1983).

30. M. K. Simon and C. C. Wang, "Differential detection of Gaussian MSK in a mobile radio environment," *IEEE Trans. Vehicular Technology*, **VT-33**, 302–320 (November 1984).

31. A. Yongacoglu, D. Makrakis, and K. Feher, "Differential detection of FMSK using decision feedback," *IEEE Trans. Communication*, **COM-36**, 641–649 (June 1988).

32. M. Hirono, T. Miki, and K. Murota, "Multilevel decision method for band-limited digital FM with limiter–discriminator detection," *IEEE Trans. Vehicular Technology*, **VT-33**, 114–122 (August 1984).

33. K. Ohono and F. Adachi, "Half-bit offset decision frequency detection of differentially encoded GMSK signals," *Electronics Letters*, **23**, 1311–1312 (November 1987).

34. S. M. Elnoubi, "Analysis of GMSK with discriminator detection in mobile radio channels," *IEEE Trans. Vehicular Technology*, **VT-35**, 71–76 (May 1986).

35. S. M. Elnoubi, "Predetection filtering on the probability of error of GMSK with discriminator detection in mobile radio channels," *IEEE Trans. Vehicular Technology*, **VT-37**, 104–107 (May 1988).

36. N. A. B. Svenson and C. E. W. Sundberg, "Performance evaluation of differential and discriminator detection of continuous phase modulation," *IEEE Trans. Vehicular Technology*, **VT-35**, 106–117 (August 1986).

37. K. Murota, "Spectrum efficiency of GMSK land mobile radio," *IEEE Trans. Vehicular Technology*, **VT-34**, 69–75 (August 1985).

38. T. Okai, F. Sugiyama, S. Asakawa, and Y. Okamura, "Narrow band constant envelope digital phase modulation system," *IECE Technical Report*, **CS79-133**, 55–62 (1979).

39. S. Asakawa and F. Sugiyama, "A compact spectrum constant envelope digital phase modulation," *IEEE Trans. Vehicular Technology*, **VT-3**, 102–111 (August 1981).

40. F. Muratore and V. Palestini, "Features and performance of 12PM3 modulation

methods for digital land mobile radio," *IEEE Journal on Selected Areas in Communications*, **SAC-5**, 906–914 (June 1987).

41. T. Maseng, "Digitally phase modulated (DPM) signals," *IEEE Trans. Communications*, **COM-33**, 911–918 (September 1985).

42. T. Aulin and C.-E. W. Sundberg, "Continuous phase modulation—part I: full response signaling," *IEEE Trans. Communications*, **COM-29**, 196–209 (March 1981).

43. T. Aulin, N. Rydbeck, and C.-E. W. Sundberg, "Continuous phase modulation—part II: partial response signaling," *IEEE Trans. Communications*, **COM-29**, 210–225 (March 1981).

44. J. B. Anderson, T. Aulin, and C. E. Sundberg, *Digital Phase Modulation*, Plenum Press, New York, 1986.

45. Y. Akaiwa, I. Takase, M. Ikoma, and N. Saegusa, "An FM modulation–demodulation scheme for narrow-band digital communication," *IECE Technical Report*, **CS79-132**, 47–54 (1979).

46. Y. Akaiwa, I. Takase, S. Kojima, M. Ikoma, and N. Saegusa, "Performance of baseband-bandlimited multilevel FM with discriminator detection for digital mobile telephony," *Trans. IECE*, **E64**, 463–469 (July 1981).

47. Y. Akaiwa and S. Kojima, "Performance of baseband bandlimited 4-level FM with coherent and differential detection," *Proc. IECE National Convention on Communication*, No. 475 (1980).

48. K. Honma, E. Murata, and Y. Tatsuzawa, "A study of digital mobile radio communication method using PLL," *IECE Technical Report*, **CS79-134** (1979).

49. K. Honma, E. Murata, and Y. Rikou, "On a method of constant envelope modulation for digital mobile radio communication," *Proc. IEEE International Conference on Communication*, 24.1.1–24.1.5 (1980).

50. K. Takagi and B. Yamamoto, "Performance of narrow band digital transmission system with a duobinary filter," *Proc. 1981 IECE National Convention*, No. 2169 (1981).

51. K. Takagi and B. Yamamoto, "Narrow band digital FM scheme using duobinary filter," *Proc. IEEE National Telecommunication Conference*, B.8.5.1–B8.5.5 (1981).

52. Y. Akaiwa and E. Okamoto, "An analysis of error rates for Nyquist- and partial response-baseband-filtered digital FM with discriminator detection," *Trans. IECE*, **J66-B**, 534–541 (April 1983).

53. F. G. Jenks, P. D. Morgan, and C. S. Warren, "Use of four level phase modulation for digital mobile radio," *IEEE Trans. Electromag. Compat.*, **EMC-14**, 113–128 (November 1972).

54. Y. Akaiwa and Y. Nagata, "A linear modulation method for digital mobile radio communication," *Proc. 1985 National Convention, IECE*, No. 2384. Y. Nagata and Y. Akaiwa, "Characteristics of a linear modulation method for digital mobile radio communications," *ibid.*, No.2385.

55. Y. Akaiwa and Y. Nagata, "Highly efficient digital mobile radio communications with a linear modulation method," *IEEE Journal on Selected Area in Communications*, **SAC-5**, 890–895 (June 1987).

56. J. A. Tarralo and G. I. Zysman, "Modulation techniques for digital cellular systems," *Proc. IEEE Vehicular Technology Conference*, 245–248 (1988).

57. C. Liu and K. Feher, "Noncoherent detection of $\pi/4$-QPSK systems in a CCI-AWGN combined interference environment," *Proc. IEEE Vehicular Technology Conference*, 83–94 (1989).

58. F. Adachi, M. Mori, and T. Ooi, "Radio channel structure for QPSK digital mobile radio," *Proc. IEEE Vehicular Technology Conference*, 220–223 (1989).

59. K. Raith, B. Hedberg, G. Larsson, and R. Kahre, "Performance of a digital cellular experiment test bed," *Proc. IEEE Vehicular Technology Conference*, 175–177 (1989).

60. S. Tomisato and H. Suzuki, "Envelope controlled digital modulation improving power efficiency of transmitter amplification—an application to trellis-coded 8PSK for mobile radio," *Trans. IEICE*, **J75-B-II**, 918–928 (December 1992).

61. P. M. Martin, A. Bateman, J. P. McGeehan, and J. D. Marvil, "The implementation of a 16-QAM mobile data system using TTIB-based fading correction techniques," *Proc. IEEE Vehicular Technology Conference*, 71–76 (1988).

62. S. Sampei and T. Sunaga, "Rayleigh fading compensation method for 16-QAM in digital land mobile radio channels," *Proc. IEEE Vehicular Technology Conference*, 640–646 (1989).

63. M. Yokoyama, "BPSK system with sounder to combat Rayleigh fading in mobile radio communications," *IEEE Trans. Vehicular Technology*, **VT-34**, 35–40 (February 1985).

64. M. K. Simon, "Dual-pilot tone calibration technique," *IEEE Trans. Vehicular Technology*, **VT-35**, 63–70 (May 1986).

65. N. Kinoshita et al., "Field experiments on 16QAM/TDMA and trellis coded 16QAM/TDMA systems for digital land mobile radio communications," *IEICE Trans. Communications*, **E77-B**, 911–920 (July 1994).

66. S. Ono, N. Kondoh, and Y. Shimazaki, "Digital cellular system with linear modulation," *Proc. IEEE Vehicular Technology Conference*, 44–49 (1989).

67. H. Tomita, "Polar loop linearlizer to $\pi/4$ shift QPSK," *Proc. Autumn National Convention of IEICE*, No. B-540 (1989).

68. M. J. Kosh and R. F. Fisher, "A high efficiency 835 MHz linear power amplifier for digital cellular telephony," *Proc. IEEE Vehicular Technology Conference*, 17–18 (1989).

69. Y. Nagata, "Linear amplification technique for digital mobile communications," *Proc. IEEE Vehicular Technology Conference*, 159–164 (1989).

70. W. Gosling, J. P. McGeehan, and P. G. Holland, "Receivers for the Wolfson single-sideband V.H.F. land mobile radio systems," *Electronics Engineer*, **49**, 231–235 (May 1979).

71. L. R. Kahn, "Single-sideband transmission by envelope elimination and restoration," *Proc. I.R.E.*, **40**, 803–806 (July 1952).

72. V. Petrovic, "Reduction of spurious emission from radio transmitters by means of modulation feedback," *IEEE Conference on Radio Spectrum Conservation Techniques*, 44–49 (September 1983).

73. V. Petrovic and W. Gosling, "Polar-loop transmitter," *Electronics Letters*, **15**, 286–288 (May 1979).

74. J. K. Cavers, "A linearizing predistorter with fast adaptation," *Proc. IEEE Vehicular Technology Conference*, 41–47 (May 1990).

75. S. Uebayashi, K. Ohno, T. Nojima, M. Murata, and Y. Yamada, "Base station equipment technologies for digital cellular systems," *NTT Review*, **4**, 55–63 (January 1992).

76. D. C. Cox, "Linear amplification with nonlinear components," *IEEE Trans. Communications*, **COM-22**, 1942–1945 (December 1974).

77. H. Miyakawa, H. Harashima, and Y. Tanaka, "A new digital modulation scheme, multimode binary CPFSK," *Proceedings of Third International Conference on Digital Satellite Communication, Kyoto*, 105–112 (November 1975).

7

OTHER TOPICS IN DIGITAL MOBILE RADIO TRANSMISSION

7.1 ANTI-MULTIPATH FADING MODULATION

A family of digital modulation methods that show immunity against frequency-selective fading have been proposed. The basic idea is to use some special and redundant pulse waveforms. A wider spectrum bandwidth results from using redundant waveforms.

$\pi/2$-TFSK ($\pi/2$ transition frequency shift keying) [1] takes phase shift characteristics as shown in Fig. 7.1. For a symbol time period, the phase shift occurs in two stages: linear and constant. The proportion of the linear and flat (constant) components is given by the time ratio t_0/T. It becomes MSK for $t_0 = T$, and $\pi/2$-shift BPSK with an NRZ pulse for $t_0 = 0$.

DSK (double shift keying) [2] shows a two-stage staircase phase transition for BPSK as shown in Fig. 7.2. PSK-RZ (phase shift keying—return to zero) [3] is a PSK with a return-to-zero pulse waveform as shown in Fig. 7.3.

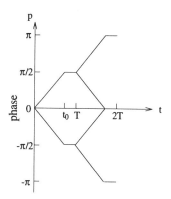

Figure 7.1. Phase transition of $\pi/2$ TFSK.

243

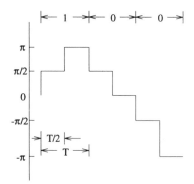

Figure 7.2. Phase transition of DSK.

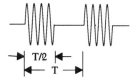

Figure 7.3. PSK-RZ signal.

Manchester coded PSK [4] uses Manchester coding. PSK-VP (phase shift keying with varied phase) [5] adopts a generalized phase pulse.

All of the above employ differential detection. They are immune against frequency-selective fading as long as the multipath delay difference is within one symbol period. This is explained in the following, taking PSK-RZ and DSK as examples. Let us assume a two-path model, consisting of a mainpath and an echo path. It is obvious that PSK-RZ never suffers from the echo path signal if the delay difference between the two paths is $0.5T$; since the main path and echo path signals never overlap, they are demodulated independently with one symbol delay differential detection. When the main and echo signals are subjected independently to fading, we can get a diversity effect. This effect is maximum for a delay difference of $0.5T$. In order to understand the effect of DSK on the two-path channel, consider a special case where the main and echo signals take equal amplitudes and opposite phases and assume a data sequence of a succession of 1s, as shown in Fig. 7.4. The received signal (sum of the main and echo signal) disappears in case of an ordinary BPSK. On the other hand, part of the DSK signal survives. The surviving DSK signal can be demodulated with $0.5T$ delay differential detection.

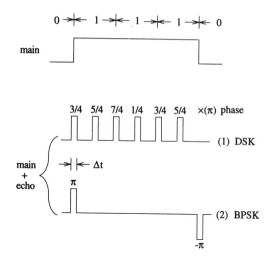

Figure 7.4. The principle of DSK.

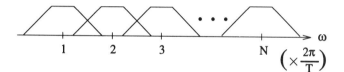

Figure 7.5. Spectrum of a multicarrier transmission system with spectrum overlap.

7.2 MULTICARRIER TRANSMISSION

In this system, digital signals are transmitted using multiple carriers with different frequencies. The input signal is serial-to-parallel converted and used to modulate the multiple carriers. If N-carriers are used, the data rate on a single carrier becomes $1/N$ times slower compared with a one-carrier system. This serves to mitigate the effect of frequency-selective fading that appears for a case of high data rate transmission.

For multicarrier systems with orthogonal frequency-division multiplexing (OFDM), spectrum-nonoverlapped systems and spectrum-overlapped systems are known. Even with spectrum overlap (Fig. 7.5) the components can be split at the demodulator ([6] and Appendix 7.3). With overlapping of the spectra, the total signal bandwidth can be decreased: an equivalent roll-off factor becomes $1/N$ times smaller than for a single modulated signal. The block diagram of the spectrum-overlapped multicarrier QAM system is shown in Fig. 7.6.

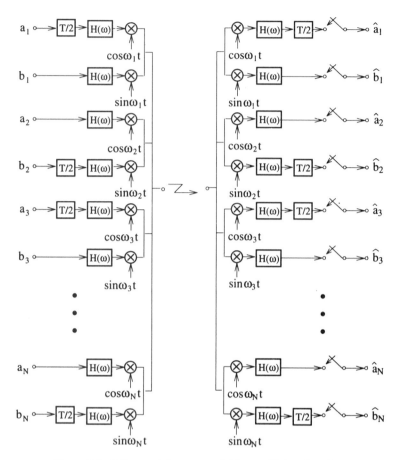

Figure 7.6. Orthogonal frequency division multiplexed system.

The modulator and demodulator of the multicarrier system are implemented effectively with the fast Fourier transform technique (Section 8.7.3).

An example of a multicarrier transmission system is a data communication system at HF or short-wavelength band [7] for mobile communications. At HF band, the frequency-selective fading becomes severe because the delay difference between multipath signals becomes large due to ionospheric propagation conditions. Channel distortion caused by frequency-selective fading is avoided with the multicarrier transmission systems.

7.3 SPREAD-SPECTRUM SYSTEM

In spread-spectrum systems the spectrum is spread by introducing additional modulation with spectrum spreading (SS) codes. If FSK is used for the

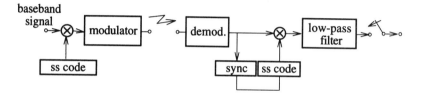

(a) direct sequence system

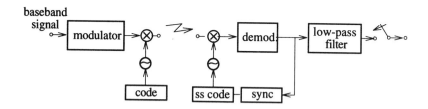

(b) frequency hopping system

Figure 7.7. Spread-spectrum communication system.

additional modulation, the spread-spectrum system is called FH (frequency hopping), and if the baseband signal or the modulated signal is multiplied with an SS code, it is called a DS (direct sequence) system (Fig. 7.7(a), (b)).

An SS code is a sequence consisting of so-called chips. Pseudo-noise (PN) sequences are used for the SS codes, whose characteristics vary depending on the purpose of the system. With multiplication of a PN (SS) code with the data signal, the spectrum spreads according to the spectrum bandwidth of the SS code: multiplying the PN code in the time domain results in a convolution integral in the frequency domain (Eq. A2.11). If we use an SS code with a length of N chips for each data symbol then the chip rate of the SS code is N times the data rate. As a result, the spectral bandwidth is increased (spread) by N times the original data spectral bandwidth. The spectrum for an FH system depends on the range of frequency hopping. If the hopping rate is slow (slow FH) compared to the data rate, its spectrum is given as a set of the modulated signal spectra for different (hopped) carrier frequency. If the hopping rate is faster (fast FH) than the data rate, its spectrum further spreads by an amount equal to a stationary signal spectrum multiplied with the number of hops per data symbol period.

The receiver for an SS system performs as follows. Let us consider first a DS system as in Fig. 7.7(a). The demodulator outputs the data signal multiplied by the SS code. Assume that the SS code is a binary polar code. Then by multiplying the demodulated signal with the SS code in synchronism with the received signal (despreading process), we have a narrow-band signal which is the transmit baseband signal. Hence, this system can be seen as an ordinary system where the pulse waveforms take the SS codes. The above receiver is a correlator receiver. From the discussion on ordinary digital transmission systems, a matched filter instead of the correlator (Section 3.3) can be employed to the SS system. The matched filter takes an impulse response of the time-inverted version of the SS code. A (matched) filter for the baseband signal must be considered separately.

From the above argument and the fact that the error rate performance (versus E_s/N_0) never depends on pulse waveform (see Section 3.3.5), it is obvious that an SS system shows the same error rate performance as an ordinary transmission system under the assumption of flat noise. If we consider the signal to noise power ratio instead of E_s/N_0 before the despreading process, its takes a low value due to the wide bandwidth of the SS signal. The receiver for an FH system consists of frequency dehopping circuits. Error rate performance is the same as in the DS system or an ordinary system.

The SS system has characteristics that are different from ordinary systems as follows:

(a)   Low Power Spectrum Density. Given a transmit power, the power spectrum density of an SS signal drops by a factor of $1/N_s$, where N_s is the spectrum spreading factor. This means that an SS signal introduces little interference into other narrow-band systems.

The low power spectral density characteristic of an SS system results in secure communications, since it becomes difficult for a third party to detect the communication by sensing a carrier level with an ordinary receiver.

(b)   Immunity Against Interference or Jamming. A narrow-band (interfering) signal is spectrum-spread at the despreading process of an SS receiver (Fig. 7.8). At the output of the receiver (narrow-band) low-pass filter the interference signal power is reduced by an amount equal to the spectrum spreading factor. Since the SS matched filter is a linear circuit (i.e., a transversal filter whose tap coefficients take chip amplitudes of the SS code), the SS matched filter receiver does not spread the spectrum of the narrow band interfering signal. However, its effect in suppressing the interference is the same as with the correlator receiver: The narrow-band signal level at the output of the matched filter is suppressed owing to the random addition in the filter.

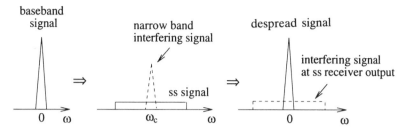

Figure 7.8. Spectra in spread-spectrum system.

(c)   *CDM and CDMA.* If SS codes with low cross correlations are used, SS signals that are transmitted on the same frequency and at the same time can be received with low interference. If orthogonal codes are used, there is no interference between the SS signals. As a result, the simultaneous use of radio spectrum at the same frequency and time between different signals, that is, CDM (code division multiplexing) is possible. If this concept is applied to multiple access systems, it is called CDMA (code division multiple access). A CDMA system requires no frequency management between users by virtue of using different SS codes as long as the interference between different code signals can be neglected.

(d)   *High Time Resolution.* In general, a narrower pulse shows a higher time resolution. Because of the wide bandwidth of an SS signal, the time difference between signals can be measured more accurately than with a narrow-band system as shown in the following. For this purpose SS codes with a sharp autocorrelation function (e.g., the delta function) are appropriate, as explained in the following. Let us analyze an SS system as shown in Fig. 7.9. At the output of the SS signal matched filter, the received signal becomes

$$r(t) = s(t)*c(t)*s(t_0 - t)$$
$$= c(t)*s(t)*s(t_0 - t), \qquad (7.1)$$

Figure 7.9. Channel impulse response measurement by matched filtered spread-spectrum system.

where, $*$ denotes the convolution integral, $s(t)$ is the SS signal, $c(t)$ is the channel impulse response and $s(t_0 - t)$ is the impulse response of the matched filter where t_0 is a time constant. Define a signal $z(t) \equiv s(t)*s(t_0 - t)$; it is expressed as

$$z(t) = \int_{-\infty}^{\infty} s(t - x)s(t_0 - x)\, dx. \tag{7.2}$$

At a time $t = t_0 + \tau$, we have

$$z(t_0 + \tau) = \int_{-\infty}^{\infty} s(y + \tau)s(y)\, dy. \tag{7.3}$$

The right-hand side of Eq. (7.3) is the autocorrelation function $R(\tau)$ of the SS code, $s(t)$. Since $r(t) = c(t)*z(t)$, if $s(t)$ shows a sharp autocorrelation, that is, $|R(\tau)| \ll 1$ for $\tau \neq 0$ [ideally $R(\tau) = \delta(\tau)$, the delta function], then we have the channel impulse response, $r(t) \approx c(t)$. The maximum sharpness of the autocorrelation function is of the order of the chip duration. Then an SS system shows a time resolution given by a pulse with the chip duration. The high time resolution characteristic can be useful for ranging distances with a radar. In some radio communication systems, measurement of the channel impulse response becomes important. An SS system can measure it with a high time resolution.

(e)   RAKE Receiver. First let us consider a conventional system rather than SS systems. A (mobile) radio channel is subjected to multipath fading. In this channel, a narrow-band signal experiences flat fading, where all frequency components of the signal drop by the same amount at the same time. As a result, signal level may drop below the threshold value for adequate communication. As the signal bandwidth becomes comparable to or wider than the coherence bandwidth of the multipath channel, the signal experiences frequency-selective fading and the signal level seldom drops below the threshold value (Section 4.4), which is an advantage for the wideband signal. On the other hand, signal waveform distortion or intersymbol interference due to the frequency-selective fading becomes a problem. Automatic channel equalization can be used to solve this problem; however, this requires sophisticated signal processing.

An SS system offers a simple technique that is effective in mitigating the multipath fading: the technique is called a RAKE receiver (Fig. 7.10) and uses a filter matched to the channel transfer characteristics. The matched filter in the RAKE receiver outputs, at the sampling instant, a signal made by coherently combining the multipath signal components (see Section 3.3.5). Since the multipath signal components are subjected to independent fading, the combined signal has a diversity gain (Section 7.4). Owing to the high time resolution, an SS system yields the channel impulse response necessary

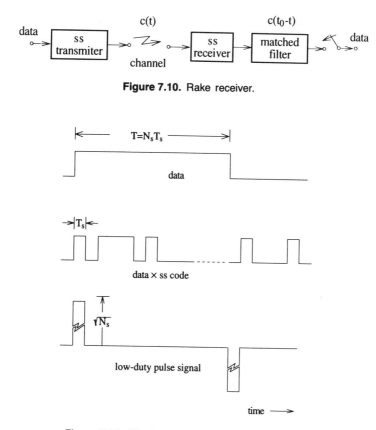

Figure 7.10. Rake receiver.

Figure 7.11. SS signal and low-duty polar signal.

for the matched filter (RAKE) receiver. Thus the SS RAKE receiver achieves the benefit of wide-band transmission with low effort.

In order to understand the SS system, it is instructive to compare an SS system with a non-SS polar signaling system under conditions of same spectrum bandwidth, same data rate, and same average transmit power (Fig. 7.11). Let us denote the data rate and the spectrum spreading factor as f_b and N_s, respectively. To get the same spectrum bandwidth, the pulse duty ratio of the non-SS polar signal must be $1/N_s$; that is, the pulse waveform is the same as the "chip" pulse waveform of the SS system. To have the same average transmit power, the pulse height of the non-SS polar signal must be $\sqrt{N_s}$ times higher than that of the SS system. Under the above conditions, the average power spectrum density of both systems is lowered by a factor N_s when compared with a full-duty polar signaling system. Error rates between the three systems are also the same when a matched filter or correlator receiver is used.

Both the wide-band systems show the same immunity against interference. The low-duty pulse polar signal has an instantaneous power that is higher by a factor of N_s when compared to a narrow-band signal, assuming the same average power. For the SS system, the narrow-band signal is spectrum spread by a factor N_s and the average power of the interfering narrower-band signal is reduced by this amount at the output of a low-pass filter placed at the spectrum despreading circuit.

Using the low-duty pulse signal, it is seen that N_s number of signals can be multiplexed in the time domain without any interference (orthogonality). In the SS system there are N_s orthogonal codes, for example, the Walsh code (Section 2.1.5), over a spectrum bandwidth expanded by a factor of N_s. Hence N_s signals can also be multiplexed for the SS system. Permitting interference between signals, more signals can be multiplexed using other codes. This is accomplished also with the low-duty pulse system by partially overlapping the pulse in time.

High time resolution with the low-duty pulse system is evident from Fig. 7.12. If the autocorrelation function of the SS code is the same as the autocorrelation of the low-duty pulse waveform, then the same results are given for channel impulse response measurement.

So far in this discussion we have not seen any differences between the SS system and the low-duty pulse system. However, we see a big difference when peak power is considered. The low-duty pulse system has a peak power N_s (spectrum spreading factor) times higher than the SS system peak power. This may become important when selecting high-power transmitters.

A slow FH system does not have high time resolution, even when the average spectrum bandwidth is as wide as for a DS system. This results because the instantaneous or a short-time bandwidth is narrow within the FH system. The FH system has an advantage over the non-FH system: effects of fading are randomized when the frequency-hopping range is wide enough compared with the correlation bandwidth of the fading channel. This is accomplished without intersymbol interference, since the instantaneous

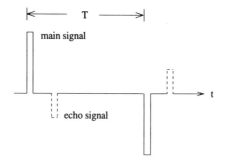

Figure 7.12. Explanation of channel impulse response for a low-duty pulse system.

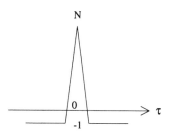

Figure 7.13. Autocorrelation function of the maximum length code.

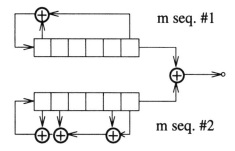

Figure 7.14. Gold code generator.

bandwidth at the slow FH system is narrow. Thus the FH technique is an effective countermeasure for shortening the faded periods of a very slow-fading channel. Without this, we must employ a space or frequency diversity technique. A slow FH system is equivalent to a frequency diversity system, where frequency branches are switched periodically according to the FH code.

SS codes are used appropriately in different systems. The maximum length sequence (m-sequency) code is applied to measure the channel, since its shows a sharp autocorrelation function as shown in Fig. 7.13. It has side lobes of $1/N$ times the peak value, where N is the length of the sequence. If the measuring distance is long, a long code is necessary. For a CDM(A) system, codes with low cross correlation are required. The Walsh codes and Gold codes have this property; the value of worst cross correlation is kept small. A Gold code generator is shown in Fig. 7.14. The code is generated by adding two specially selected (preferred pair) m-sequences.

Synchronization of SS codes at a correlation receiver is a central issue in the system design. The signal to noise power ratio is quite low when the synchronization is lost, making it difficult to reestablish the SS code synchronization.

Further description of SS systems is beyond the scope of this book. The reader is referred to [8,9] for more details.

7.4 DIVERSITY TRANSMISSION SYSTEM

Diversity techniques mitigate fading by using multiple received signals which experienced different fading conditions. This system is based on the fact that it is unlikely that all signals experience simultaneously bad fading conditions.

A fading channel causes three kinds of transmission errors: (i) thermal noise error due to level fading, (ii) irreducible error due to random FM effects, and (iii) errors due to waveform distortion caused by frequency-selective fading. The term irreducible errors means that the error rates cannot be reduced even if average SNR is arbitrarily increased. A diversity system is effective for reducing the three kinds of errors. The effect of a diversity system on the thermal noise error is easily understood since the signal level dropping probability is diminished with a diversity system. The random FM effect and the waveform distortion become significant when the signal level drops. Thus we can understand that the random FM effect and the waveform distortion caused by the fading channel are mitigated in the diversity system by preventing the signal level from dropping to a low value.

A detailed description of diversity systems is beyond the scope of this book. Instead, a brief introductory discussion is given followed by a description of the error rate analysis and some diversity systems used for digital mobile radio communications. For comprehensive treatments, refer to [10,11].

Many different diversity systems are known: space diversity with multiple antennas, polarization diversity using differently polarized waves, frequency diversity with multiple frequencies, time diversity by transmission of the same signal in different times, and angle diversity using directive antennas aimed in different directions.

With respect to the signal combining methods, we have maximum ratio combining, equal gain combining, and selection (switching) combining. These are further classified into RF, IF, and post-demodulation combining. RF and IF (or pre-demodulation) combining require synchronization so that the modulated signals are in phase with each other. Without synchronization, the diversity effect is lost for maximum ratio or equal gain combining and switching noise appears for the selection diversity. This problem is absent with post-demodulation combining; however the technique requires two or more receivers.

Another category of classification of diversity communications includes transmitter diversity and receiver diversity, using multiple transmitters and receivers, respectively.

Space diversity is classified into micro-diversity and macro- (or site) diversity, according to whether antennas are closely spaced of the order of a wavelength or separated wide enough to cope with the topographical conditions (e.g., buildings, road and terrain). The former is effective for fast fading where signal fades in a distance of the order of a half-wavelength. The latter is effective for shadowing, where the signal fades due to the topographical obstructions (Section 4.2).

All of the above systems are introduced intentionally to mitigate fading; they can be called explicit diversity. The "implicit diversity" effect is obtained unintentionally when an automatic equalizer is used in frequency-selective fading channels. In a frequency-selective fading channel, signals are received subject to different fading conditions and different propagation delays depending on the details of multipath propagation. An automatic equalizer (Section 7.5), introduced to equalize the channel distortion due to the multipath fading, uses time diversity in the channel equalizing process by combining signals from different paths. The same effect is obtained by a spread-spectrum RAKE receiver (Section 7.3).

7.4.1 Probability Density Function of SNR for Diversity Systems

First we discuss error rates due to the thermal noise in fading channels with diversity systems. The average error rate due to the thermal noise is given as

$$\langle P_e \rangle = \int_0^\infty p_e(\gamma) p(\gamma) \, d\gamma, \tag{7.4}$$

where $p_e(\gamma)$ is error rate for a given SNR (signal to noise power ratio) denoted by γ and $p(\gamma)$ is the probability density function of γ. Without a diversity system, $p(\gamma)$ is assumed here to take an exponential distribution (Eq. 5.153) that is given by the Rayleigh distribution.

Assume the same mean signal power and mean noise power for all branches. Then mean SNR is the same for all branches. The local mean SNR for the ith branch, γ_i, is given as

$$p(\gamma_i) = \frac{1}{\gamma_0} e^{-\gamma_i/\gamma_0} \qquad (0 \le \gamma_i \le \infty), \tag{7.5}$$

where γ_0 is the average SNR.

The probability that γ_i takes values less than γ is

$$P(\gamma_i \le \gamma) = \int_0^\gamma p(\gamma_i) \, d\gamma_i$$
$$= 1 - e^{-\gamma/\gamma_0}. \tag{7.6}$$

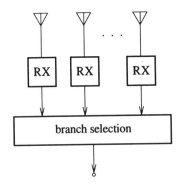

demodulator for RF or IF signal and
decision device for baseband signal

Figure 7.15. Space diversity selection combining system.

Selection Diversity. Let us consider an M-branch space diversity system with the selection combining method shown in Fig. 7.15. Assume that the received signal levels are independent of each other. The branch with the highest γ_i is selected at each instant. Denote the highest γ_i in the selected branch as γ at an instant. The other branches have γ_i less than γ. Thus the probability that this situation occurs, that is, the selection diversity system output SNR yielding γ, becomes

$$P_M(\gamma_1, \gamma_2, \ldots, \gamma_M \le \gamma) = (1 - e^{-\gamma/\gamma_0})^M. \tag{7.7}$$

Differentiating Eq. (7.7) with respect to γ, we have the probability density function as

$$p(\gamma) = \frac{d}{d\gamma}(1 - e^{-\gamma/\gamma_0})^M \tag{7.8a}$$

$$= \frac{M}{\gamma_0}e^{-\gamma/\gamma_0}(1 - e^{-\gamma/\gamma_0})^{M-1} \tag{7.8b}$$

$$\approx M\frac{\gamma^{M-1}}{\gamma_0^M} \quad (\gamma \ll \gamma_0). \tag{7.8c}$$

Maximal Ratio Combining. The maximal ratio combining system is shown in Fig. 7.16. Received signals at each branch are weighted by a factor of each branch SNR and coherently summed. This system gives the best error rate

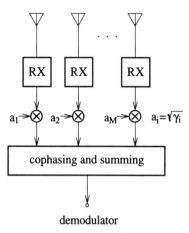

Figure 7.16. Maximal ratio combining system.

performance. The probability density function of γ at the output of the combiner becomes [10],

$$p(\gamma) = \frac{1}{(M-1)!} \frac{\gamma^{M-1}}{\gamma_0^M} e^{-\gamma/\gamma_0} \tag{7.9a}$$

$$\approx \frac{1}{(M-1)!} \frac{\gamma^{M-1}}{\gamma_0^M} \qquad (\gamma \ll \gamma_0). \tag{7.9b}$$

The ratio between Eq. (7.9b) and Eq. (7.8c) becomes $1/M!$. This means that the average error rate with the maximal ratio combiner is $1/M!$ times smaller than that for selection combining.

Equal Gain Combining. The equal gain combining system is given by taking the same weighting factor, $a_i = a_0$ for all i, in the maximal ratio combining system. Unfortunately, the probability density function for the equal gain combining is not obtained in closed form. Approximate expression for $\gamma \ll \gamma_0$ is given [10] as

$$p(\gamma) \approx \frac{2^{M-1} M^M}{(2M-1)!} \frac{\gamma^{M-1}}{\gamma_0^M} \qquad (\gamma \ll \gamma_0). \tag{7.10}$$

For $M \gg 1$, maximal ratio combining has advantage by 1.3 dB over equal gain combining. We can guess intuitively that the degree of the effect of equal gain combining on SNR distribution improvement is between those for maximum ratio combining and selection combining.

Effects of Branch Correlation [11,12]. If the branch SNRs are correlated, the diversity effect is diminished. For an extreme case of complete correlation, we have no diversity effect. Here, the results of analyses for a two-branch system ($M = 2$) are cited as follows.

Selection Diversity. The distribution function is given as

$$P_2(\gamma) = 1 - e^{-\gamma/\gamma_0}\left[1 - Q(a,b) + Q(b,a)\right], \tag{7.11}$$

where

$$Q(a,b) = \int_b^\infty e^{-(a^2+x^2)/2} I_0(ax)\, x\, dx,$$

$$b = \sqrt{\frac{2\gamma}{\gamma_0(1-\rho^2)}}, \qquad a = b\rho,$$

where $I_0(x)$ is the zeroth order modified Bessel function of the first kind and ρ denotes the magnitude of the complex cross covariance of fading Gaussian processes. It is shown that the normalized envelope covariance of the two branch signals is very near ρ^2. The average error rate is approximately determined by probabilities for $\gamma \ll \gamma_0$. For $\gamma \ll \gamma_0$, Eq. (7.11) is approximated as

$$P_2(\gamma) \approx \frac{\gamma^2}{\gamma_0^2(1-\rho^2)}. \tag{7.12}$$

The probability density function is given as

$$p_2(\gamma) \approx \frac{d}{d\gamma} P_2(\gamma).$$

Using Eq. (7.12) we have

$$p_2(\gamma) \approx \frac{2\gamma}{\gamma_0^2(1-\rho^2)}. \tag{7.13}$$

Maximal Ratio Combining. The distribution function is given as

$$P_2(\gamma) = 1 - \frac{1}{2\rho}\left[(1+\rho)e^{-\gamma/\gamma_0(1+\rho)} - (1-\rho)e^{-\gamma/\gamma_0(1-\rho)}\right]. \tag{7.14a}$$

For $\gamma \ll \gamma_0$,

$$P_2(\gamma) \approx \frac{\gamma^2}{2\gamma_0^2(1-\rho^2)}. \tag{7.14b}$$

The probability density function becomes

$$p_2(\gamma) = 1 - \frac{1}{2\rho\gamma_0}(e^{-\gamma/\gamma_0(1+\rho)} - e^{-\gamma/\gamma_0(1-\rho)}). \qquad (7.15a)$$

For $\gamma \ll \gamma_0$,

$$p_2(\gamma) \approx \frac{\gamma}{\gamma_0^2(1-\rho^2)}. \qquad (7.15b)$$

Comparing Eqs. (7.13) and (7.15b) and considering Eq. (7.4), we can see that selection diversity has an average (approximate) error rate twice that of the maximal ratio combining system. The effect of the correlation between the two branch signals is reflected in the term $\gamma_0^2(1-\rho^2)$ in Eqs. (7.13) and (7.15b). Hence the effect of the correlation is equivalent to reducing the average SNR by $\sqrt{1-\rho^2}$. For example, SNR is reduced by 0.02 dB, 0.14 dB, and 0.62 dB for $\rho = 0.1$, 0.25, and 0.5, respectively.

7.4.2 Average Error Rate for Diversity Systems

7.4.2.1 *Error Rates due To Additive Gaussian Noise.* Error rates for

a given bit energy to noise power density ratio $E_b/N_0 (\equiv \gamma)$ are given in the following. For BPSK and QPSK with coherent detection [Eqs. (5.43a) and (5.49)], we have

$$P_e(\gamma) = \tfrac{1}{2}\text{erfc}(\sqrt{\alpha\gamma}) \qquad \text{(coherent detection)}, \qquad (7.16a)$$

where the parameter α is introduced for convenience as a degradation factor. For ASK and FSK with noncoherent detection, and BPSK with differential detection [Eqs. (5.63) and (5.76)], we have

$$P_e(\gamma) = \tfrac{1}{2}e^{-\alpha\gamma} \qquad \text{(noncoherent detection)}, \qquad (7.16b)$$

where $\alpha = 1/2$ for ASK and FSK and $\alpha = 1$ for BPSK with ideal performance.

The average error rate is calculated as

$$\langle P_e \rangle = \int_0^\infty \tfrac{1}{2}\text{erfc}(\sqrt{\alpha\gamma})p(\gamma)\,d\gamma \qquad \text{(coherent detection)} \qquad (7.17a)$$

and

$$\langle P_e \rangle = \int_0^\infty \tfrac{1}{2}e^{-\alpha\gamma}p(\gamma)\,d\gamma \qquad \text{(noncoherent detection)}. \qquad (7.17b)$$

Maximal Ratio Combining. From Eqs. (7.4) and (7.9), average error rate is given by

$$\langle P_e \rangle = \int_0^\infty P_e(\gamma) \frac{\gamma^{M-1}}{(M-1)!} \frac{1}{\gamma_0^M} e^{-\gamma/\gamma_0} d\gamma. \tag{7.18}$$

For noncoherent detection, using Eq. (7.16b) and integrating Eq. (7.18) by parts, we have

$$\langle P_e \rangle = \frac{1}{2} \frac{1}{(1+\alpha\gamma_0)^M} \qquad \text{(noncoherent detection)}. \tag{7.19a}$$

For low error rates,

$$\langle P_e \rangle \approx \frac{1}{2} \frac{1}{(\alpha\gamma_0)^M} \qquad (\alpha\gamma_0 \gg 1). \tag{7.19b}$$

The same result as this approximate formula can be obtained using the approximate probability density function (Eq. 7.9b) for low γ as

$$\langle P_e \rangle \approx \int_0^\infty \frac{1}{2} e^{-\alpha\gamma} \frac{1}{(M-1)!} \frac{\gamma^{M-1}}{\gamma_0^M} d\gamma \qquad (\gamma \ll \gamma_0) \tag{7.20a}$$

$$= \frac{1}{2} \frac{1}{(\alpha\gamma_0)^M}. \tag{7.20b}$$

For coherent detection,

$$\langle P_e \rangle = \int_0^\infty \frac{1}{2} \operatorname{erfc}(\sqrt{\alpha\gamma}) \frac{\gamma^{M-1}}{(M-1)!} \frac{1}{\gamma_0^M} e^{-\gamma/\gamma_0} d\gamma. \tag{7.21}$$

For $M \geq 2$ we have (Appendix 7.1)

$$\langle P_e \rangle = \frac{1}{2} - \frac{1}{2} \frac{1}{\sqrt{1+1/\alpha\gamma_0}} \left(1 + \sum_{m=1}^{M-1} \frac{(2m-1)!!/(2m)!!}{(1+\alpha\gamma_0)^m} \right)$$
$$\text{(coherent detection)}. \tag{7.22}$$

For $M = 2$,

$$\langle P_e \rangle = \frac{1}{2} - \frac{1}{2} \frac{1}{\sqrt{1+1/\alpha\gamma_0}} \left(1 + \frac{1}{2} \frac{1}{1+\alpha\gamma_0} \right) \tag{7.23a}$$

$$\approx \frac{3}{16} \frac{1}{(\alpha\gamma_0)^2} \qquad (\alpha\gamma_0 \gg 1). \tag{7.23b}$$

If we use the approximate probability density function, we have (Appendix 7.2)

$$\langle P_e \rangle \approx \int_0^\infty \frac{1}{2} \mathrm{erfc}(\sqrt{\alpha\gamma}) \frac{1}{(M-1)!} \frac{\gamma^{M-1}}{\gamma_0^M} d\gamma \qquad (\gamma \ll \gamma_0) \qquad (7.24a)$$

$$= \frac{1}{2} \frac{(2M-1)!!}{(2M)!!} \frac{1}{(\alpha\gamma_0)^M}. \qquad (7.24b)$$

For $M = 2$, Eq. (7.24b) yields the same result as Eq. (7.23b).

Selection Diversity. Using Eqs. (7.4) and (7.8a), we have

$$\langle P_e \rangle = \int_0^\infty P_e(\gamma) \frac{d}{d\gamma} (1 - e^{-\gamma/\gamma_0})^M d\gamma. \qquad (7.25)$$

Integrating by parts, P_e becomes

$$\langle P_e \rangle = [P_e(\gamma)(1 - e^{-\gamma/\gamma_0})^M]_0^\infty - \int_0^\infty \left(\frac{d}{d\gamma} P_e(\gamma) \right) (1 - e^{-\gamma/\gamma_0})^M d\gamma.$$

Considering $P_e(\infty) = 0$, and expanding $(1 - e^{-\gamma/\gamma_0})^M$ as

$$(1 - e^{-\gamma/\gamma_0})^M = \sum_{k=0}^M \binom{M}{k} (-1)^k e^{-k\gamma/\gamma_0},$$

where

$$\binom{M}{k} = \frac{M!}{(M-k)!\,k!},$$

we have

$$\langle P_e \rangle = \sum_{k=0}^M \binom{M}{k} (-1)^k \int_0^\infty \left(\frac{d}{d\gamma} P_e(\gamma) \right) e^{-k\gamma/\gamma_0} d\gamma. \qquad (7.26)$$

For noncoherent detection, using Eqs. (7.16b) and (7.26), we have

$$\langle P_e \rangle = \frac{1}{2} \sum_{k=0}^M \binom{M}{k} (-1)^k \int_0^\infty \alpha e^{-(\alpha + k/\gamma_0)\gamma} d\gamma$$

$$= \frac{1}{2} \sum_{k=0}^M \binom{M}{k} (-1)^k \frac{1}{1 + k/\alpha\gamma_0} \qquad \text{(noncoherent detection)}. \qquad (7.27)$$

For $M = 2$,

$$\langle P_e \rangle = \frac{1}{2} - \frac{1}{1 + 1/\alpha\gamma_0} + \frac{1}{2}\frac{1}{1 + 2/\alpha\gamma_0} \tag{7.28a}$$

$$\approx \frac{1}{(\alpha\gamma_0)^2} \qquad (\alpha\gamma \gg 1). \tag{7.28b}$$

If we use the approximate probability distribution (Eq. 7.8c),

$$\langle P_e \rangle \approx \int_0^\infty \frac{1}{2} e^{-\alpha\gamma} M \frac{\gamma^{M-1}}{\gamma_0^M} \, d\gamma \qquad (\gamma \ll \gamma_0). \tag{7.29a}$$

Comparing this with Eq. (7.20a), we have

$$\langle P_e \rangle \approx \frac{1}{2} \frac{M!}{(\alpha\gamma_0)^M} \qquad (\gamma \ll \gamma_0). \tag{7.29b}$$

This result for $M = 2$ is the same as Eq. (7.28b).
For coherent detection,

$$\frac{d}{d\gamma} P_e(\gamma) = -\frac{1}{2\sqrt{\pi}} \sqrt{\frac{\alpha}{\gamma}} e^{-\alpha\gamma}.$$

Inserting this into Eq. (7.16) and using the relation

$$\int_0^\infty \frac{1}{\sqrt{\gamma}} e^{-\beta\gamma} \, d\gamma = \sqrt{\frac{\pi}{\beta}}$$

we have

$$\langle P_e \rangle = \frac{1}{2} \sum_{k=0}^M \binom{M}{k} (-1)^k \frac{1}{\sqrt{1 + k/\alpha\gamma_0}} \qquad \text{(coherent detection).} \tag{7.30}$$

For $M = 2$,

$$\langle P_e \rangle = \frac{1}{2} - \frac{1}{\sqrt{1 + 1/\alpha\gamma_0}} + \frac{1}{2}\frac{1}{\sqrt{1 + 2/\alpha\gamma_0}} \tag{7.31a}$$

$$\approx \frac{3}{8}\frac{1}{(\alpha\gamma_0)^2}. \tag{7.31b}$$

If we use the approximate probability distribution function, then

$$\langle P_e \rangle \approx \int_0^\infty \frac{1}{2} \text{erfc}(\sqrt{\alpha\gamma}) M \frac{\gamma^{M-1}}{\gamma_0^M} \, d\gamma \qquad (\gamma \ll \gamma_0). \tag{7.32a}$$

Comparing this with Eq. (7.24a), we get

$$\langle P_e \rangle \approx \frac{M!}{2} \frac{(2M-1)!!}{(2M)!!} \frac{1}{(\alpha \gamma_0)^M} \quad \text{(coherent detection).} \quad (7.32b)$$

For $M = 2$, we have the same result as Eq. (7.31b).

Equal Gain Combining. Since no exact expression for the probability density function γ is known, we use the approximate probability density function given by Eq. (7.10). Comparing Eq. (7.10) with Eq. (7.9b) and using Eqs. (7.20b) and (7.24b), we get

$$\langle P_e \rangle \approx \begin{cases} \dfrac{1}{2} \dfrac{(2M)^M M!}{(2M)!} \dfrac{1}{(\alpha \gamma_0)^M} & \text{(noncoherent detection)} \\[3mm] \dfrac{1}{2} \dfrac{(2M)^M M!(2M-1)!!}{(2M)!\,(2M)!!} \dfrac{1}{(\alpha \gamma_0)^M} & \text{(coherent detection).} \end{cases} \quad (7.33)$$

For $M = 2$,

$$\langle P_e \rangle \approx \begin{cases} \dfrac{2}{3} \dfrac{1}{(\alpha \gamma_0)^2} & \text{(noncoherent detection)} \\[3mm] \dfrac{1}{4} \dfrac{1}{(\alpha \gamma_0)^2} & \text{(coherent detection).} \end{cases} \quad (7.34)$$

Using Eqs. (7.19a), (7.22), (7.27), (7.30), and (7.33), bit error rates are calculated as shown in Figs. 7.17(a) and (b).

The post-detection combining diversity system for QPSK and $\pi/4$ shifted QPSK with differential detection is extensively analyzed by Adachi and Oono [12] (Section 5.5.5). When square root raised cosine roll-off Nyquist filters are used for the transmit and receive filters, a simple result is given as [12]

$$\langle P_e \rangle \approx K_M \left(\frac{1}{\gamma_0} \right)^M, \quad (7.35)$$

where γ_0 is the bit energy to noise power density ratio, and K_M is given as

$$K_M = \begin{cases} \dfrac{(2M-1)!!}{2} & \text{for selection combining,} \\[3mm] \dfrac{1}{2} \dfrac{M^M}{M!} & \text{for equal gain combining,} \\[3mm] \dfrac{1}{2} \dfrac{(2M-1)!!}{M!} & \text{for maximal ratio combining.} \end{cases}$$

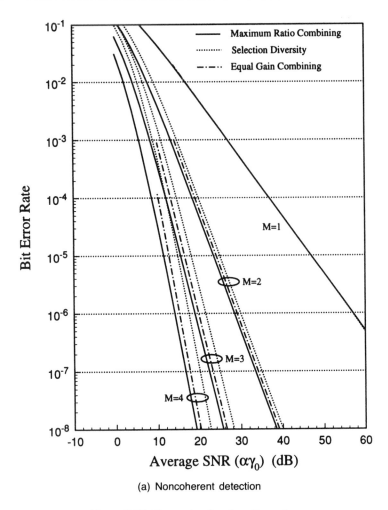

(a) Noncoherent detection

Figure 7.17. Error rates for diversity systems.

7.4.2.2 Irreducible Error Rates due to Random FM Effect. In contrast to the error rate analysis for the thermal noise effects, it is difficult to evaluate the effects of diversity on the random FM effect and waveform distortion resulting from fading channels. This is because the various parameters of the system, such as modulation/demodulation method, pulse waveforms, and the diversity system, must be considered.

The two-branch post-detection combining system for BPSK with differential detection is analyzed in [11]. The result is cited here as

$$\langle P_e \rangle = \left(\frac{1 + \gamma_0[1 - J_0(2\pi f_m T)]}{2(1 + \gamma_0)} \right)^2 \left(\frac{2(\gamma_0 + 1) + \gamma_0 J_0(2\pi f_m T)}{1 + \gamma_0} \right), \qquad (7.36)$$

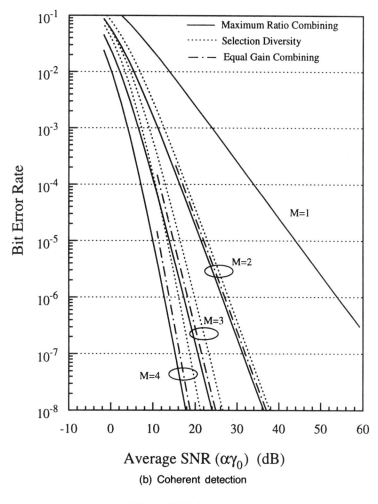

Figure 7.17. (continued).

where f_m is the maximum Doppler frequency and T is the symbol duration. When we let the average signal to noise power ratio γ_0 to be infinite, the irreducible error rate is given as

$$\langle P_e \rangle_{\text{ir}} \approx \tfrac{1}{4}[1 - J_0(2\pi f_m T)]^2 [2 + J_0(2\pi f_m T)]. \qquad (7.37)$$

This is further simplified for $f_m T \ll 1$:

$$\langle P_e \rangle_{\text{ir}} \approx \tfrac{3}{4}(\pi f_m T)^4, \qquad (7.38)$$

where $J_0(x) \approx 1 - (x/2)^2$ for $x \ll 1$ is used.

If this result is compared with that for the single-branch system given as

$$\langle P_e \rangle_{ir} \approx \tfrac{1}{2}(\pi f_m T)^2, \tag{7.39}$$

we can see that the irreducible error rate, which is the second order of a small value $f_m T$, is reduced to fourth order of a small value $f_m T$ with the two-branch diversity.

If we assume slow fading ($f_m T \to 0$), then the error rate corresponds to the thermal noise effect alone. The error rates are improved with diversity as $\langle P_e \rangle_{th} \propto \gamma_0^{-M}$, where M is the number of branches.

Results for a post-detection combining diversity system for QPSK and $\pi/4$ shifted QPSK with differential detector are given [12] as

$$\langle P_e \rangle_{ir} \approx K_M (2\pi f_{rms} T)^{2M} \tag{7.40}$$

Also for this case, the irreducible error rates is reduced as the Mth order of $(f_{rms} T)^2$, similarly to Eqs. (7.38) and (7.39).

The irreducible error rate of a post-detection diversity system for digital FM with frequency detection is given as [11]

$$\langle P_e \rangle_{ir} \approx \frac{(2M-1)!}{(M-1)!^2} \left(\frac{f_m}{2\sqrt{2}\Delta f} \right)^{2M}, \tag{7.41}$$

where Δf is the peak frequency deviation. Using the modulation index m, Δf is expressed as $m/2T$. Thus $\langle P_e \rangle_{ir}$ is proportional to $[(f_m T)^2]^M$ for given a modulation index.

7.4.2.3 *Error Rates due to Frequency Selective Fading.* For the raised cosine roll-off filtered ($\pi/4$ shifted) QPSK, error rates under frequency selective fading are given in Fig. 5.34(d).

7.4.2.4 *Error Rates due to Cochannel Interference.* For the raised cosine roll-off filtered ($\pi/4$ shifted) QPSK, approximate error rates are given in Fig. 5.34(c).

7.4.3 Multiple Transmitter Diversity System

In this system, a signal is transmitted simultaneously from multiple base stations that are located far away from each other. Thus it is a macro-diversity system. This technique is used in some commercial paging systems. For paging systems, a crucial issue is expansion of the service area with low transmit power. For this purpose, a multiple base station transmit system is appropriate. In addition, since it is a macro-diversity system, it can mitigate the effects of shadowing as well as of fast fading in areas where the zones of the base station overlap.

However, attention must be paid to interference between the signals from different base stations. At least bit timings are synchronized with each other. Some digital modulation techniques have been proposed to avoid interference effects and to obtain diversity effects [13–15]. Digital FM is exclusively used for paging systems, since transmit power efficiency is the main concern for the high-power transmitter and since the receivers must be simple and small. The multiple base station simultaneous transmission systems proposed so far use different baseband signals between the transmitters. The limiter–discriminator detector is used for the receiver. The integrate-and-dump filter is also used since the transmit baseband signal is the NRZ rectangular pulse (Section 3.3.3).

The diversity effect is analyzed by Adachi [15] (Copyright © IEEE 1979) as follows. Consider a two-transmitter system. We denote the transmitted signals as

$$u_1(t) = \cos[\omega_0 t + \varphi_1(t)], \tag{7.42a}$$

$$u_2(t) = \cos[\omega_0 t + \varphi_2(t)], \tag{7.42b}$$

where ω_0 is the carrier frequency and $\varphi_i(i = 1, 2)$ denote the baseband phase signal. Assuming the Rayleigh fading, the received signals are expressed as

$$v_i(t) = R_i(t) \cos[\omega_0 t + \varphi_i(t) + \theta_i(t)] \qquad (i = 1, 2), \tag{7.43}$$

where $R_i(t)$ is the envelope, which is Rayleigh distributed, and the phase $\theta_i(t)$ has a uniform distribution between 0 and 2π (Section 4.3).

At a receiver, the limiter–discriminator output signal $v(t)$ relative to ω_0 is given, when $R_1(t) > R_2(t)$, as

$$v(t) = \dot{\varphi}_1(t) + \dot{\theta}_1(t) - \frac{d}{dt} \left(\tan^{-1} \frac{\alpha(t) \sin \psi(t)}{1 + \alpha(t) \cos \psi(t)} \right), \tag{7.44}$$

where the dot denotes the time derivative and

$$\alpha(t) = \frac{R_2(t)}{R_1(t)}, \tag{7.45}$$

$$\psi(t) = \varphi(t) - \varphi_2(t) + \theta_1(t) - \theta_2(t), \tag{7.46}$$

and $\dot{\psi}$ is the beat frequency signal. When $R_1(t) < R_2(t)$ we should interchange the subscript in Eqs. (7.44) to (7.46). The third term on the right-hand side of Eq. (7.44) represents the interference between the two received signals. This term must disappear for no interference between the two signals. Since the fading is relatively slow compared with the bit rate $1/T$, $R_i(t)$ and $\theta_i(t)$ remain constant for a bit duration T. By Fourier series expansion, Eq. (7.44) can be rewritten as

$$\nu(t) = \dot{\varphi}_1(t) + \dot{\theta}_1(t) - \frac{d}{dt}\left(\sum_{m=1}^{\infty}(-1)^m\frac{\alpha^m}{m}\sin[m\psi(t)]\right). \qquad (7.47)$$

$W(nT)$, the output of the integrate-and-dump filter at the sampling time nT $(n = 0, \pm 1, \pm 2, \ldots)$ or equivalently, phase shift during T, becomes

$$W(nT) \approx \varphi_1(nT) - \varphi_1(nT - T)$$
$$+ \left[\sum_{m=1}^{\infty}(-1)^m\frac{2\alpha^m}{m}\cos\left(\frac{m}{2}\psi_\sigma(nT)\right)\sin\left(\frac{m}{2}\Delta\psi(nT)\right)\right], \quad (7.48)$$

where

$$\Delta\psi(nT) = \psi(nT) - \psi(nT - T),$$

$$\psi_\sigma(nT) = \psi(nT) + \psi(nT - T).$$

In order for the interference term to disappear, $\Delta\psi(nT)$ should be

$$\Delta\psi(nT) = \pm 2\pi k, \qquad k = 1, 2, 3, \qquad (7.49)$$

The output signal is then

$$W(nT) \approx \begin{cases} \varphi_1(nT) - \varphi_1(nT - T), & R_1(nT) > R_2(nT) \\ \varphi_2(nT) - \varphi_2(nT - T), & R_1(nT) < R_2(nT), \end{cases} \qquad (7.50)$$

This result shows that selection diversity effect is obtained.

A method for getting the condition in Eq. (7.49) is phase sweeping or carrier frequency offset [13]. In this method, we set

$$\varphi_1(t) = \frac{\omega_s t}{2} + \frac{\pi\beta}{T}\int_{-\infty}^{t}s(\tau)\,d\tau, \qquad (7.51a)$$

$$\varphi_2(t) = \frac{-\omega_s t}{2} + \frac{\pi\beta}{T}\int_{-\infty}^{t}s(\tau)\,d\tau, \qquad (7.51b)$$

where ω_s denotes the offset frequency and $s(t)$ is a transmit binary NRZ rectangular signal that takes the levels $+1$ or -1. β is the modulation index given by $\beta = 2\Delta f_d T$, where Δf_d denotes the frequency deviation. From Eq.

(7.49) and the above equations we get

$$\frac{\omega_s}{2\pi} = \frac{l}{T} \qquad (l = 1, 2, 3, \ldots). \tag{7.52}$$

Another method [15] is to use different modulation indices such as

$$\varphi_1(t) = \frac{\pi\beta_1}{T} \int_{-\infty}^{\infty} s(t) \, dt, \tag{7.53a}$$

$$\varphi_2(t) = \frac{\pi\beta_2}{T} \int_{-\infty}^{\infty} s(t) \, dt. \tag{7.53b}$$

From Eq. (7.49) and the above equation with $\int_{-\infty}^{\infty} s(t) \, dt = T$, we have

$$\Delta\beta \equiv \beta_2 - \beta_1 = 2l \qquad (l = 1, 2, 3, \ldots). \tag{7.54}$$

The other method [14] is to add time-varying signals, that meet the conditions of Eq. (7.49).

The average received power may be different between the signals from the two base stations. In this case, the argument in Section 7.4.2 cannot be applied. The error rates are analyzed as follows.

Since error rates are not given in closed form for digital FM with limiter–discriminator detection, the following approximate expression is used:

$$P_e(R) \approx \frac{1}{2} \exp\left[-\alpha\left(\frac{T}{2N_0}\right) R^2\right], \tag{7.55}$$

where R is the received signal envelope, N_0 is the noise power density at the input to the limiter–discriminator, and α denotes a constant which expresses the appropriate effect of the modulation index and degradation in the receiver error rate performance. Define $P_e(R_1, R_2)$ as

$$P_e(R_1, R_2) \approx \begin{cases} \frac{1}{2} \exp[-\alpha(T/2N_0) R_1^2], & R_1 > R_2 \\ \frac{1}{2} \exp[-\alpha(T/2N_0) R_2^2], & R_1 < R_2. \end{cases} \tag{7.56}$$

The average error rate is given using Eq. (7.56) and the joint probability density function $p(R_1, R_2)$ as

$$\langle P_e \rangle = \int_0^\infty dR_1 \int_0^{R_1} P_e(R_1, R_2) p(R_1, R_2) \, dR_2$$
$$+ \int_0^\infty dR_2 \int_0^{R_2} P_e(R_1, R_2) p(R_1, R_2) \, dR_1. \tag{7.57}$$

$p(R_1, R_2)$ is given in [10] as

$$p(R_1, R_2) = \frac{R_1 R_2}{\sigma_1^2 \sigma_2^2 (1 - \rho^2)} \left[I_0 \left(\frac{\rho R_1 R_2}{\sigma_1 \sigma_2 (1 - \rho^2)} \right) \right]$$

$$\times \exp\left[-\frac{1}{1 - \rho^2} \left(\frac{R_1^2}{2\sigma_1^2} + \frac{R_2^2}{2\sigma_2^2} \right) \right], \tag{7.58}$$

where σ_1^2 and σ_2^2 are average powers of the received fading signals. The term ρ^2 (appearing in Section 7.4.1) is a constant that is very near the value of the normalized envelope correlation of the two-branch signals. $I_0(\cdot)$ is the zeroth-order modified Bessel function of the first kind, which is defined by

$$I_0(z) = \frac{1}{2\pi} \int_{-\pi}^{\pi} \exp[-z \cos \theta] \, d\theta. \tag{7.59}$$

Substituting Eqs. (7.56) and (7.58) into (7.57) and letting $R_2/R_1 = t$, and integrating Eq. (7.57) with respect to R_1, θ, and t, we have

$$\langle P_e \rangle = \frac{1}{4(1 + \alpha_1 \gamma_1)} \left(1 - \frac{1 - (\gamma_1/\gamma_2) + \alpha_1 \gamma_1 (1 - \rho^2)}{\{[1 + (\gamma_1/\gamma_2) + \alpha_1 \gamma_1 (1 - \rho^2)]^2 - 4\rho^2 (\gamma_1/\gamma_2)\}^{1/2}} \right)$$

$$+ \frac{1}{4(1 + \alpha_1 \gamma_2)} \left(1 - \frac{1 - (\gamma_2/\gamma_1) + \alpha_2 \gamma_2 (1 - \rho^2)}{\{[1 + (\gamma_2/\gamma_1) + \alpha_2 \gamma_2 (1 - \rho^2)]^2 - 4\rho^2 (\gamma_2/\gamma_1)\}^{1/2}} \right), \tag{7.60}$$

where $\gamma_1 (\equiv \sigma_1^2 T/N_0)$ and $\gamma_2 (\equiv \sigma_2^2 T/N_0)$ are the average received signal bit energy to noise power density ratios.

For $\gamma_1, \gamma_2 \gg 1$,

$$\langle P_e \rangle \approx \frac{1}{4\gamma_1 \gamma_2 (1 - \rho^2)} \left(\frac{1}{\alpha_1^2} + \frac{1}{\alpha_2^2} \right), \tag{7.61}$$

which reduces to Eq. (7.19b) with $M = 2$ and $\gamma_1 = \gamma_2 = \gamma_0$, $\alpha_1 = \alpha_2 = \alpha$, and $\rho = 0$.

7.4.4 Antenna Selection Diversity System

Antenna selection diversity systems have advantages over the others because they require no cophasing circuit for RF or IF combining and no dual receiver for post-detection diversity. However, they generate switching noise, since the received signals at the antenna are not cophased. The switching noise appears as click noise at the demodulator output. In analog systems, the effect of click noise can be mitigated to some degree by use of blanking or

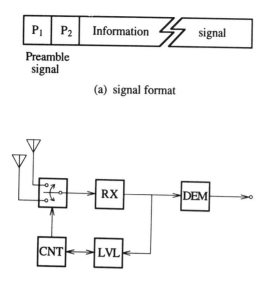

(a) signal format

(b) Antenna selection diversity receiver.

Figure 7.18. Antenna selection diversity receiver.

sample-and-hold techniques [11]. In digital systems, such techniques are not effective.

A method which avoids the switching noise for digital systems was proposed by Afrasteh and Chukurov [16] and Akaiwa [17], independently. A block diagram of this system is shown in Fig. 7.18. The idea for this method originated in noticing that the digital signals are usually sent in the form of a data block or frame, which consists of a preamble signal and an information signal. The antenna selection is made at the preamble signal period (P_1 and P_2 in Fig. 7.18(a)). The selection is held during the information signal period. The two antennas are switched to measure each of the received signal levels. The higher signal level antenna is then selected. Although the switching noise appears at the preamble signal period, its effect can be ignored since the preamble signal is a periodic signal for clock or carrier recovery purposes. When the technique is applied to a subscriber (mobile) receiver in a TDMA system, the antenna selection operation can be made at the time slot just before the dedicated time slot. The result is that no switching noise appears in the (preamble) signal.

The proposed method is useful for portable radio terminals, where the equipment needs to be small. The idea of the method was applied in [17] for a transmit diversity system using different frequencies for the up-link and down-link: a diversity antenna is not required for a subscriber radio. In this system the base station sends post-amble signals (P_1' and P_2') from the

Figure 7.19. Signal format for antenna selection feedback diversity system.

antennas in turn, following the information signal, as shown in Fig. 7.19. The subscriber receiver examines which part of the post-amble signal has the higher level and sends back the results of the examination to the base station. The base station selects the reported antenna to transmit the preamble and information signals in the next time slot. Thus, this method is a type of feedback diversity scheme.

If the base station adopts the diversity system, then both down-link and up-link diversity transmission can be realized. Diversity circuits including antennas may only be required at the base station. When the same frequency is used for the up-link and down-link, then the above technique of transmitting the post-amble signal is not necessary. Since the correlation of the up-link and down-link is high, the base station can use the antenna that was selected at the last receiving period for the next transmission. The performance of the proposed system is inferior to that of the other systems since the selected antenna is held for a burst period. The degradation in performance becomes significant in severely fast-fading conditions.

After the proposal of this antenna selection diversity system, a timely theoretical analysis on the performance of the system was published by Barnard and Pauw [18] as follows. As we discussed in Section 7.4.1, the average error rate can be calculated if the probability density function of the received signal power to noise power ratio γ is given. Denote the signal amplitude at the output of the selector as z. The bit energy to noise power ratio γ is given as

$$\gamma = \frac{z^2 T}{N_0}, \tag{7.62}$$

where T is the bit duration and N_0 is the noise power density. Taking the ensemble average of Eq. (7.62), we have the average bit energy to noise power density ratio as

$$\langle \gamma \rangle (\equiv \gamma_0) = \frac{\langle z^2 \rangle T}{N_0}. \tag{7.63}$$

$\langle z^2 \rangle$ is given as

$$\langle z^2 \rangle = \langle x^2 \rangle + \langle y^2 \rangle = 2R(0) \qquad (\langle x^2 \rangle = \langle y^2 \rangle = R(0)), \tag{7.64}$$

where x and y are the in-phase and quadrature-phase components, respectively, of z. $R(\tau)$ is the autocorrelation function of the quadrature components x and y, which are Rayleigh-faded signals. From Eqs. (7.62), (7.63), and (7.64), we get

$$\gamma = \frac{z^2}{2R(0)}\,\gamma_0. \tag{7.65}$$

γ_0 is assumed to take the same value for each branch. z and hence γ are time varying due to fading.

Assume that a selection is made at $\tau = 0$ and let the M branch amplitudes be $r_1, r_2, \ldots, r_M$. Let branch i be selected, then

$$z(\tau) = r_i(\tau), \tag{7.66}$$

where $r_i(0) > r_j(0)$ for all $j(\neq i)$ $(1, 2, \ldots, M)$, and τ is the time elapsed after the selection was made.

The cumulative distribution function of $z(\tau)$ becomes

$$F[z(\tau)] = \text{Prob}[r_1(\tau) \le z(\tau), r_1(0) > r_2(0), r_1(0) > r_3(0), \ldots, r_1(0) > r_M(0)]$$
$$+ \cdots + \text{Prob}[r_M(\tau) \le z(\tau), r_M(0) > r_1(0), r_M(0) > r_2(0), \ldots, r_M(0) > r_{M-1}(0)]. \tag{7.67}$$

If we assume an identical distribution of the M branch amplitudes, then

$$F[z(\tau)] = M\,\text{Prob}[r_1(\tau) \le z(\tau), r_1(0) > r_2(0), r_1(0) > r_3(0), \ldots, r_1(0) > r_M(0)]$$
$$= M \int_0^{z(\tau)} \int_0^\infty \int_0^{r_1(0)} \int_0^{r_1(0)} \cdots \int_0^{r_1(0)} f_{r_1(\tau), r_1(0), r_2(0), \ldots, r_M(0)}$$
$$\times\, dr_M(0) \cdots dr_2(0)\, dr_1(0)\, dr_1(\tau). \tag{7.68}$$

Assuming independent fading between the branches, the probability density function of $z(\tau)$ is given in [18] as

$$f_{z(\tau)} = \sum_{k=0}^{M-1} \binom{M}{k+1}(-1)^k \frac{z(\tau)}{P_k(\tau)} \exp\!\left(\frac{-z^2(\tau)}{2P_k(\tau)}\right), \tag{7.69}$$

where

$$P_k(\tau) = \frac{R^2(0) - R^2(\tau)\,k/(k+1)}{R(0)}. \tag{7.70}$$

Departing from [18], we assume (Eq. 4.34),

$$R(\tau) = R_{xx}(\tau) = R_{yy}(\tau) = b_0 J_0(\omega_m \tau), \tag{7.71}$$

where ω_m is the maximum Doppler frequency. Notice that $f_{z(\tau)}$ is the sum of the Rayleigh distribution functions.

Transforming the variable z to γ in Eq. (7.69), we have

$$f_{\gamma(\tau)} = \sum_{k=0}^{M-1} \binom{M}{k+1}(-1)^k \frac{R(0)}{P_k(\tau)\,\gamma_0} \exp\left(\frac{-R(0)\,\gamma(\tau)}{P_k(\tau)\,\gamma_0}\right). \tag{7.72}$$

Denoting the error rate for a given $\gamma(\tau)$ as $P_e[\gamma(\tau)]$,

$$P_e(\tau) = \int_0^\infty P_e[\gamma(\tau)]f_{\gamma(\tau)}d\gamma(\tau). \tag{7.73}$$

The average error rate for our system is then

$$\overline{P}_e = \frac{1}{T_d}\int_0^{T_d} P_e(\tau)\,d\tau, \tag{7.74}$$

where T_d is the information time length after a selection is made.

Let us assume that the error rate for a given γ is

$$P_e(\gamma) = \begin{cases} \frac{1}{2}\mathrm{erfc}(\sqrt{\alpha\gamma}) & \text{(coherent detection)} \\ \frac{1}{2}e^{-\alpha\gamma} & \text{(noncoherent detection),} \end{cases} \quad \begin{matrix} (7.75) \\ (7.76) \end{matrix}$$

where α is a degradation factor. Using Eqs. (7.72), (7.73), (7.75), and (7.76), we have

$$P_e(\tau) = \sum_{k=0}^{M-1} \binom{M}{k+1}(-1)^k g_k(\tau), \tag{7.77}$$

where

$$g_k(\tau) = \begin{cases} \frac{1}{2}\left[1 - \left(1 + \dfrac{R(0)}{P_k(\tau)\alpha\gamma_0}\right)^{-1/2}\right] & \text{(coherent detection)} \\ \dfrac{1}{2}\dfrac{1}{1 + P_k(\tau)\alpha\gamma_0/R(0)} & \text{(noncoherent detection).} \end{cases} \quad \begin{matrix} (7.78) \\ \\ (7.79) \end{matrix}$$

If $R(\tau) = 0$ then $P_e(\tau)$ yields error rates of a system without diversity. On the other hand, if $R(\tau) = R(0)$ then ideal selection diversity is obtained; we can confirm that Eq. (7.77) with Eqs. (7.78) and (7.79) reduce to Eqs. (7.30) and (7.27), respectively, for the above cases. Error rates calculated with Eqs. (7.74), (7.75), (7.76), and (7.77) are given in Fig. 7.20.

The performance of the antenna selection diversity method was improved

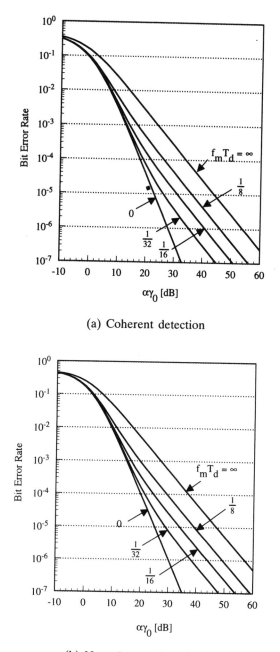

(a) Coherent detection

(b) Noncoherent detection

Figure 7.20. Error rate performance of a 2-branch antenna selection diversity system, where antenna selection is made at the preamble part of the frame and Rayleigh fading is considered. T_d is the frame length and f_m is the maximum Doppler frequency.

in [42] by introducing a selection algorithm based on prediction of received signal levels.

7.5   ADAPTIVE AUTOMATIC EQUALIZER

As the signal bandwidth becomes comparable to the coherence bandwidth of the fading mobile radio channel (frequency-selective fading), the signal suffers from waveform distortion. The automatic equalizer is known to remove distortion at the receiver. Automatic equalizer technologies are well developed and are applied to data transmission through public telephone channels. Many of the techniques can be applied to mobile radio channels. However, the mobile radio has a special requirement for the equalizer: adapting to the fast-fading channel. Research and development on equalizers is continuing at the time of writing. Here, a short introductory description of automatic equalizers and applications to mobile radio channels is given. For comprehensive understanding, refer to [19].

The most common example of signal distortion in radio channels is the so-called "ghost" on a TV picture. This phenomenon occurs when echo signals with significant power and delay are received after the main signal. There are ways to resolve the ghost problem: (i) improve the receiving antenna system, for example, its directivity, (ii) enter a CATV system, and (iii) use an echo canceler. The echo canceler is a type of equalizer, which we will discuss. Digital signal transmission has an advantage over analog signal transmission in view of equalizing the channel because the digital signal takes discrete values, as will be shown in the following.

7.5.1   Linear Equalizer [19,20]

A distorted signal can be equalized by using a filter with a transfer function that is the inverse of the channel transfer function. For example, consider a two-path channel with impulse response $h(t)$ given as

$$h(t) = A_0 \delta(t) + A_1 \delta(t - t_1). \qquad (7.80)$$

The transfer function becomes

$$H(\omega) = A_0 + A_1 e^{-j\omega t_1}, \qquad (7.81)$$

and for the inverse filter $H^{-1}(\omega) \equiv 1/H(\omega)$.

Since digital signal processing is generally adopted for automatic equalizers, it is convenient to use a discrete-time (sampled) representation of the received signal. The received signal sampled with sampling period of T is given as

$$r(nT) = s(t)*h(t)|_{t=nT}$$

$$= A_0 s(nT) + A_1 s(nT - t_1) \qquad (n = 0, \pm 1, \pm 2, \ldots), \qquad (7.82)$$

where $s(t)$ denotes transmit signal and the second term of Eq. (7.80) represents the echo signal. For simplicity, we assume $t_1 = mT$ ($m = 0, 1, 2, \ldots$) in the following. Denote a T-time delay element by z^{-1} (z-transform). The channel transfer function is then expressed as

$$H(z) = A_0 + A_1 z^{-1}, \qquad (7.83)$$

where $z = e^{j\omega T}$.

The transfer function of the inverse filter is

$$H(z)^{-1} = \frac{1}{A_0 + A_1 z^{-1}}$$

$$= \frac{1}{A_0} \frac{1}{1 + (A_1/A_0) z^{-1}}. \qquad (7.84)$$

The above transfer function is realized with the feedback circuit in Fig. 7.21. We can expand Eq. (7.84) as

$$H(z)^{-1} = \frac{1}{A_0} \sum_{m=0}^{\infty} \alpha^m z^{-m}, \qquad (7.85)$$

where $\alpha = -A_1/A_0$. This transfer function can be realized by the circuit known as the transversal filter, as in Fig. 7.22. The impulse response

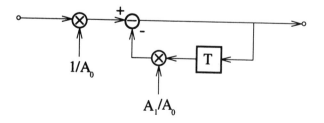

Figure 7.21. Feedback equalizer for a two-path channel.

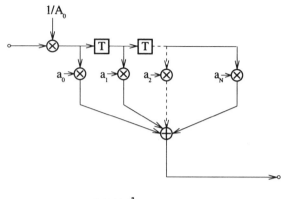

$$a_n = (-A_1/A_0)^n$$

Figure 7.22. Transversal equalizer.

converges if $|\alpha z| = |\alpha| < 1$, otherwise it diverges. These results correspond to whether the feedback circuit is stable or not. We rewrite Eq. (7.84) as

$$H(z)^{-1} = \frac{1}{A_1} \frac{z}{1 + (A_0/A_1)z} \qquad (7.86)$$

$$= \frac{1}{A_1} z \sum_{m=0}^{\infty} \beta^m z^m, \qquad (7.87)$$

where $\beta = -A_0/A_1$. The circuit is stable when $|\beta| < 1$. For the transversal filter with delay element z instead of z^{-1}, the time is inverted in the transversal filter.

When the transversal equalizer is implemented we must truncate the length of the filter. To show the effect of the truncation, an example of the impulse response of the system (channel plus equalizer) is shown in Fig. 7.23.

We can generalize the equalizer with finite length as shown in Fig. 7.24. This structure is called a linear equalizer since it is realized by a linear circuit. The transfer function becomes

$$H(z) = \frac{N(z^{-1})}{D(z^{-1})} = \frac{\displaystyle\sum_{m=0}^{M} \beta_m z^{-m}}{1 + \displaystyle\sum_{n=1}^{N} \alpha_n z^{-n}}. \qquad (7.88)$$

For stability of the circuit, the equation $D(z^{-1}) = 0$ should have roots, all of whose absolute values are less than 1 in the z-plane. The task of designing a linear equalizer becomes one of assigning coefficients α_n and β_m to get the

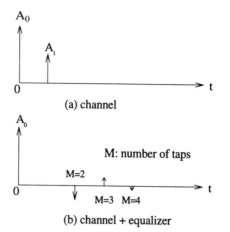

Figure 7.23. Effect of transversal equalizer for impulse response $(A_1/A_0 = 0.5)$.

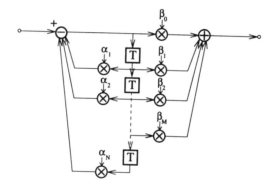

Figure 7.24. Linear equalizer.

best performance under the constraints of stable operation and size of the equalizer. Because of the instability problem, linear equalizers exclusively use the transversal filter structure. The criterion for equalizer performance should be error rate. However, this is a difficult problem since the error rate is a highly nonlinear function of the tap coefficients. Therefore the error, that is the difference between the desired signal and the equalized one, is considered as a performance criterion.

For digital transmission systems we are interested in the signal errors at the sampling or decision-making instants. In this case the equalizer can be operated on the basis of one sample per symbol (symbol rate sampling equalization). However, it is shown that symbol rate sampling equalization is sensitive to the sample timing: the performance of the equalizer deteriorates with timing errors. Hence, joint optimization for the sample timing and equalization must be carried out. Double sampling per symbol resolves this problem [20].

7.5.2 Performance Criteria for Equalization

Let us consider a digital transmission system that has no intersymbol interference when channel distortion is absent. The sampled signal becomes $a_k (k = 0, 1, 2, \ldots)$, where a_k takes a value from a discrete set of levels corresponding to the transmitted data. Intersymbol interference deviates the sample signal from a_k due to channel distortion. We denote $y(t)$ as an input signal to an equalizer.

$$y(t) = \sum_k a_k h(t - kT), \qquad (7.89)$$

where $h(t)$ denotes the impulse response for the tandem combination of a transmit filter, the channel, and a receive filter. The error or the intersymbol interference at $t = kT$ is given by $y(kT) - a_k$.

Let us consider a transversal equalizer. The output signal of the equalizer becomes

$$z(kT) = \sum_{n=-N}^{N} y(kT - nT') w_n \qquad (T' \leq T), \qquad (7.90)$$

where w_n are the tap coefficients, and T' is the sampling period for the equalizer. Hereafter we use the notation such as $z_k = z(kT)$ and $y_k = y(kT)$. The error is given by

$$e_k = z_k - a_k. \qquad (7.91)$$

We have two criteria to measure the error: mean squared error $\langle e_k^2 \rangle$, and the maximum absolute error max $|e_k|$. We must adjust w_n to minimize $\langle e_k^2 \rangle$ or max $|e_k|$.

7.5.2.1 Mean Square Error Criterion and Mean Square Algorithm. Assuming symbol rate equalization ($T' = T$), we have

$$\langle e_k^2 \rangle = \left\langle \left(\sum_{n=-N}^{N} y_{k-n} w_n - a_k \right)^2 \right\rangle. \qquad (7.92)$$

In order to minimize $\langle e_k^2 \rangle$ we require the derivatives of $\langle e_k^2 \rangle$ with respect to w_n to be zero. We then have $2N + 1$ linear equations given as

$$\frac{\partial}{\partial w_n} \langle e_k^2 \rangle = \left\langle y_{k-n} \left(\sum_{m=-N}^{N} y_{k-m} w_m - a_k \right) \right\rangle = 0 \qquad (n = -N, -N+1, \ldots, N). \tag{7.93}$$

The above equations can be rewritten in matrix form as

$$\left(\begin{array}{c} Y_{mn} \end{array} \right) \left(\begin{array}{c} w_{-N} \\ w_{-N+1} \\ \vdots \\ w_N \end{array} \right) = \left(\begin{array}{c} c_{-N} \\ c_{-N+1} \\ \vdots \\ c_N \end{array} \right), \tag{7.94}$$

where the matrix elements are given as

$$Y_{mn} = \langle y_{k-m} y_{k-n} \rangle \tag{7.95}$$

and

$$C_m = \langle y_{k-m} a_k \rangle. \tag{7.96}$$

Multiplying Eq. (7.94) by the inverse matrix Y_{mn}^{-1}, we have the optimum tap coefficients given as

$$\left(\begin{array}{c} w_{-N} \\ w_{-N+1} \\ \vdots \\ w_N \end{array} \right) = \left(\begin{array}{c} Y_{mn} \end{array} \right)^{-1} \left(\begin{array}{c} c_{-N} \\ c_{-N+1} \\ \vdots \\ c_N \end{array} \right). \tag{7.97}$$

In the above argument, we must know a_k, which are the transmitted data. For this purpose, a given sequence of a_k, called a test signal, reference signal or training signal, is transmitted prior to the information signal. In practice, real-time matrix inversion is difficult. Alternatively, an iterative method may be applied. Notice that $\partial/\partial w_m \langle e_k^2 \rangle$ tells us whether the mean square error $\langle e_k^2 \rangle$ increases ($\partial/\partial w_n \langle e_k^2 \rangle > 0$) or decreases ($\partial/\partial w_n \langle e_k^2 \rangle < 0$) when the tap coefficient w_n is changed to value away from the optimum value. Thus we can know whether we should increase or decrease w_n to get closer to the optimum value. The tap coefficients are updated as

$$w_n(j+1) = w_n(j) - \Delta \frac{\partial}{\partial w_n} \langle e_k^2(j) \rangle \qquad (j = 0, 1, 2, \ldots), \tag{7.98}$$

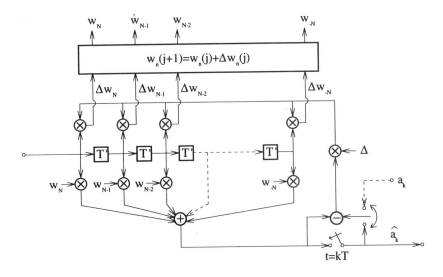

Figure 7.25. Least mean square equalizer.

where j denotes the jth iteration and Δ is a small constant. This method is known as the steepest-descent method. Using the expression

$$\frac{\partial}{\partial w_n} \langle e_k^2 \rangle = \left\langle y_{k-n} \left(\sum_{m=-N}^{N} y_{k-m} w_m - a_k \right) \right\rangle$$

$$= \langle y_{k-n} e_k \rangle \tag{7.99}$$

we have the least mean squared automatic equalizer as in Fig. 7.25. In the training mode of operation the reference signal a_k is used, while in the information signal transmit mode of operation the decided data $\hat{a}_k$ are used. If the channel characteristics never change, the tap coefficients w_n are fixed after the training is finished. For time-varying channels such as mobile radio channels, the equalizer adapts to the time-varying channel with an adaptive algorithm using the decided data.

7.5.2.2 Peak Distortion Criteria and Zero-Forcing Algorithm. Peak distortion is defined as

$$C_1 = \max_{\{a_k\}} |z_k - a_k|$$

$$= \max_{\{a_k\}} \left| \sum_{n=-N}^{N} y_{k-n} w_n - a_k \right|$$

$$= \max_{\{a_k\}} \left| \sum_{n} a_{k-n} t_n - a_k \right|, \tag{7.100}$$

where

$$t_n = \sum_{i=-N}^{N} h_{n-i} w_i$$

and $h_n = h(nT)$. t_n is the sampled impulse response of the total system including the equalizer. For simplicity, assume max $|a_k| = 1$ and $t_0 = 1$. We then have

$$C_1 = \sum_{n(\neq 0)} |t_n|. \tag{7.101}$$

C_1 is minimum $(= 0)$ if $t_n(n \neq 0)$ are forced to be zero. $t_n = 0(n \neq 0)$ means we have no intersymbol interference at the output of the equalizer. It is shown in [20] that if $\Sigma_{i(\neq 0)} |h_i| < |h_0|$, $t_n = 0$ $(n \neq 0)$ can be achieved by adjusting the tap coefficient w_n so that

$$\langle a_{k-n}(z_k - a_k) \rangle = 0 \qquad (n = -N, \ldots, N). \tag{7.102}$$

Since a_k are random values, we rewrite Eq. (7.102) as

$$\left\langle a_{k-n} \left(\sum_m a_{k-m} t_m - a_k \right) \right\rangle = \langle a_{k-n}^2 \rangle t_n \qquad (n \neq 0).$$

The recursive algorithm to achieve Eq. (7.102) is

$$w_n(j+1) = w_n(j) + \Delta a_{k-n}(z_k - a_k) \qquad (n = -N, \ldots, N), \tag{7.103}$$

where Δ is a small constant. An equalizer with the zero-forcing algorithm is shown in Fig. 7.26. The zero-forcing algorithm yields $\langle a_{k-n} e_k \rangle = 0$, which is different from the least mean square algorithm which yields $\langle y_{k-n} e_k \rangle = 0$.

The linear equalizer with either the LMS or zero-forcing algorithm is an inverse filter for the channel as long as noise is absent. (If we consider the noise, the difference between the two algorithms is seen: the LMS algorithm tends to find optimum tap coefficients, compromising between the effects of intersymbol interference and noise power increase, while the zero-forcing algorithm never takes account of noise). Thus, the equalization of the channel is equivalent to prediction of the channel characteristics. From this viewpoint, the linear equalizer principle plays an important role in adaptive equalization. Linear equalizers are analyzed rigorously in [20].

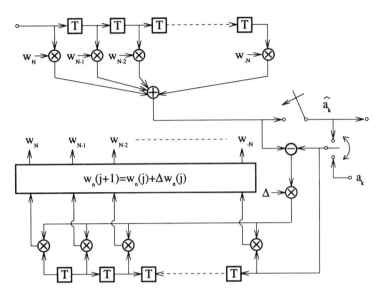

Figure 7.26. Zero-forcing equalizer.

7.5.3   Decision Feedback Equalizer [19]

Figure 7.27 depicts the decision feedback equalizer. The equalizer output signal, z_k can be expressed as

$$z_k = \sum_{n=-N_1}^{0} y_{k-n} w_n + \sum_{n=1}^{N_2} \hat{a}_{k-n} w_n. \tag{7.104}$$

The decision feedback equalizer differs from the linear feedback equalizer in that decided data $\hat{a}_k$ are fed back instead of z_k. Since a nonlinear element, viz., decision circuit, is introduced in the feedback path, we have no instability problem.

For a performance criterion, we can use either the least mean squared error or the peak distortion. The mean square error criterion is used in most systems. The target function is then $\langle |e_k (= z_k - \hat{a}_k)|^2 \rangle$. The analysis of the system is difficult when taking decision errors into consideration. We therefore assume no decision errors: $\hat{a}_k = a_k$. Furthermore, we assume random data. Using Eq. (7.104) and making the derivative of $\langle |e_k|^2 \rangle$ with respect to the feedback coefficients w_n zero, we can show that the feedback coefficients w_n are expressed in terms of the feed-forward coefficients as

$$w_n = - \sum_{k=-N_1}^{0} w_k h_{n-k}, \qquad n = 1, 2, \ldots, N_2. \tag{7.105}$$

If we assume $h_i = 0$ for $i < 0$ or $i > L$ then $w_n = 0$ for $n > L$.

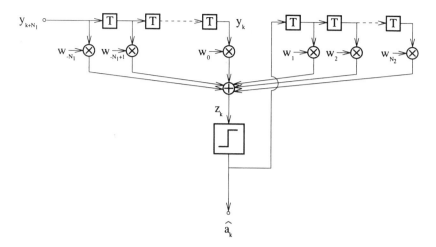

Figure 7.27. Decision feedback equalizer.

Making the derivative of $\langle |e_k|^2 \rangle$ with respect to the feedforward coefficients zero and using Eq. (7.105), we have

$$\sum_{k=-N_1}^{0} \varphi_{nk} w_k = h^*_{-n} \qquad (n = -N_1, \ldots, 1, 0), \qquad (7.106)$$

where

$$\varphi_{nk} = \sum_{m=0}^{-1} h^*_m h_{n+m-k} \qquad (n, k = -N_1, \ldots, 1, 0). \qquad (7.107)$$

If we use the steepest descent method, the tap coefficients w_n are updated in the feed-forward sections as

$$w_n(j+1) = w_n(j) + \Delta e_j y_{j-n} \qquad (n = -N_1, \ldots, -2, -1, 0) \quad (7.108)$$

and in the feedback sections as

$$w_n(j+1) = w_n(j) + \Delta e_j \hat{a}_{j-n} \qquad (n = 1, 2, \ldots, N_2), \qquad (7.109)$$

where Δ is a small constant.

7.5.4 The Viterbi Equalizer

The maximum-likelihood sequence estimation (MLSE) receiver gives the optimum performance for a digital communication system where intersymbol interference exists. The Viterbi algorithm is exclusively used in MLSE

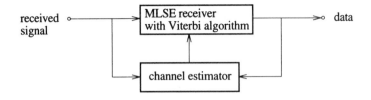

Figure 7.28. Viterbi equalizer.

receiver. We can apply this method to frequency-selective fading channels. For this purpose, an estimate of the time-varying channel impulse response is required for the design of the matched filter. The Viterbi equalizer is shown in Fig. 7.28.

The Viterbi equalizer compensates for the channel distortion in a way that is different in the usual sense of equalizers: it never produces a circuit with inverse transfer function of the channel, but produces replica signals at a receiver. Nevertheless, it is the optimum receiver as long as the channel impulse response is known. Like the DFE, nonlinear operations are made on the received signal, so it is also a nonlinear equalizer.

7.5.5 Adaptation and Prediction Algorithm

Besides the zero-forcing and least mean squared error algorithms, some other methods are known, such as recursive least-squares or Kalman algorithms, the fast Kalman, and lattice algorithms. Most of the equalizers assume a training sequence for updating the tap coefficients. A category of equalizers that require no training sequence is known as blind equalizers [22]. Description of these topics is beyond the scope of this book. Readers can refer to [19], [20], [23] and [24] for more detail.

7.5.6 Applications to a Mobile Radio Channel

From field experiments [25] the highest data rate in a land mobile radio channel without an equalizer was measured to be less than 100 kbps. Adaptive equalizers for application to mobile radio channels were studied corresponding to a proposal for a TDMA mobile telephone system with transmission bit rate of several hundred kbps.

Decision feedback equalizers were studied by Raith, Stjernvall, and Uddenfeldt [26] in frequency-selective fading channels by computer simulation. It was shown that a transmission rate of 300 kbps was feasible with use of an equalizer. It is claimed that the implicit diversity effect obtained by the equalizer as well as some other factors prove advantageous for a TDMA system over a FDMA system. Field tests with the equalizers at transmission rates of 170 and 340 kbps were successful [27]. A decision feedback equalizer with the Kalman algorithm for fast adaptation is described in [28]. Decision feedback equalization together with adjacent channel interference cancellation is discussed in [29]. To cope with both the minimum and nonminimum phase shift conditions, a decision feedback equalizer that has a time-reversal function has been proposed in [30] and [31].

A receiver for digital PM proposed maximum-likelihood sequence estimation (MLSE) [32]; thus, it can be applied to a frequency-selective fading channel. The performance of a MLSE Viterbi receiver for fading mobile channels was investigated in [33–39]. For the radio signals for the Pan-European digital mobile telephone system, it was shown the Viterbi equalizer works well for a multipath channel with echo delays on the order of $20\,\mu s$ and vehicle speeds of 200 km/h [36].

Frequency offset is a practical problem for receivers. Frequency offset compensation techniques for the MLSE receiver are described in [36, 37]. Another practical problem with the MLSE receiver is sample timing control. For this purpose a technique based on double sampling was proposed [38]. The speed of adaptation during a data transmission period must be fast, especially for a long burst length. This is the case for the signals in the US and Japanese digital cellar systems. The MLSE receiver can cope with a maximum Doppler frequency of 80 Hz [38,39] in digital cellular systems.

Most equalizers assume a linear channel: the transmitted signal, which may be linearly modulated or not, must be passed through a linear channel. Some channels are not linear, such as an amplitude limiter, a differential detector, or a frequency discriminator. An equalizer for a frequency-selective fading channel with a limiter–differential detection system has been proposed [40]. The principle of this method is based on the learning of intersymbol interference with a training sequence. Corresponding to past decided data and input signals, intersymbol interference is estimated and subtracted from the signal to be decided.

7.6 ERROR CONTROL TECHNIQUES

Transmission errors occur due to noise and distortion. We have two types of errors: random errors and burst errors. Random errors may be caused by thermal noise. Burst errors are generated during a fade in the transmission channel. Depending on the type of error, different approaches become useful, as shown later.

Interleaving is a technique for changing burst errors into random error. In this method a block of data is stored in a two-dimensional table as shown in Fig. 7.29. At the transmitter, data is written in the horizontal direction and read and transmitted in the vertical direction. At the receiver, data are written and read in the opposite manner. The burst errors are then separated from each other by the range of the horizontal length of the table. The delay

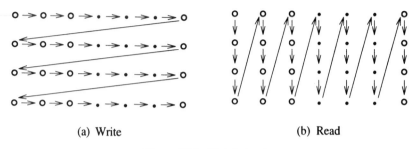

(a) Write   (b) Read

Figure 7.29. Bit interleaving.

in signal transmission is the penalty with this method. This delay may not be tolerable when interleaving is performed to cope with long burst errors caused by a very slowly fading channel. In these circumstances, a diversity transmission or frequency hopping system becomes useful to reduce the length of burst errors.

We can detect transmission errors by adding a redundant signal (check bits) to the information signal. Upon detecting a transmission error, we have two ways to control the error. One is called ARQ (automatic repeat request) and uses feedback control to request a retransmission of the corrupted data. The other, called FEC (feed-forward error correction), uses feed-forward control to control the errors. In this section, a brief description of error control techniques is given. The reader can refer to other books [43, 44] for further discussion.

The most primitive code used for error detection is a parity check code. A check bit is added so that the modulo 2 addition of the coded signal bits is zero (Fig. 7.30); denote the information bits as $(a_1, a_2, \ldots, a_n)$, then the check bit c is given by $\Sigma_{i=1}^{n} a_i$ (mod 2). We can detect a single error or an odd number of errors by checking that $\Sigma_{i=1}^{n+1} b_i$ equals 0 or not for the received bits b_i; that is, the parity of the received bits including the check bit. We can apply parity check codes to the horizontal and vertical directions of a

a_1	a_2	a_3	c
0	0	0	0
0	0	1	1
0	1	0	1
0	1	1	0
1	0	0	1
1	0	1	0
1	1	0	0
1	1	1	1

Figure 7.30. Parity check coding.

a_{11}	a_{12}	a_{13}	c_1
a_{21}	a_{22}	a_{23}	c_2
a_{31}	a_{32}	a_{33}	c_3
b_1	b_2	b_3	c_4

Figure 7.31. Horizontal and vertical parity check code.

two-dimensional array of data signals (Fig. 7.31). In this case, a single error can be detected and corrected.

By using more check bits with sophisticated coding algorithms, better performance from error correction and/or detection methods can be achieved. Error-correcting codes are classified into two important group: block codes and convolutional codes. In block codes, a block of k data bits is encoded to produce a code word of $n(>k)$ bits. In convolutional codes, a stream of data bits is encoded successively by modulo-2 convolutions: a data bit affects the encoded code word for a length of N bits, called the constraint length. (For example, see Fig. 7.34; $N = 3$ in this case.)

An important category of error-correcting codes is linear block code. A code word c of length n bits can be expressed as an n-dimensional vector, whose components are 1 or 0; $c = (c_1, c_2, \ldots c_n)$. For a data word d of length k, we denote $d = (d_1, d_2, \ldots, d_k)$. The number of check bits is $m = n - k$. The ratio k/n is known as the code rate or the code efficiency.

In an n-dimensional space, code words may take 2^k different points which are a subset of 2^n points in the n-dimensional space. Error-correcting codes are designed so that the minimum number of difference bits between any two code words, called the Hamming distance, is greater than a given value (Fig. 7.32). If the minimum distance is $2t + 1$, then t errors can be corrected or $2t$ errors can be detected.

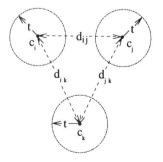

Figure 7.32. Code words with the minimum distance of $2t + 1$.

A family of codes, where a code word consists of data bits and check bits produced by a linear combination of the data bits, is known as a systematic code.

7.6.1 Linear Block Codes [45]

From *Modern Digital and Analogue Communication Systems*, Second Edition, by B. P. Lathi. Copyright. Used by permission from Oxford University Press, Inc.

A linear (systematic) block code is given as

$$
\begin{aligned}
c_1 &= d_1, \\
c_2 &= d_2, \\
&\vdots \\
c_k &= d_k, \\
c_{k+1} &= h_{11}d_1 \oplus h_{12}d_2 \cdots \oplus h_{1k}d_k \\
c_{k+2} &= h_{21}d_1 \oplus h_{22}d_2 \cdots \oplus h_{2k}d_k \\
&\cdots\cdots\cdots\cdots\cdots\cdots\cdots\cdots \\
c_n &= h_{m1}d_1 \oplus h_{m2}d_2 \cdots \oplus h_{mk}d_k,
\end{aligned}
\tag{7.110}
$$

where $\oplus$ denotes modulo 2 addition.

Using matrix notation,

$$
c = dG \tag{7.111}
$$

where

$$
G = \begin{bmatrix}
100\cdots 0 & h_{11}h_{21}\cdots h_{m1} \\
010\cdots 0 & h_{12}h_{22}\cdots h_{m1} \\
\cdots\cdots & \cdots\cdots\cdots \\
000\cdots 1 & h_{1k}h_{2k}\cdots h_{mk}
\end{bmatrix}
$$

$$
\begin{array}{cc}
I_k(k \times k) & P(k \times m)
\end{array}
$$

$$
= [I_k, P].
$$

The matrix G is called the *generator matrix*.

The code vector is expressed as

$$
c = [d, c_p], \tag{7.112}
$$

where $c_p = dP$ is a parity check vector.

Decoding. A received word r is given as

$$
r = c \oplus e,
$$

where e is an error vector. Define a matrix H^T as

$$H^T = \begin{bmatrix} P \\ I_m \end{bmatrix}, \tag{7.113}$$

where I_m is the identity matrix of order $m \times m$ $(m = n - k)$.
The matrix H^T has the property

$$cH^T = [d, c_p] \begin{bmatrix} P \\ I_m \end{bmatrix}$$
$$= dP \oplus c_p$$
$$= c_p \oplus c_p = 0. \tag{7.114}$$

The transpose of the matrix H^T, $H = [P^T I_m]$ is called the parity-check matrix. We can calculate a vector, called the *syndrome*, as

$$s = rH^T$$
$$= (c \oplus e) H^T$$
$$= cH^T \oplus eH^T$$
$$= eH^T. \tag{7.115}$$

The vectors s have $m(= n - k)$ dimensions; s is zero if there is no error $(e = 0)$ in the received vector r. Thus, we can decide that there are errors if $s \neq 0$. After calculating syndrome s, we proceed to the next step of finding the error vector e. If e is known, we have the corrected word as $c = r \oplus e$. However, the error vector cannot be uniquely determined from the syndrome s. There may be multiple combinations of error vectors and data vectors producing the same syndrome. The best strategy, in the sense that the error probability for the corrected word is minimum, is to choose the error vector with the minimum number of 1s as elements (minimum-weight vector). For this error-correction method, we can prepare a table of 2^m pairs of minimum-weight error vectors and their corresponding syndromes.

7.6.2 Cyclic Codes [45]

Cyclic codes are a subclass of linear block codes. In cyclic codes a shift of a code becomes another code. From this property of cyclic codes, treatment becomes simple: encoding/decoding is performed with a shift register and a mathematical expression in polynomial form can be used. The code word with n bits is expressed as

$$c(x) = c_1 x^{n-1} + c_2 x^{n-2} + \ldots c_n. \tag{7.116}$$

The coefficients express the elements of the code and take either 0 or 1. Addition of the coefficients is modulo 2 and multiplication obeys the usual integer multiplication. We denote the data polynomial and the generator polynomial by $d(x)$ and $g(x)$, respectively. The systematic cyclic codes are generated as

$$c(x) = x^{n-k}d(x) + \rho(x), \tag{7.117}$$

where $\rho(x)$ is the remainder from dividing $x^{n-k}d(x)$ by $g(x)$, that is,

$$\rho(x) = \text{Rem}\frac{x^{n-k}d(x)}{g(x)}. \tag{7.118}$$

Since $f(x) + f(x) = 0$, the code word $c(x)$ is a multiple of the generator polynomial $g(x)$.

Denote the error polynomial by $e(x)$; then the received signal polynomial $s(x)$ becomes

$$s(x) = \text{Rem}\frac{r(x)}{g(x)}$$

$$= \text{Rem}\frac{c(x) + e(x)}{g(x)}$$

$$= \text{Rem}\frac{e(x)}{g(x)} \tag{7.119}$$

If there is no error or a code word is received that may be different from the transmitted one, we have $s(x) = 0$. For deciding on an error polynomial $e(x)$ from $s(x)$, we choose the one (the minimum-weight vector) that has the minimum number of 1s as its coefficients.

A cyclic code can be generated by a feedback shift register as shown in Fig. 7.33. The switch S_1 is at the positions P_1. k data bits followed by $n-k$ successive 0s are fed into the shift register. As the kth data bit cleans out of the last register (the content of the registers at this moment corresponds to the remainder), the switch is thrown to the position P_2 and the $m(= n-k)$ parity check bits are output. The feedback shift register performs the division $d(x)x^{n-k}/g(x)$. Hence, the syndrome $s(x)$ can be obtained by the same feedback shift register.

The ability to correct/detect errors is given by the generator polynomial. The code with a minimum distance $2t + 1$ can detect up to $2t$ errors. A discussion of the burst error-detecting capability of a cyclic code follows. A burst error of length L bits is expressed as

$$e(x) = x^i b(x) \qquad (i = 0, 1, 2, \ldots, n - L),$$

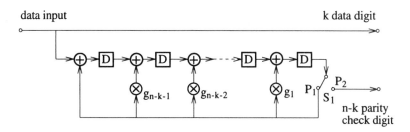

data input   k data digit

Figure 7.33. Encoder for cyclic codes.

where $b(x)$ is a polynomial expressing the error pattern.

$$b(x) = x^{L-1} + b_2 x^{L-2} + \cdots 1.$$

Assume a generator polynomial $g(x)$ of degree $m = n - k$. We can detect any burst error of length $L \le m$, since $e(x)$ is not a factor of $g(x)$ in this case. A burst error of length $L > m$ can be detected with some probability. We have 2^{L-2} error patterns. The number of error patterns that may be a factor of $g(x)$ is 1 for $L = m + 1$ and 2^{L-m-2} for $L > m + 1$. The probability that we cannot detect the error becomes

$$P_m = \begin{cases} 1/2^{m-1}, & L = m + 1 \\ 1/2^m, & L > m + 1. \end{cases} \tag{7.120}$$

Standardized error-detecting codes are given by CCITT. One of which is given as

$$g(x) = x^{16} + x^{12} + x^5 + 1$$
$$= (x + 1)(x^{15} + x^{14} + x^{13} + x^{12} + x^4 + x^3 + x^2 + 1). \tag{7.121}$$

It can be shown that the minimum distance of this code is $d_{\min} = 4$. It can detect up to 3 errors. Since $g(x)$ has a factor of $x + 1$, and $g(-1) = g(1) = c(1) = 0$, any odd number of errors can be detected. Since the degree of $g(x)$ is 16, we can detect any burst error with length of less than or equal to 16. The probability that a burst error of length greater than 16 cannot be detected is given by Eq. (7.120).

7.6.3 Convolutional Codes

It is difficult to describe convolutional codes using a mathematical expression; therefore we discuss some specific examples. A convolutional encoder with a code rate of 1/2 is given in Fig. 7.34. The encoder outputs two bits for every one input bit. The output bits are determined from the input bit and the two

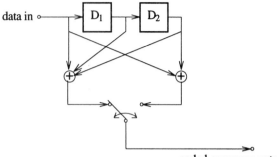

Figure 7.34. Encoder for a convolutional code.

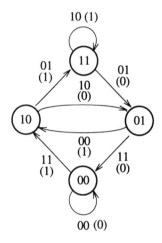

Figure 7.35. State diagram of the encoder in Fig. 7.32.

previous input bits stored in the shift register. Thus, this is an 8-state machine. Although it has eight states, its input–output relation is uniquely described by the 4-state diagram shown in Fig. 7.35, where state 10, for example, corresponds to the registers D_1 and D_2 containing 1 and 0, respectively. Once an input bit is received, the state changes or remains the same, outputting two bits as shown on the branches. The input bit is shown in parentheses.

Since the input data affect the three successor code words, decoding with the maximum-likelihood sequence estimator or the Viterbi algorithm discussed in Section 3.3.8 is useful.

Figures 7.36 (a) and (b) show examples of the decoding process assuming

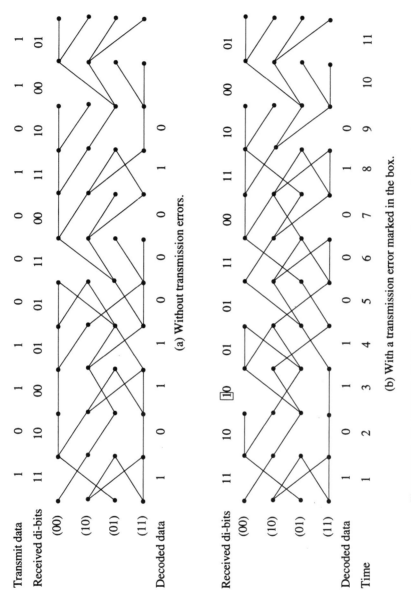

Figure 7.36. An example of Viterbi decoding process of a convolutional code.

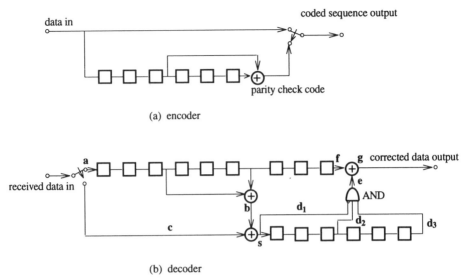

(a) encoder

(b) decoder

Figure 7.37. An encoder/decoder for the Hagelberger code.

the same transmit data sequence. The Hamming distance is used for the branch metric. In the case of no transmission errors (Fig. 7.36 (a)), path merge occurs at times 6 and 7. We can decide the data before each merge. Even if the error occurs (Fig. 7.36 (b)), data are correctly decoded. However, the decision on data is delayed since the path merge is delayed due to the error.

Another example of convolutional codes is the Hagelberger code, which is intended for burst error correction. An encoder/decoder is shown in Fig. 7.37. The decoder is simply implemented with shift register circuits. The receiver shift register is the same as that for the encoder, so the sequences *b* and *c* in Fig. 7.37 are the same and, hence, the syndrome *s* becomes 0 when there is no error. This code can correct a burst error with length of up to 6 bits, as long as the burst error is isolated from other errors by 19 bits. The error-correcting process is demonstrated by an example in Table 7.1.

7.6.4 ARQ (Automatic Repeat Request)

A system employing an ARQ scheme requests a retransmission of the data received in error. There are three ARQ strategies: (i) Stop and Wait, (ii) Go Back N and (iii) Selective Repeat. In the Stop and Wait ARQ system, the receiver sends an ACK signal to acknowledge a correct reception, and sends a NAK signal to request a retransmission of a coded signal block detected in error. The efficiency of transmission is low, since confirmation of the received signal is made for every block of data.

In the Go Back N ARQ system, coded data block is transmitted

TABLE 7.1   Error correction process of the coder/decoder in Fig. 7.37. Transmit data are a sequence of 0s. The burst error sequence is assumed to be 111001.

a	· · · 0 0 1 1 0 0 · ·
c	· · · 0 0 1 0 1 0 · ·
b	· · · 0 0 1 1 0 1 1 0 0 · · ·
s	· · · 0 0 1 0 1 1 1 0 1 1 0 0 · · ·
$\bar{d}_1$	1 1 1 1 1 0 1 0 0 0 1 0 0 1 1 1 1 1 1
d_2	· · · 0 0 1 0 1 1 1 0 1 1 0 0 0 · · ·
d_3	· · · 0 0 1 0 1 1 1 0 0 1 1 0 0 · · ·
e	· · · 0 0 0 0 0 0 1 1 0 0 0 0 0 0 · · ·
f	· · · 0 0 1 1 0 0 0 · · ·
g	· · · 0 0 0 0 0 0 0 · · ·

continuously. If the sender receives a NAK signal requesting a retransmission then the transmitter goes back by N data blocks and restarts the transmission from that data block. The time delay between the transmission of a data block and the reception of a NAK signal must be less than the time for the transmission of N data blocks. The idle time is shorter for this system than for the Stop and Wait ARQ system. However, it still wastes time, since N data blocks, which may include blocks without error, are retransmitted for every retransmission.

In the Selective Repeat ARQ system, data blocks are sent continuously and only the blocks in error are retransmitted. Thus, this system has the highest efficiency. The management of the order of data blocks and the buffer is more complicated for this system compared to the other systems.

There are other techniques, known as hybrid ARQ systems, that use a combination of ARQ and forward error correction. For further discussion, refer to [41].

7.6.5 Application to Mobile Radio Channels

Mobile radio channels are characterized by burst errors and high average error rates caused by fading. Error control techniques can be used depending on the requirements of the application. Error detection is a must for data services.

Burst error-correcting codes or interleaving techniques are used to cope

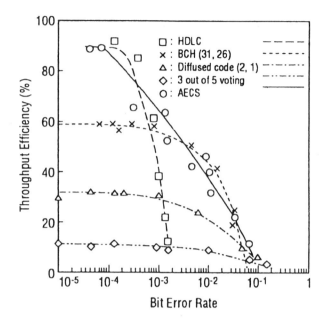

Figure 7.38. Simulated throughput efficiency vs. bit error rate for coded data transmission in analog cellular channels. Courtesy of OKI Electric Co. Ltd.

with burst errors. These techniques are useless for a very long burst of errors. In this case, an ARQ system works effectively.

A data transmission system which automatically changes its mode of transmission (AECS: adaptive error control system) is developed in [46] for application to analog cellular systems. A transmission rate of 4800 kbps is possible with a voice band data modem (modulator/demodulator) (CCITT standard V.27 ter.). The error rate performance is shown in Fig. 7.38. For error rates less than 10^{-4}, selective repeat ARQ without error correction (HDLC) is used in view of transmission efficiency. For error rates of around 10^{-3}, a 3-error correcting BCH code ($n = 31$, $k = 26$) is used with interleaving. For errors rates around 10^{-2}, a burst error-correcting code (diffused code) with a code rate of 1/2 is used. For high error rates such as 10^{-1}, five successive transmissions of a data block are made. At the receiver, majority voting decoding is performed. Thus the code rate with the above modes of error correction varies as $26/31 = 0.84$, 0.5, and 0.2.

7.7 TRELLIS-CODED MODULATION [47,48]

Trellis-coded modulation is a combined coding and modulation technique: coding and modulation are integrated to achieve better error rate perfor-

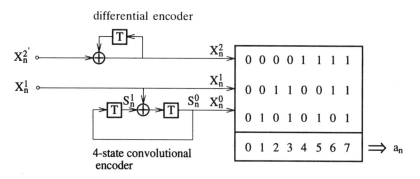

Figure 7.39. An encoder for the 4-state 8PSK code [47]. (Copyright © IEEE 1987).

mance without compromising bandwidth. To understand the principle of this technique, it is helpful to consider a system where coding and modulation are introduced independently. Let us compare a QPSK system without coding and an 8PSK system with a coding rate of 2/3. For the same data rate, both systems have the same spectral bandwidth. In the 8PSK receiver, demodulation and symbol-by-symbol decision-making are performed, followed by the error-correction process. The 8PSK system yields no significant advantage in error rate performance over the QPSK system without coding. This is because the coding gain is lost due to the higher error probability in the 8PSK system.

In trellis-coded modulation, coding and modulation are combined to produce the minimum error rates provided that the signals are detected as a sequence: the maximum-likelihood sequence estimation. The criterion is to minimize Euclidean distance between coded and modulated signal sequences. For signal sequences with the minimum distance, called the free distance, d_{free}, and the uncorrelated Gaussian noise with variance σ^2, error rates are approximately given as

$$P_e \simeq N_{\text{free}} Q[d_{\text{free}}/2\sigma)] \qquad (P_e \ll 1), \qquad (7.122)$$

where

$$Q(x) = \frac{1}{\sqrt{2\pi}} \int_x^\infty \exp(-y^2/2)\, dy$$

and N_{free} denotes the average number of nearest-neighbor signal sequences with distance d_{free} that diverge at any state from a transmitted signal sequence and remerge with it after one or more transitions.

As an example of a trellis-coded modulation system, an encoder for an 8PSK code [47] is shown in Fig. 7.39. A bit of the di-bit input signal is fed

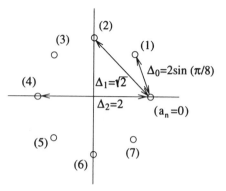

Figure 7.40. 8PSK signal.

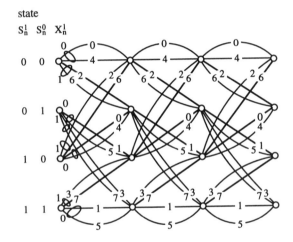

Figure 7.41. Trellis diagram for 4-state trellis-coded 8PSK [47]. (Copyright © IEEE 1987.)

to a 4-state convolutional encoder with coding rate of 1/2. Thus the total coding rate is 2/3. The code word a_n is assigned a phase as in Fig. 7.40. The 4-state trellis diagram for the trellis-coded 8PSK system is shown in Fig. 7.41. The 8PSK signals are assigned to the transitions in the 4-state trellis according to the following rules:

1. Parallel transitions occur with signals in the subsets (0,4), (1,5), (2,6), or (3,7). The distance between signals in this subset is $\Delta_2 = 2$.
2. Four transitions originating from or merging in one state are from signals in the subsets (0,4,2,6) or (1,5,3,7). The distance between the signal in this subset is at least $\Delta_1 = \sqrt{2}$.

Any two signal sequences in the trellis of Fig. 7.41 originating from one state and remerging in another state after more than one transition have a squared distance of at least $\Delta_1^2 + \Delta_0^2 + \Delta_1^2 = \Delta_0^2 + \Delta_2^2$ between them. Since the squared distance between the parallel paths is Δ_2^2, the free distance d_{free} in the 4-state 8PSK becomes $d_{\text{free}} = 2$. If we compare this with the free distance $d_{\text{free}} = \sqrt{2}$ for uncoded QPSK, a 3 dB gain is obtained for the trellis-coded 8PSK system over the uncoded QPSK. More gains are achieved with other codes. Trellis-coded modulation is applied to other modulation schemes such as QAM.

Trellis coding assumes coherent detection and maximum-likelihood sequence estimation, for example, the Viterbi algorithm.

Applications of trellis-coded modulation to fading channels are discussed in [49] and [50]. A trellis-coded 16QAM system has been investigated for applications in TDMA systems for mobile radio communications [51]. A trellis-coded 8PSK system with envelope control (for power efficiency in the mobile radio transmitter amplifier) is described in [52].

7.8 ADAPTIVE INTERFERENCE CANCELLATION

In radio communication systems, the received signal is sometimes interfered with by other signals. The interference may be intentional or unintentional. Cancellation of the intentional interference ("jamming") in communication or radar systems is important in military applications. Even in nonmilitary systems, interference cancellation becomes important since it results in a more efficient use of the spectrum, which is important in cellular systems. Interference cancellation techniques for mobile radio communications are currently a hot topic.

Interference cancellation must be adaptive since both the interfering signal and the desired signal are time-varying in mobile radio communications. There are two categories of adaptive interference cancellation techniques. One uses adaptive antenna arrays [53] and the other uses signal processing techniques to suppress the interfering signal.

Let us consider adaptive antennas first. Adaptive antennas are used to induce a null point in the directive pattern in the direction of the interfering signal. Consider, for example, an antenna array system consisting of two antennas that have an omnidirectional directive pattern in horizontal plane as shown in Fig. 7.42. The received signals are weighted with complex coefficients and summed.

Let us assume a plane wave for simplicity. The signals at antennas 1 and 2 become the same except for a phase difference $\Delta\varphi$ that is $\Delta\varphi = (2\pi d/\lambda)\cos\theta$; d is the distance between the antennas and λ is the wavelength of the carrier. The effect the modulating signal has on the received signal is neglected: this is valid when the modulating signal changes only little in the arrival time difference between the signals. Let the complex

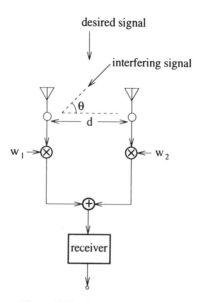

Figure 7.42. An antenna array.

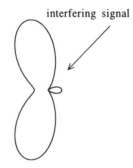

Figure 7.43. Directive pattern of an antenna array.

weighting factors w_1 and w_2 satisfy $|w_1| = |w_2| = 1$, and $\angle w_1 - \angle w_2 = \Delta\theta$; then the directive pattern of the antenna array or the relative power of the summed signal as a function of θ becomes

$$G(\theta) = \cos^2\left(\frac{\pi d}{\lambda}\cos\theta + \frac{\Delta\theta}{2}\right). \tag{7.123}$$

A null point is produced in the direction of θ_0 by letting $\Delta\theta = -(2\pi d/\lambda)\cos\theta_0 \pm \pi$. A directive pattern is shown in Fig. 7.43 for $d = \lambda/2$ and $\Delta\theta = (1 - 1/\sqrt{2})\pi$.

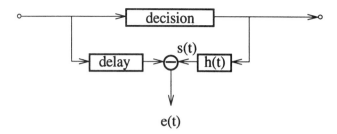

Figure 7.44. Generation of an error signal for digital communication system.

In order for the antenna array to be adaptive to the interference, the weighting factors w_1 and w_2 are automatically adjusted. A well-known method is the least mean squared algorithm. The weighting factors are adjusted to minimize the average least mean square error

$$\langle |\varepsilon(t)|^2 \rangle = \left\langle \left| \sum_{i=1}^{n} w_i r_i(t) - s(t) \right|^2 \right\rangle.$$

$r_i(t)$ is the received signal at antenna i and $s(t)$ is the reference signal or the desired signal.

In radar systems the reference signal is given; it is a transmitted and reflected signal. In analog communication systems, the reference signal must be transmitted periodically. In digital communication systems, the reference signal can be generated from the received signal as shown in Fig. 7.44; $h(t)$ is the impulse response of the overall channel including transmit and receive filters. The capability of digital systems to generate the reference signal is due to discrete-level signal transmission. To get the optimum weighting factors, recursive methods can be applied as discussed in Section 7.5 for the automatic equalizer. A diversity receiver combined with the least mean squared algorithm acts as an adaptive interference canceler if the interference levels become significant and as a maximal ratio combining diversity system if the interference is absent [86,87].

The antenna array system cannot cancel an interfering signal if it is in the same direction as the desired signal. The other category of adaptive interference cancellation can cope with this case. Figure 7.45 shows an adaptive interference canceler based on MLSE (maximum likelihood estimation) [54] without an array antenna. The MLSE or the Viterbi algorithm is used for estimating both the desired and the interfering signal. Adaptive equalizers are also assumed in this system. The adaptation algorithm used in the equalizers is the recursive least squares method [55]. The reference signals for the desired and interfering signals are produced using the

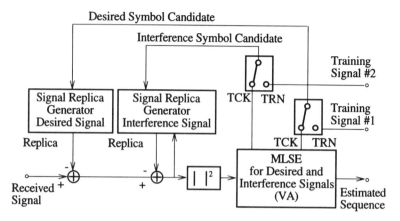

Figure 7.45. Adaptive interference canceler using RLS-MLSE [54]. (Copyright © IEICE 1994.)

candidates for desired and interfering signals and the channel impulse response. They are subtracted from the received signal to output an error signal. The squared error signal is the target function for the MLSE. Training signals are assumed for initial acquisition. The adaptive interference canceler/equalizer is further combined with a diversity receiving system. Good results are obtained with computer simulation. Trellis-coded modulation is introduced to this adaptive interference canceler in [56]. Computer simulation shows better results than that without trellis-coded modulation.

For suppressing the effect of the interfering signal, forward error correction using information about interference level is proposed [57]. The interference signal is periodically measured in the absence of signal transmission.

Adaptive interference cancellation is discussed in the following for spread-spectrum code division multiple-access (CDMA) systems. A primitive multiuser detector for a CDMA system consists of a bank of matched filters and a threshold device as shown in Fig. 7.46. The performance of the receiver suffers from the cross correlation between the spectrum spreading codes for different users. The optimum multiuser receiver can be obtained with the maximum likelihood sequence detector. The complexity of the receiver, however, increases exponentially with the number of users. Therefore, some suboptimum receivers are proposed to reduce the receiver complexity [58–62]. A multistage adaptive interference canceler is shown in Fig. 7.47. The interference from users #1 through #$(N-1)$ is subtracted successively from the input signal to get the desired signal (user #N). Tentative decisions on the data are used to produce replica signals of the interfering signals. Knowledge about the other users' spectrum-spreading codes is assumed in this receiver. Adaptive interference canceling receivers for SS-CDMA

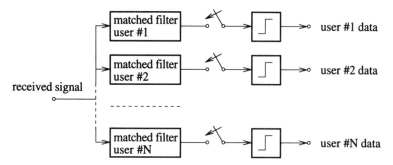

Figure 7.46. A multiuser detector for a CDMA system.

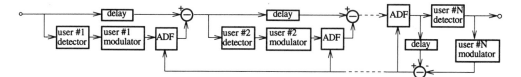

Figure 7.47. A multistage adaptive interference canceling receiver for an SS-CDMA system. ADF denotes adaptive digital filter [58]. (Copyright © IEEE 1990.)

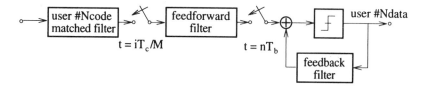

Figure 7.48. Interference canceling SS-CDMA receiver without knowledge of other user codes. T_c and T_b are the chip and data duration, respectively. M is an integer.

systems (Fig. 7.48) are discussed in [63] and [64] for systems where other users' codes are unknown. The received signal is fed to a filter matched to the desired (user #N) signal SS code. The output of the matched filter is sampled at a fractional time interval of the chip rate. The interference from other users is suppressed with the adaptive feed-forward filter using the orthogonal property of the codes. The feedback filter equalizes the intersymbol interference. Any algorithm such as the minimum mean squared error method can be used to adapt the filter. Improvements in adaptation speed in fading channels are reported in [65] and [66].

Some people would consider that antenna arrays are different from diversity systems. However, as seen from Figs. 7.42 and 7.16, an antenna array has the same structure as a cophased diversity combining system. If we consider a diversity combining system, similar to an antenna array, the diversity system gives a directivity pattern that is determined from the weighting factor. (The directivity pattern for the receiver is the same as for a transmitter, because of the reciprocity of a linear circuit.) The difference between an antenna array and a diversity combining system is only superficial, and consists in the assumption of arrival signals. Array antennas usually assume a desired signal and a small number of interfering signals, all of which are plane waves; on the other hand, diversity antennas assume a desired signal that is a sum of many plane wave signals which are independent of each other. Consequently the correlation of received signals between antennas is high for an antenna array and low for a diversity system. These situations correspond to the environments for mobile terminals and base stations.

Using a proper combining algorithm such as the least mean squared error algorithm, an antenna array and a diversity system work in the same manner. For example, consider a situation where an echo signal with a time delay arrives at a receiver in a direction different from the main signal; then the echo signal may be used or canceled depending on the time delay in both the antenna array and the diversity system.

7.9 VOICE CODING [67]

The most straightforward system for voice coding is the usual analog-to-digital conversion at periodic sampling instants. In this system the sampled signals are quantized into discrete levels spaced at a given distance. The difference between the input signal level and the quantized level is called quantization error or quantization noise. The quantization noise power is calculated as follows. Quantization noise is assumed to be independent and uniformly distributed in the range $-\Delta/2$ to $\Delta/2$ within the frequency range of $-f_s/2$ to $f_s/2$, where Δ is the level spacing and f_s is the sampling frequency. The expected value of the noise power N is then

$$N = \int_{-\Delta/2}^{\Delta/2} x^2 p(x)\, dx = \frac{\Delta^2}{12}, \tag{7.124}$$

where the probability density function $p(x) = 1/\Delta$ for $|x| \le \Delta/2$ and $p(x) = 0$ for $|x| \ge \Delta/2$ is used. Let us consider the case of quantizing a sinusoidal signal of peak value A_p using a 2^n-level quantizer. The average power S of the signal is expressed using Δ as $S = \frac{1}{2}A_p^2 = \frac{1}{2}(2^{n-1}\Delta)^2$. The average signal to quantization noise power ratio becomes

$$\frac{S}{N} = \frac{\frac{1}{2}(2^{2(n-1)}\Delta^2)}{\Delta^2/12}$$

$$= \frac{3}{2} \cdot 2^{2n} \tag{7.125a}$$

$$\approx 6n + 1.8 \text{ [dB]} \tag{7.125b}$$

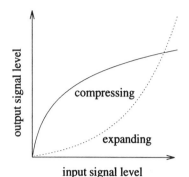

Figure 7.49. Compression and expansion of a signal.

7.9.1 PCM (Pulse Code Modulation)

When we decrease the signal level in the above coding system, the signal to noise power ratio becomes small for a fixed n and Δ. This scheme is not appropriate for voice signal transmission, since the voice signal level is not always at a constant value. To solve this problem, the PCM system employs a technique that compresses the input signal to the quantizer as shown schematically in Fig. 7.49. The input signal is compressed instantaneously depending on its level. The effect of compressing is equivalent to decreasing Δ for lower-level signals. Thus, the signal to noise power ratio becomes less dependent on the input signal level. The signal is expanded at the receiver to recover the original, uncompressed signal.

The CCITT has standardized two compression laws [68], the μ-law:

$$y = \frac{\text{sgn}(x)}{\ln(1+\mu)} \ln\left(1 + m\left|\frac{x}{x_p}\right|\right), \qquad \left|\frac{x}{x_p}\right| \le 1, \qquad (7.126a)$$

and the A-law

$$y = \begin{cases} \dfrac{A}{1+\ln A}\dfrac{x}{x_p}, & \left|\dfrac{x}{x_p}\right| \le \dfrac{1}{A} \\[2ex] \dfrac{\text{sgn}(x)}{1+\ln A}\left(1 + \ln A\left|\dfrac{x}{x_p}\right|\right), & \dfrac{1}{A} \le \left|\dfrac{x}{x_p}\right| \le 1, \end{cases} \qquad (7.126b)$$

where x_p is the maximum input signal level and $\mu = 100$ and 255 and $A = 87.6$ are the standardized values. The sampling frequency is 8 kHz and there are 2^8 quantization levels. The voice coding rate is then 64 kbps. PCM is seldom used for mobile radio communications because its transmission rate is high, requiring a wide bandwidth.

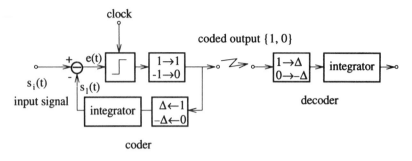

Figure 7.50. Delta modulation system.

7.9.2 Delta Modulation

The coder and decoder for a delta modulation system are shown in Fig. 7.50. The local signal $s_l(t)$, which is generated using previously decoded signal, is subtracted from the input signal $s_i(t)$ to produce an error signal. The polarity of the error signal is detected and a '1' or '0' signal is assigned as the coded output signal (i.e., one-bit quantization). The decoder is the same as the feedback circuit, called a local decoder, in the encoder. The local decoder tracks the input signal by adding a step size of Δ or $-\Delta$ to its value, that is, integrating at every clock time (Fig. 7.50). The integrator in the local decoder is actually replaced with a more general low-pass filter. The filter may be called a predictor, which predicts the next signal from the previous signals. The minimum error in tracking is $\pm\Delta$, and using a smaller Δ corresponds to a higher signal to noise ratio. However, tracking speeds become slower for a smaller Δ, and a larger error is produced for rapidly changing input signals, as shown in Fig. 7.51. The tracking speed is raised by increasing the coding clock frequency.

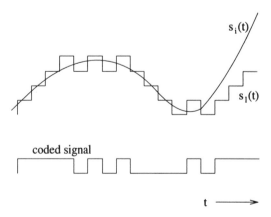

Figure 7.51. Adaptive delta modulation encoder.

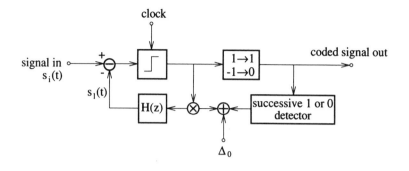

Figure 7.52. Waveforms for delta modulation system.

The adaptive delta modulation system changes its step size in order to track the rapidly changing input signal without increasing the coding frequency. The block diagram of the adaptive delta modulation encoder is shown in Fig. 7.52. A general filter is assumed for the local decoding filter. If a successive 1 or 0 detector detects a given number of successive occurrences of 1s or 0s, it outputs a value to be added to the given step size Δ_0, increasing the total step size. The decoder at the receiver is the same as the local decoder of the encoding circuit. The receiver decoder outputs the replica signal $s_l(t)$ (assuming no transmission errors) using the received data signal and knowing $H(z)$, Δ_0, and the structure of the successive 1 or 0 detector. ADM (adaptive delta modulation) is used for mobile radio communications at coding speed of 10–30 kbps.

7.9.3 Adaptive Differential Pulse Code Modulation

The ADPCM system uses a multilevel quantizer and an adaptive prediction filter. Figure 7.53 shows the ADPCM system standardized by the CCITT. The sampling frequency is 8 kHz and there are 2^4 quantization levels, resulting in a coding rate of 32 kbps.

The quantizer and predictor are shown in more detail in Fig. 7.54. If the quantization error, $e(n) - e'(n)$, is small, the signal $\tilde{x}(n) = \hat{x}(n) + e'(n)$, which is obtained at the receiver output, is close to the input signal $x(n)$ so long as there are no transmission errors. In Fig. 7.53, if we assume analog transmission by removing the quantizer, encoder, and decoder and connecting their inputs and outputs, exactly the same signal as $x(n)$ is obtained at the receiver output. This means that the transfer function of the transmit circuit is the inverse transfer function of the receiver circuit. The receive feedback filter is called the synthesis filter, which is also used in the feedback path of the transmit filter. The number of bits for quantization can be

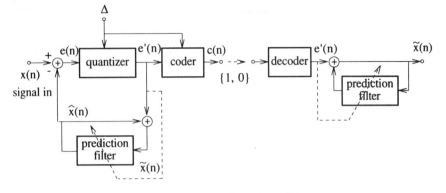

Figure 7.53. Adaptive differential PCM system.

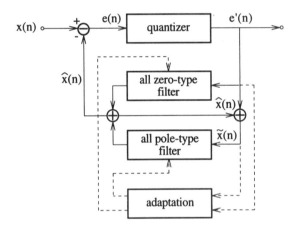

Figure 7.54. Detailed block diagram of ADPCM predictor.

decreased, since, due to prediction, the dynamic range of $e(n)$ is reduced compared with that of $x(n)$.

Since the adaptation is done using the variable $e'(n)$, which can be obtained from the transmitted signal sequence $c(n)$, we do not need to send any extra information to control the receiver adaptive filter.

The 32 kbps ADPCM is adopted for the European and Japanese digital cordless telephone system.

7.9.4 Adaptive Predictive Coding

If more sophisticated prediction is used, such as APC (adaptive predictive coding), coding efficiency can be increased. This scheme is based on the periodic structure of voice signals. An example of a voice signal waveform

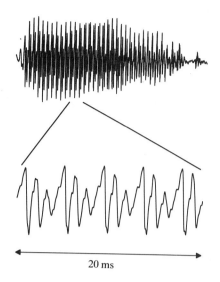

Figure 7.55. Example of a voice signal waveform.

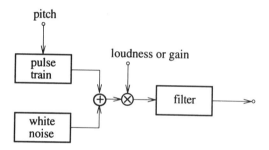

Figure 7.56. Model for generation of a voice signal.

is shown in Fig. 7.55. It is the Japanese pronunciation of 'SAI' by a female college student. We can see some regularities in the waveform due to the mechanism of voice signal generation. A voice signal generation model is shown in Fig. 7.56. Periodic pulses and white noise are used as the sound sources. The former are generated as to model the vibration of vocal cords to produce voiced sound. The vibration frequency, or pitch, corresponds to the repetition of peak pulses in Fig. 7.55. The white noise is generated for the fricative (or unvoiced) sounds. The filter represents the vocal tract from the sound source input to the output, in other words, the mouth, lips, and nose.

The regularity of the waveform does not change for several tens of milliseconds. Thus, the regular waveform in some time period can be expressed by a small set of parameters, such as the pitch, loudness, and transfer function of the vocal tract, or coefficients of the equivalent short-term prediction filter.

The coefficients of the short-term prediction filter are obtained with LPC (linear predictive coding) analysis. In short, the LPC analysis determines the transfer function of the vocal tract using of the short-term voice signal as a test signal. A more detailed description of linear predictive coding analysis is beyond the scope of this book; refer to [67]–[70] for further information. Adaptive predictive coding is used for most of the efficient voice coding systems.

The long-term or pitch prediction is made so as to minimize the squared pitch prediction error. Assume a first-order digital filter; then the error is defined as

$$E = \sum_{n=0}^{N-1} [e(n) - \beta e(n - T)]^2, \tag{7.127}$$

where $e(n)$ is the short-term prediction error given by LPC, β is a coefficient, T is the pitch duration, and N is the number of samples of the signal in a block. β and T are found so as to minimize E.

Figure 7.57 shows an APC (adaptive predictive coding) system. The major difference of this coding from the ADPCM system consists in the way of estimation and transmission of the adaptive parameters. In the APC system, the input signal is stored in a buffer for a time period of, say, 20 ms; then the short-term parameters are analyzed by LPC, followed by long-term estimation. The quantization step size is also determined using the buffered signal. These parameters, the side information, are multiplexed and transmitted with the coded signal.

The parameter estimation and coding are carried out independently in the systems so far described. In the following systems, sometimes called hybrid coding systems, combined analysis of the parameter estimation and coding is made using a cost function for the minimum waveform error between input signal and locally decoded signal.

7.9.5 Multipulse Coding

Figure 7.58 shows a block diagram of the multipulse coding system. The estimation of short-term and/or long-term parameters is made using the buffered signal for a given size of block. For the sound source input to the synthesis filter, many pulses with different amplitude and position are generated to achieve the minimum cost function,

$$E = \sum_{n=0}^{N-1} \left[\left(x(n) - \sum_{j=1}^{k} g_j h(n - m_j) \right) * w(n) \right]^2, \tag{7.128}$$

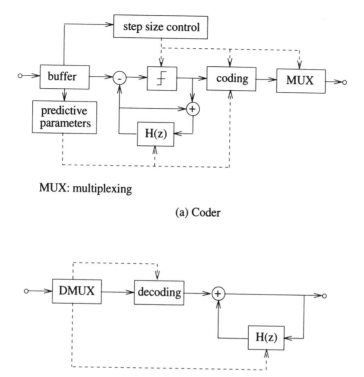

MUX: multiplexing

(a) Coder

DMUX: demultiplexing

(b) Decoder

Figure 7.57. Adaptive predictive coding system.

where $x(n)$ are samples of the input signal, g_j and m_j are the amplitude and position of jth pulse, $h(n)$ are samples of the impulse response of the synthesis filter, and $w(n)$ are samples of the impulse response of the perceptual weighting factor. The symbol $*$ denotes convolution. The perceptual weighting factor is used to improve voice quality as heard by the human ear. The coding (analysis) is carried out by feeding back the locally decoded (synthesis) signal and is called the A-b-S (analysis-by-synthesis) method. The key technical issue in multipulse coding is determining the multipulse and its efficient coding for good voice quality. Multipulse coding requires a high volume of signal processing. To reduce the amount of signal processing, intervals of pulses are constrained to given values (regular pulses) in the pan-European (GSM) digital cellular speech coding: RPE-LPC/LTP (regular pulse excited LPC with long-term prediction) [71].

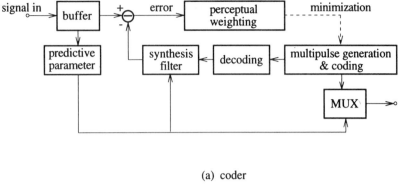

(a) coder

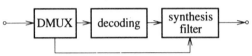

(b) decoder

Figure 7.58. Multipulse coding system.

First, the LPC predicted error is calculated as

$$e(n) = x(n) - \sum_{i=1}^{P} a_i x(n - i),$$

where a_i and P are the coefficients and order of the LPC filter. The pitch parameters, β and M, are given by minimizing,

$$E(\beta, M) = \sum_{n=0}^{N-1} [e(n) - \beta v(n - M)]^2, \tag{7.129}$$

where N is the subframe length (say, 5 ms, i.e., $N = 40$), β is the pitch coefficient, M is the pitch duration, and $v(n)$ is past subframe sound source expressed with the regular pulses. Using the given β and M, the regular pulses $x_m(i)$ are made by sampling $d(n)$, defined by $d(n) \equiv e(n) - \beta v(n - M)$, at some interval, say 4,

$$x_m(i) = d(m + 4 \cdot i) \qquad (m = 0, 1, 2, 3), \tag{7.130}$$

where m denotes the initial phase of a regular pulse. The initial phase is determined by max $\sum_{i=0}^{N_{sub}} x_m^2(i)$, where N_{sub} (say 40) is the subframe length. The amplitudes of the regular pulses are coded with adaptive PCM.

The transmitted information comprises the LPC predictive parameters for

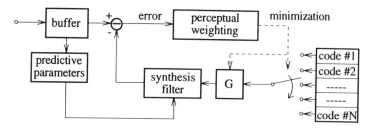

Figure 7.59. CELP (code excited linear predictive) coding.

each frame length (20 ms): the pitch duration M, the pitch coefficient β, and the amplitudes and the initial phase for each subframe length (40 samples = 5 ms).

7.9.6   Code Excited LPC (CELP)

CELP is different from multipulse coding because it uses prepared waveforms instead of multipulse waveforms as the source signal input to the synthesis filter (Fig. 7.59). The selection of one of the waveforms stored in the codebook and the gain factor is made to minimize the perceptually weighted error power. The index for a waveform and the gain factor are transmitted together with parameters of the synthesis filter. Assigning an index to a waveform of a given number of samples corresponds to vector quantization. Using vector quantization, the average number of bits per sample may become less than 1.

The long-term or pitch prediction filter in the synthesis filter can be adaptively controlled with the error signal. For this case, the synthesis filter is given in Fig. 7.60. The pitch predictor consists of a first-order filter. The delay and coefficient β are adjusted to minimize the perceptually weighted error power. A subframe data block of say 5 ms is stored and delayed in the shift register. The registor is called the adaptive codebook.

VSELP (vector sum excited LPC) is proposed to decrease the computational complexity for the optimum code search and memory for the codebook. In addition, it has immunity against transmission errors. Multiple codebooks are used for the sound source in the VSELP as shown in Fig. 7.61. Each codebook is implemented as shown in Fig. 7.62. Consider that the number of sound sources is 2^M (indexed by i as $i = 0, 1, 2, \ldots, 2^M - 1$). We prepare M different basis vectors, which are determined by experiment and standardized. A sound source vector is generated as

$$u_i(n) = \sum_{m=1}^{M} \theta_{im} v_m(n), \qquad i = 0, 1, 2, \ldots, 2^M - 1, \qquad (7.131)$$

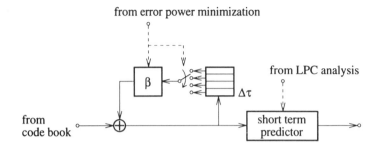

Figure 7.60. CELP with adaptive codebook (pitch prediction).

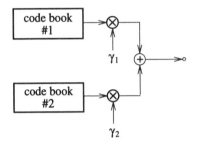

Figure 7.61. Sound source codebook for VSELP.

where θ_{im} takes 1 or -1 corresponding to sound source index i. Thus, 2^M sound sources are produced with stored basis vectors. The index of M bits for each codebook is transmitted at each subframe. The receiver constructs a sound source with Eq. (7.131) and the circuit of Fig. 7.58b. The transmitted analog signal is obtained by inputting the sound source to the synthesis filter (Fig. 7.59).

Consider that a transmission error occurs in the M bits: the polarity of one of the M basis vectors is reversed. The generated sound is not much different from the original sound. This explains why VSELP has immunity against transmission errors.

The use of two codebooks helps to reduce the amount of signal processing required. Description of this topic is beyond the scope of this book.

VSELP was selected as the full rate standard, of which the coding rate is 13 kbps (7.9 kbps for voice coding and the rest for error control) and 11.2 kbps (6.7 kbps for voice coding and the rest for error control) for American and Japanese digital cellular telephone systems, respectively [72,73].

For the half-rate standard in Japanese digital cellular telephone systems, PSI-CELP (pitch synchronous/innovation CELP) [74] was adopted. The

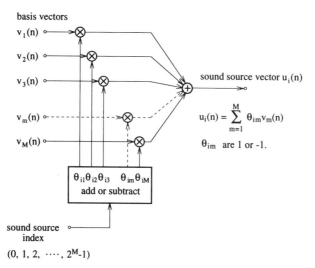

basis vectors

$v_1(n)$

$v_2(n)$

$v_3(n)$

$v_m(n)$

$v_M(n)$

sound source vector $u_i(n)$

$$u_i(n) = \sum_{m=1}^{M} \theta_{im} v_m(n)$$

θ_{im} are 1 or -1.

$\theta_{i1} \theta_{i2} \theta_{i3}$ $\theta_{im} \theta_{iM}$
add or subtract

sound source index

$(0, 1, 2, \cdots, 2^M-1)$

Figure 7.62. Implementation of sound source generator for VSELP.

coding rate is 5.6 kbps with 3.45 kbps for voice coding and 2.15 kbps for error control. The PSI-CELP coder is shown in Fig. 7.63. The lengths of the frame and subframe are 40 ms and 10 ms, respectively. The pitch prediction, predicted error, and codebook gain are coded in a subframe. The PSI-CELP has three kinds of codebooks for quantization of source signal: adaptive (for pitch prediction), fixed, and stochastic codebooks. The adaptive codebook and the fixed codebook are selected for voiced and unvoiced signals, respectively. The most significant point of the PSI-CELP is that the stochastic codes are synchronized to the pitch.

7.9.7 LPC Vocoder

In general the voice decoders reconstruct the transmit voice signal by applying the source signal to the (synthesis) filter. Similarly to the other linear predictive coding schemes, the LPC vocoder uses the LPC synthesis filter. In this system, the input signal to the synthesis filter is simplified as shown in Fig. 7.64.

At the encoder, the LPC short-term analysis, the pitch analysis, and voiced or unvoiced signal decision are carried out. The unvoiced signal and voiced signal are represented by white noise and a periodic pulse train, respectively. The pitch frequency, voiced or unvoiced signal indicator, the gain factor g_n or g_p, and synthesis filter parameters are transmitted. Since the coding rate is as low as 2.4 kbps, the voice quality is not sufficient for use in public communications at the time of writing.

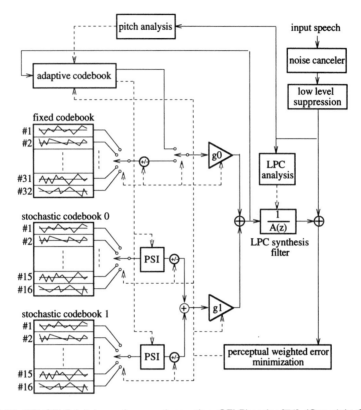

Figure 7.63. PSI-CELP (pitch synchronous innovation–CELP) coder [74]. (Copyright © IEICE 1993.)

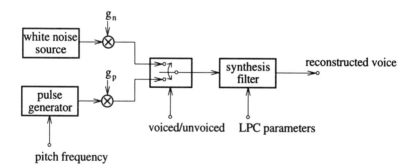

Figure 7.64. Decoder for LPC vocoder.

7.9.8   Application to Mobile Radio Communications

When we apply voice coding to mobile radio communications, the voice quality, coding rate, coding/decoding delay, and required signal processing volume become important issues. The voice quality must be considered with as well as without transmission errors. The coding rate affects the spectrum efficiency and its importance depends on the system. Voice coding methods currently used in public mobile radio communication ranges from 32 kbps ADPCM for digital cordless telephones to 5.6 kbps PSI-CELP for the Japanese half-rate digital cellular system. When coding/decoding delays become long, echo suppression/cancellation must be introduced. The signal processing requirement is being reduced by virtue of improvements in micro-semiconductor technologies.

Burst error as well as random error in mobile radio channels causes degradation in decoded voice quality. The degree of degradation depends on the voice coding schemes. (Adaptive) Delta modulation is immune to transmission error: the acceptable average bit error rates are around 10^{-2} for a relatively low coding rate, say of 16 kbps.

The transmission errors produce annoying click noise in PCM or ADPCM systems. Techniques have been proposed to reduce the effects of click noise [75–77]. Two kinds of techniques are known: noise blanking and replacement. In the former method, the signal is killed during the click noise period. In the latter method, the noisy signal period is replaced by the previous period. In these techniques smooth continuation of the voice signal is desirable. The click noise is detected using error detection coding and/or by differentiating the signal, detecting abrupt change in the signal level. For ADPCM a technique is proposed in which correlation between the coefficients of the prediction filter for voice signal is used: when an error is detected, the coefficients are replaced with those of the previous signal period.

For a highly efficient (i.e., low coding rate) coding system, transmission errors cause intolerable deterioration in voice quality. In this system, error control is a must. The impact of errors on the voice quality depends strongly on the bits assigned for the synthesis filter parameters and sound source. For example, Fig. 7.65 shows how the signal to noise power ratio depends on errors in the different bits assigned for PSI-CELP. The delay parameters used for the pitch prediction and the pitch synchronization are fairly sensitive to transmission errors. To protect the bits with different sensitivity from transmission errors, error correction codes with different degree of error-correcting capability for respective information-bearing bits are used [78].

For a long burst error resulting from a long duration of fading, forward error correction is useless. The error signals are replaced with the previous one, using an error flag given by the CRC (Cyclic Redundancy Check) error detection. If the error burst becomes very large, we have no option but to blank the signal.

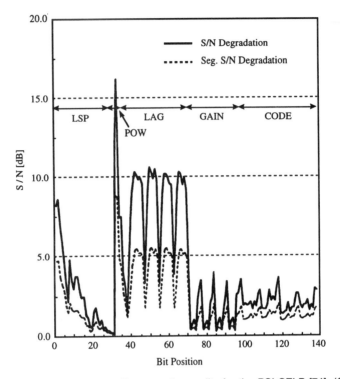

Figure 7.65. Transmission error effect on voice quality for the PSI-CELP [74]. (Copyright © IEICE 1993.)

Departing from the topic of voice coding, voice activity detection and comfort noise insertion are issues in the VOX (voice-operated transmission) system, where radio signals are transmitted only when the voice signal exists and are turned off during silent periods. Voice activity detection includes methods for measuring the input signal level and level crossing rate, measuring LPC predicted error signal power [79], and for use of the prediction coefficients [77]. With the VOX system, the transmit sound is completely blanked for silent periods. The listener hears no sound, which causes the listener concern that the channel might be disconnected. To remove this, "comfort" noise is generated during silent periods at the receiver [80].

APPENDIX 7.1 AVERAGE ERROR RATE FOR MAXIMAL RATIO COMBINER WITH COHERENT DETECTOR

Average error rate is given by Eq. (7.21) as

$$\langle P_e \rangle = \int_0^\infty \frac{1}{2} \operatorname{erfc}(\sqrt{\alpha\gamma}) \frac{\gamma^{M-1}}{(M-1)!} \frac{1}{\gamma_0^M} e^{-\gamma/\gamma_0} d\gamma. \tag{A7.1}$$

Integrating by part we have

$$\langle P_e \rangle = [\tfrac{1}{2}\mathrm{erfc}(\sqrt{\alpha\gamma})\,I_M(\gamma)]_0^\infty - \int_0^\infty \frac{d}{d\gamma}(\tfrac{1}{2}\mathrm{erfc}(\sqrt{\alpha\gamma}))\,I_M(\gamma)\,d\gamma, \quad (A7.2)$$

where

$$I_M(\gamma) = \int \frac{\gamma^{M-1}}{(M-1)!\,\gamma_0^M} e^{-\gamma/\gamma_0}\,d\gamma \qquad (A7.3a)$$

$$= -\sum_{m=0}^{M-1} \frac{\gamma^m}{m!\,\gamma_0^m} e^{-\gamma/\gamma_0}. \qquad (A7.3b)$$

We can find

$$[\tfrac{1}{2}\mathrm{erfc}(\sqrt{\alpha\gamma})\,I_M(\gamma)]_0^\infty = \tfrac{1}{2} \qquad (M \geq 1) \qquad (A7.4)$$

using

$$\frac{d}{d\gamma}(\tfrac{1}{2}\mathrm{erfc}(\sqrt{\alpha\gamma})) = -\frac{1}{2}\sqrt{\frac{\alpha}{\pi}}\frac{1}{\sqrt{\gamma}}e^{-\alpha\gamma}. \qquad (A7.5)$$

The second term on the right-hand side of Eq. (A7.2) is given using Eqs. (A7.3b) and (A7.5) as

$$\int_0^\infty \frac{d}{d\gamma}(\tfrac{1}{2}\mathrm{erfc}(\sqrt{\alpha\gamma}))\,I_M(\gamma\,d\gamma = -\frac{1}{2}\sqrt{\frac{\alpha}{\pi}}\sum_{m=0}^{M-1} J_m, \qquad (A7.6)$$

where

$$J_m = \int_0^\infty \frac{\gamma^m}{m!\,\gamma_0^m}\frac{1}{\sqrt{\gamma}}e^{-(\alpha+1/\gamma_0)\gamma}\,d\gamma. \qquad (A7.7)$$

Integrating Eq. (A7.7) by part for $m \geq 1$,

$$J_m = \frac{(2m-1)/2m}{\beta\gamma_0} J_{m-1}, \qquad (A7.8)$$

where $\beta = \alpha + 1/\gamma_0$.

Using the relation

$$\int_0^\infty \frac{1}{\sqrt{\gamma}}e^{-\beta\gamma}\,d\gamma = \sqrt{\frac{\pi}{\beta}},$$

we get $J_0 = \sqrt{\pi/\beta}$, and we have

$$J_m = \frac{(2m-1)!!}{(2m)!!} \frac{1}{(\beta\gamma_0)^m} \sqrt{\frac{\pi}{\beta}}. \tag{A7.9}$$

With use of Eqs. (A7.9), (A7.6), and (A7.4), Eq. (A7.2) yields

$$\langle P_e \rangle = \frac{1}{2} - \frac{1}{2} \sqrt{\frac{\alpha}{\pi}} \left(J_0 + \sum_{m=1}^{M-1} J_m \right)$$

$$= \frac{1}{2} - \frac{1}{2} \frac{1}{\sqrt{1 + 1/\alpha\gamma_0}} \left(1 + \sum_{m=1}^{M-1} \frac{(2m-1)!!/(2m)!!}{(1+\alpha\gamma_0)^m} \right). \tag{A7.10}$$

APPENDIX 7.2   AVERAGE ERROR RATE OF MAXIMAL RATIO COMBINING SYSTEM WITH COHERENT DETECTOR, WITH USE OF APPROXIMATE PROBABILITY DENSITY FUNCTION

The error rate is given by

$$\langle P_e \rangle \approx \int_0^\infty \frac{1}{2} \mathrm{erfc}(\sqrt{\alpha\gamma}) \frac{1}{(M-1)!} \frac{\gamma^{M-1}}{\gamma_0^M} d\gamma. \tag{A7.11}$$

Integrating by part we have

$$\langle P_e \rangle = \left[\frac{1}{2} \mathrm{erfc}(\sqrt{\alpha\gamma}) \frac{1}{M!} \frac{\gamma^M}{\gamma_0^M} \right]_0^\infty - \int_0^\infty \frac{d}{d\gamma} \left(\frac{1}{2} \mathrm{erfc}(\sqrt{\alpha\gamma}) \right) \frac{1}{M!} \frac{\gamma^M}{\gamma_0^M} d\gamma$$

$$= \frac{1}{2} \sqrt{\frac{\alpha}{\pi}} \int_0^\infty \frac{1}{\sqrt{\gamma}} e^{-\alpha\gamma} \frac{1}{M!} \frac{\gamma^M}{\gamma_0^M} d\gamma. \tag{A7.12}$$

Comparing this with Eqs. (A7.7) and (A7.9), we get

$$\langle P_e \rangle = \frac{1}{2} \frac{(2M-1)!!}{(2M)!!} \frac{1}{(\alpha\gamma_0)^M}. \tag{A7.13}$$

APPENDIX 7.3   ORTHOGONAL FREQUENCY DIVISION MULTIPLEXING

Consider two modulated signals expressed in a complex form,

$$z_1(t) = [a_1 x_1(t) + jb_1 y_1(t)] e^{j\omega_1 t}, \tag{A7.14a}$$

$$z_2(t) = [a_2 x_2(t) + jb_2 y_2(t)] e^{j\omega_2 t}, \tag{A7.14b}$$

where a_i, b_i are information-bearing signals, $x_i(t)$ and $y_i(t)$ are waveforms, and ω_i are carrier frequencies for $i = 1, 2$. All of these are real-valued. Actual signals are given by the real part of $z_i(t)$. The two signals are summed and transmitted through a channel. We will find (orthogonal) conditions so that the two signals can be separately received. Demodulation of the signal includes frequency translation into a baseband signal and filtering. Letting $z(t) = z_1(t) + z_2(t)$, the terms including a_i and b_i are given by

$$\text{Re}\left[\int_{-\infty}^{\infty} z(t) e^{-j\omega_i t} x_i(t)\, dt\right] \quad \text{and} \quad \text{Im}\left[\int_{-\infty}^{\infty} z(t) e^{-j\omega_i t} y_i(t)\, dt\right],$$

respectively, where we assume a correlation receiver instead of filtering (correlation and matched filtering are equivalent; see Section 3.3).

The conditions for any a_i and b_i to be separated at a receiver without interchannel interference are

$$\int_{-\infty}^{\infty} x_1(t) x_2(t) \cos \Delta\omega t\, dt = 0, \tag{A7.15a}$$

$$\int_{-\infty}^{\infty} y_1(t) y_2(t) \cos \Delta\omega t\, dt = 0, \tag{A7.15b}$$

$$\int_{-\infty}^{\infty} x_1(t) y_2(t) \sin \Delta\omega t\, dt = 0, \tag{A7.15c}$$

$$\int_{-\infty}^{\infty} y_1(t) x_2(t) \sin \Delta\omega t\, dt = 0, \tag{A7.15d}$$

where $\Delta\omega = \omega_2 - \omega_1$.

From Eq. (A7.15a) we have

$$X_1(\Delta\omega) * X_2(\Delta\omega) + X_1(-\Delta\omega) * X_2(-\Delta\omega)$$
$$= X_1(\Delta\omega) * X_2(\Delta\omega) + X_1(\Delta\omega) * X_2(\Delta\omega)$$
$$= 2\text{Re}[X_1(\Delta\omega) * X_2(\Delta\omega)] = 0, \tag{A7.16a}$$

where $x_i(t) \leftrightarrow X_i(\omega)$, $*$ denotes the convolutional integral [Eqs. (2.25) or (2.26)], and $X(-\omega) = X_1^*(\omega)$ (Eq. 2.8) for real $x(t)$ is used.

Similarly we have

$$Y_1(\Delta\omega) * Y_2(\Delta\omega) + Y_1(-\Delta\omega) * Y_2(-\Delta\omega)$$
$$= 2\text{Re}[Y_1(\Delta\omega) * Y_2(\Delta\omega)] = 0, \tag{A7.16b}$$

$$X_1(\Delta\omega) * Y_2(\Delta\omega) - X_1^*(\Delta\omega) * Y_2^*(\Delta\omega)$$
$$= 2\text{Im}[X_1(\Delta\omega) * Y_2(\Delta\omega)] = 0, \tag{A7.16c}$$

$$Y_1(\Delta\omega)*X_2(\Delta\omega) - Y_1^*(\Delta\omega)*X_2^*(\Delta\omega)$$

$$= 2\text{Im}[Y_1(\Delta\omega)*X_2(\Delta\omega)] = 0. \tag{A7.16d}$$

We analyze the orthogonal conditions for different cases.

Band-limited System

Without Spectrum Overlap. Assume that signals are band-limited as

$$X_i(\omega) = 0 \quad \text{and} \quad Y_i(\omega) = 0 \quad \text{for } |\omega| \geq \omega_{im} \quad (i = 1, 2).$$

Then Eqs. (A7.16a) to (A7.16d) hold for $\Delta\omega \geq \omega_{1m} + \omega_{2m}$. This is a case where spectra do not overlap between different carrier signals.

With Spectrum Overlap. Consider digital signals with symbol duration of T and band-limitation such that

$$X_i(\omega) = 0, \quad Y_i(\omega) = 0 \quad \text{for } |\omega| \geq 2\pi/T \tag{A7.17}$$

and square root of the Nyquist-I roll-off characteristics. This is the optimum transmission system under white Gaussian noise (Section 3.3). If the roll-off factor is zero (rectangular transfer function), the spectra never overlap with each other for $\Delta\omega = 2\pi/T$, and orthogonal frequency division multiplexing is possible for any QAM system. We will show that for the same channel spacing, $\Delta\omega = 2\pi/T$, introduction of the nonzero roll-off factor allows spectrum overlapping [6].

If we let

$$x_2(t) = x_1\left(t - \frac{T}{2}\right) \tag{A7.18a}$$

and

$$y_2(t) = y_1\left(t - \frac{T}{2}\right) \tag{A7.18b}$$

and assume that $x_i(t)$ and $y_i(t)$ are even function of t, then using notation $t' = t - T/4$ and $\Delta\omega = 2\pi/T$, we can see Eqs. (A7.15a) and (A7.15b) are valid as

$$\int_{-\infty}^{\infty} x_1(t)x_2(t)\cos\Delta\omega t\, dt = -\int_{-\infty}^{\infty} x_1\left(t' + \frac{T}{4}\right)x_1\left(t' - \frac{T}{4}\right)\sin\Delta\omega t'\, dt' = 0 \tag{A7.19a}$$

and

$$\int_{-\infty}^{\infty} y_1(t) y_2(t) \cos \Delta \omega t \, dt = - \int_{-\infty}^{\infty} y_1\left(t' + \frac{T}{4}\right) y_1\left(t' - \frac{T}{4}\right) \sin \Delta \omega t' \, dt' = 0.$$

(A7.19b)

If we let

$$x_1(t) = y_2(t),$$ (A7.20a)

and

$$y_1(t) = x_2(t),$$ (A7.20b)

we can similarly confirm that Eqs. (A7.15c) and (A7.15d) hold.

Thus the two signals are orthogonally frequency division multiplexed. Since $x_i(t)$ and $y_i(t)$ are band-limited within $|\omega| \leq 2\pi/T$, N channels are orthogonally multiplexed as shown in Fig. 7.5. The configuration of the transmitter and receiver is shown in Fig. 7.6. Since $X_i(\omega)$ and $Y_i(\omega)$ have the square root of the Nyquist-I characteristics, the system shows no intersymbol interference as well as no interchannel interference. The assumption that $x_i(t)$ and $y_i(t)$ are even function of t is not valid, since it violates causality. In practice, we can shift the time origin by t_0 as, for example, $x_i(\tau) = x_i(-\tau)$ where $\tau = t' - t_0$ (Section 2.1).

Non-band-limited System

Assume that all $x_i(t)$ and $y_i(t)$ $(i = 1, 2)$ are the same; then if $X_i(\omega) * Y_i(\omega)$ is zero for $\omega = n\omega_0$ $(n = \pm 1, \pm 2, \ldots)$, N channels can be orthogonally frequency division multiplexed with spacing of ω_0, since Eqs. (A7.16a) to (A7.16d) are valid.

Let

$$G(\omega) = X_i(\omega) * X_i(\omega),$$ (A7.21)

and we require that

$$G(n\omega_0) = \begin{cases} 1, & n = 0 \\ 0, & n \neq 0. \end{cases}$$ (A7.22)

Equation (A7.22) can be rewritten as

$$G(n\omega_0) = G(\omega) \sum_{n=-\infty}^{\infty} \delta(\omega - n\omega_0) = \delta(\omega).$$ (A7.23)

Taking the inverse Fourier transform of the above equation, we get

$$g(t) * \frac{1}{\omega_0} \sum_{n=-\infty}^{\infty} \delta(t - nT) = \frac{1}{2\pi}, \tag{A7.24}$$

where

$$g(t) = x_i^2(t) \leftrightarrow G(\omega). \tag{A7.25}$$

Equation (A7.24) becomes

$$\sum_{n=-\infty}^{\infty} x_i^2(t - nT) = \frac{1}{T}. \tag{A7.26}$$

The waveforms $x_i(t)$ do not need to be band-limited. The waveforms that satisfy Eq. (A7.26) are given, for example, by

$$x_i(t) = \begin{cases} 1, & 0 \leq t \leq T \\ 0, & \text{otherwise}, \end{cases}$$

or

$$x_i(t) = \begin{cases} \cos\left(\frac{\pi}{2T}t\right), & -T \leq t \leq T \\ 0, & \text{otherwise}. \end{cases}$$

For further description of orthogonally frequency division multiplexed systems, refer to [81]–[85].

REFERENCES

1. J. Horikoshi, "Error performance considerations of $\pi/2$-TFSK under the multipath interfering environment." *Trans. IECE*, **E67**, 40–46 (January 1984).
2. S. Yoshida, S. Ariyavisitakul, F. Ikegami, and T. Takeuchi, "A novel anti-multipath modulation technique DSK," *Proc. IEEE Globecom*, 36.4.1–36.4.5 (1985).
3. S. Yoshida, S. Ariyavisitakul, F. Ikegami and T. Takeuchi, "A power-efficient linear digital modulation and its application to anti-multipath modulation PSK-RZ scheme," *Proc. IEEE Vehicular Technology Conference, Tampa*, 66–71 (June 1987).
4. S. Yoshida and F. Ikegami, "Anti-multipath modulation technique— Manchester-coded DPSK and its generalization," *IEICE, Technical Report*, **CS86-47**, (1986).

5. H. Takai, "BER performance of anti-multipath modulation PSK-VP and its optimum phase-waveform," *Proc. IEEE Vehicular Technology Conference*, 412–419 (1990).

6. B. R. Saltberg, "Performance of an efficient parallel data transmission system," *IEEE Trans. Commun. Technol.*, **COM-15**, 805–811 (December 1967).

7. B. Hirosaki and H. Aoyagi, "A highly efficient HF modem with adaptive fading control algorithm," *Proc. IEEE Globecom*, 48.3.1–48.3.5 (1984).

8. R. C. Dixon, *Spread Spectrum System*, 2nd ed., Wiley-Interscience, 2nd ed., New York, (1984).

9. M. K. Simon, J. K. Omura, R. A. Scholtz, and B. K. Levitt, *Spread Spectrum Communications*, vols. I, II, III, Computer Science Press, Rockville, Md. (1985)

10. M. Schwartz, W. R. Benett, and S. Stein, *Communication Systems and Techniques*, McGraw-Hill, New York (1966).

11. W. C. Jakes (Ed.), *Microwave Mobile Communications*, Wiley, New York (1974).

12. F. Adachi and K. Oono, "BER performance of QDPSK with postdetection diversity reception in mobile radio channels," *IEEE Trans. Vehicular Technology*, **40**, 237–249 (February 1991).

13. T. Hattori and K. Hirade, "Multitransmitter digital signal transmission by using offset frequency strategy in a land mobile telephone system," *IEEE Trans. Vehicular Technology*, **VT-27**, 231–238 (November 1978).

14. T. Hattori and S. Ogose, "A new modulation scheme for multitransmitter simulcast digital mobile radio communication," *Proc. IEEE Vehicular Technology Conference*, 83–88 (March 1979).

15. F. Adachi, "Transmitter diversity for a digital FM paging system," *IEEE Trans. Vehicular Technology*, **VT-28**, 333–338 (November 1979).

16. A. Afrasteh and D. Chukurov, "Performance of a novel selection technique in an experimental TDMA system for digital portable radio communications," *Proc. IEEE Globecom*, 810–814 (November 1988).

17. Y. Akaiwa, "Antenna selection diversity for framed digital signal transmission in mobile radio channel," *Proc. IEEE Vehicular Technology Conference*, 470–473 (May 1989).

18. J. H. Barnard and C. K. Pauw, "Probability of error for selection diversity as a function of dwell time," *IEEE Trans. Communications*, **COM-37**, 800–803 (August 1989).

19. J. G. Proakis, "Digital Communications," 3rd ed., McGraw-Hill, New York (1983).

20. Y. Sato, *Theory of Linear Equalization, Adaptive Digital Signal Processing*, Maruzen, Tokyo (1990).

21. R. W. Lucky, "Automatic equalization for digital communication," *BSTJ*, **44**, 547–588 (April 1965).

22. Y. Sato, "Blind equalization and blind sequence estimation," *IEICE Trans. Communications*, **E77-B**, 545–556 (May 1994).

23. B. Widrow and S. D. Stearns, *Adaptive Signal Processing*, Prentice-Hall, Englewood Cliffs, N.J. (1985).

24. S. Haykin, *Introduction to Adaptive Filters*, Macmillan, New York (1984).

25. M. Hata and T. Miki, "Performance of MSK high-speed digital transmission in land mobile radio channels," *Proc. Globecom '84*, 518–524 (November 1984).

26. K. Raith, J.-E. Stjernvall, and J. Uddenfeldt, "Multi-path equalization for digital cellular radio operating at 300 kbits/s," *Proc. IEEE Vehicular Technology Conference*, 268–272 (1986).

27. J.-E. Stjernvall, B. Hedberg, and S. Ekemark, "Radio test performance of a narrowband TDMA system," *Proc. IEEE Vehicular Technology Conference*, 293–299 (1987).

28. M. Nakajima and S. Sampei, "Performance of a decision feedback equalizer under frequency selective fading in land mobile communications," *Trans. IEICE*, **72-B-II**, 515–523 (October 1989).

29. K. Honma, M. Uesugi, and K. Tsubaki, Adaptive equalization in TDMA digital mobile radio," *Trans. IEICE*, **J72-B-II**, 587–594 (November 1989).

30. S. Ariyavisitakul, "Performance bounds for a decision feedback equalizer with a time-reversal structure," *Proc. Fourth Nordic Seminar on Digital Mobile Radio Communications*, 10.a (June 1990).

31. A. Higashi and H. Suzuki, "Dual-mode equalization for digital mobile radio," *Trans. IEICE*, **J74-B-II**, 91–100 (March 1991).

32. T. Maseng, "Digitally modulated (DPM) signals," *IEEE Trans. Communications*, **COM-33**, 911–918 (September 1985).

33. R. D'Avella, L. Moreno, and M. Sant'Agostino, "Adaptive equalization in TDMA mobile radio systems," *Proc. IEEE Vehicular Technology Conference*, 385–392 (May 1987).

34. L. Moreno and R. D'Arella, "Maximum likelihood adaptive techniques in the digital mobile radio environment," *Proc. Int. Conf. Digital Land Mobile Radio Communications*, 227–236 (June 1987).

35. J. E. Stjernvall, B. Hedberg, K. Raith, T. Backstrom, and R. Löfdahl, "Radio test performance of a narrowband TDMA system-DMS90," *Proc. Int. Conf. Digital Land Mobile Radio Communications*, 310–318 (June 1987).

36. R. D'Avella, L. Moreno, and M. Sant'Agostino, "An adaptive MLSE receiver for TDMA digital mobile radio," *IEEE, Journal on Selected Areas in Communications*, **7**, 122–129 (January 1989).

37. K. Okanoue, Y. Nagata, and Y. Furuya, "An adaptive MLSE receiver with carrier frequency estimator for TDMA digital mobile radio," *Proc. Fourth Nordic Seminar on Digital Mobile Radio Communications*, 10.2 (June 1990).

38. K. Okanoue, A. Ushirokawa, H. Tomita, and Y. Furuya, "New MLSE receiver free from sample timing and input level controls," *Proc. IEEE Vehicular Technology Conference*, 408–411 (May 1993).

39. G. Larsson, B. Gudmundson, and K. Raith, "Receiver performance for the north American digital cellular system," *Proc. IEEE Vehicular Technology Conference*, 1–6 (May 1991).

40. T. Kohama, H. Kondoh, and Y. Akaiwa, "An adaptive equalizer for frequency-selective mobile radio channels with noncoherent demodulation," *Proc. IEEE Vehicular Technology Conference*, 770–775 (May 1990).

41. D. Bertsekas and R. Gallager, *Data Networks*, 2nd ed., Prentice-Hall, Englewood Cliffs, N.J. (1992).

42. Y. Yamao and Y. Nagao, "Predictive antenna selection diversity (PASD) for TDMA mobile radio," *Trans. IEICE*, **E77-B**, 641–646 (May 1994).

43. W. Wesley Peterson and E. J. Weldon, Jr., *Error-Correcting Codes*, 2nd ed., MIT Press, Cambridge, Mass. (1972).

44. S. Lin and D. J. Costello, Jr., *Error Control Coding*, Prentice-Hall, Englewood Cliffs, N.J. (1983).

45. B. P. Lathi, *Modern Digital and Analog Communication Systems*, Holt, Rinehart and Winston, New York (1983).

46. T. Sato, M. Kawabe, T. Kato, and A. Fukusawa, "Composition of robust error control scheme using adaptive coding and its effect on data communication," *Trans. IEICE*, **J72-B-I**, 438–445 (May 1989).

47. G. Ungerboeck, "Trellis-coded modulation with redundant signal sets, Part I: Introduction," *IEEE Communications Magazine*, **25**, 5–11 (February 1987).

48. G. Ungerboeck, "Trellis-coded modulation with redundant signal sets, Part II: State of the art," *IEEE Communications Magazine*, **25**, 12–21 (February 1987).

49. D. Divsalar and M. K. Simon, "The design of trellis coded MPSK for fading channels: Performance criteria," *IEEE Trans. Communications*, **36**, 1004–1012 (September 1988).

50. D. Divsalar and M. K. Simon, "Set partitioning for optimum code design," *IEEE Trans. Communications*, **36**, 1013–1021 (September 1988).

51. S. Sampei and Y. Kamio, "Performance of trellis coded 16QAM/TDMA system for land mobile communications," *Trans. IEICE*, **J73-B-II**, 630–638 (November 1990).

52. S. Tomisato and H. Suzuki, "Envelope controlled digital modulation improving power efficiency of transmitter amplification—Application to trellis-coded 8PSK for mobile radio," *Trans. IEICE*, **J75-B-II**, 918–928 (December 1992).

53. R. T. Compton, Jr., *Adaptive Antennas—Concept and Performance*, Prentice-Hall, Englewood Cliffs, N.J. (1988).

54. H. Yoshino, K. Fukawa, and H. Suzuki, "Adaptive interference canceller based upon RLS-MLSE," *Trans. IEICE*, **J77-B-II**, 74–84 (February 1994).

55. K. Fukawa and H. Suzuki, "Recursive least squares adaptive algorithm maximum likelihood sequence estimation: RLS-MLSE-an application of maximum likelihood estimation theory to mobile radio," *Trans. IEICE*, **J76-B-II**, 202–214 (April 1993).

56. H. Murata, S. Yoshida, and T. Takeuchi, "Trellis-coded co-channel interference canceller for mobile communication," *Technical Report of IEICE*, **RCS93-75**, 39–46 (November 1993).

57. K. Hamaguchi and H. Sasaoka, "Block coded FH-16QAM/TDMA system," *Technical Report of IEICE*, **RCS93-90**, 33–39 (February 1994).

58. R. Kohno, H. Imai, M. Hatori, and S. Pasupathy, "Combination of an adaptive array antenna and a canceller of interference for direct-sequence spread-spectrum multiple-access system," *IEEE Journal on Selected Areas in Communications*, **8**, 675–682 (May 1990).

59. R. Kohno, H. Imai, M. Hatori, and S. Pasupathy, "An adaptive canceller of cochannel interference for spread-spectrum multiple-access communication networks in a power line," *IEEE Journal on Selected Areas in Communications*, **8**, 691–699 (May 1990).

60. Z. Xie, R. T. Short, and C. T. Rushforth, "A family of suboptimum detectors for coherent multiuser communications," *IEEE Journal on Selected Areas in Communications*, **8**, 683–690 (May 1990).

61. M. K. Varanasi and B. Aazhang, "Multistage detection in asynchronous code-division multiple-access communications," *IEEE Trans. Communications*, **COM-38**, 509–519 (April 1990).

62. Y. C. Yoon, R. Kohno, and H. Imai, "A spread-spectrum multiaccess system with cochannel interference cancellation for multipath fading channels," *IEEE Journal on Selected Areas in Communications*, **11**, 1067–1075 (September 1993).

63. M. Abdulrahman, D. D. Falconer, and A. U. H. Sheikh, "Equalization for interference cancellation in spread spectrum multiple access systems," *Proc. IEEE Vehicular Technology Conference*, 71–74 (May 1992).

64. U. Madhow and M. Honig, "Minimum mean squared error interference suppression for direct-sequence spread-spectrum code division multiple-access," *Proc. First International Conference on Universal Personal Communication*, 273–277 (September 1992).

65. S. Yoshida, A. Ushirokawa, S. Yanagi, and Y. Furuya, "DS/CDMA adaptive interference canceller in mobile radio environments," *Technical Report of IEICE*, **RCS93-76**, 47–54 (November 1993).

66. K. Fukawa and H. Suzuki, "A reception scheme utilizing matched filter with interference canncelling—optimum detection for DS-CDMA mobile communication," *Proc. Spring Conference of IEICE*, 1.467–1.468 (March 1994).

67. K. Ozawa and T. Miyano, *Efficient Voice Codings for Digital Mobile Communication*, Triceps, Tokyo (1992).

68. B. P. Lathi, *Modern Digital and Analog Communication Systems*, Holt, Rinehart and Winston, New York (1983).

69. J. L. Flanagan, *Speech Analysis and Perception*, 2nd ed., Springer-Verlag, New York (1972).

70. N. S. Jayant and Peter Noll, *Digital Coding of Waveforms: Principles and Applications to Speech and Video*, Prentice-Hall, Englewood Cliffs, N.J. (1984).

71. P. Vary et al., "Speech codes for the European mobile ratio system," *Proc. ICASSP*, 227–230 (1988).

72. I. A. Gerson and M. A. Jasiuk, "Vector sum excited linear prediction (VSELP) speech coding at 8 kbps," *IEEE, Proc. ICASSP*, 461–464 (1990).

73. I. A. Gerson, M. A. Jasiuk, M. J. McLaughlin, and E. H. Winter, "Combined speech and channel coding at 11.2 kbps," *Signal Processing V: Theories and Applications*, L. Torres, E. Masgrau and M. A. Lagnus (eds.), 1339–1342, Elsevier Science Publishers B.V., Amsterdam (1990).

74. T. Ohya, H. Suda, and T. Miki, "Pitch synchronous innovation CELP (PSI-CELP)—PDC half-rate speech CODEC," *IEICE Technical Report*, **RCS 93-78**,

63–70 (November 1993). Also R. V. Cox, "Speech coding standards," in *Speech Coding and Synthesis*, W. B Kleijn and K. K. Paliwal (eds.), 49–78, Elsevier Science B.V., Amsterdam (1995).

75. O. Nakamura, A. Dobashi, K. Seki, S. Kubota, and S. Kato, "Improved ADPCM voice transmission for TDMA-TDD systems," *Proc. Vehicular Technology Conference*, 301–304 (May 1993).

76. S. Kubota, A. Dobashi, T. Hasumi, M. Suzuki, and S. Kato, "Improved ADPCM voice transmission employing click noise detection scheme for TDMA-TDD systems," *Proc. Fourth Int. Conf. Personal, Indoor and Mobile Radio Communications*, 613–617 (September 1993).

77. I. Matsumoto, S. Sasaki, M. Horaguchi, and K. Urabe, "Enhancement of speech coding for digital cordless telephone systems," *Proc. Fourth Int. Conf. Personal, Indoor and Mobile Radio Communications*, 618–621 (September 1993).

78. H. Suda and T. Miki, "An error protected 16 kbits/s voice transmission for land mobile radio," *IEEE Journal on Selected Areas in Communications*, **SAC-6**, 346–352 (1988).

79. GSM Recommendation 06.32, *Voice activity detection*.

80. GSM Recommendation 06.12, *Comfort noise aspects for full rate speech traffic channels*.

81. R. W. Chang, "Synthesis of band-limited orthogonal signals for multichannel data transmission," *Bell System Tech. J.*, **45**, 1775–1796 (December 1966).

82. D. A. Shnidman, "A generalized Nyquist criterion and an optimum linear receiver for a pulse modulation system," *Bell System Tech. J.*, **46**, 2163–2177 (November 1967).

83. B. Hirosaki, "An orthogonally multiplexed QAM system using the discrete Fourier Transform," *IEEE Trans. Communications*, **COM-29**, 982–989 (July 1981).

84. L. J. Cimini, Jr., "Analysis and simulation of a digital mobile channel using orthogonal frequency division multiplexing," *IEEE Trans. Communications*, **COM-33**, 665–675 (July 1985).

85. M. Alard and R. Lassalle, "Principles of modulation and channel coding for digital broadcasting for mobile receivers," *EBU Review—Technical*, No. 224, 168–190 (August 1987).

86. J. H. Winter, "Optimum combining in digital mobile radio with cochannel interference," *IEEE Journal on Selected Areas in Communication*, **SAC-2**, 538–539 (July 1984).

87. H. Suzuki, "Signal transmission characteristics of diversity reception with least-squares combining—Relationship between desired signal combining and interference cancelling," *Trans. IEICE*, **J75-B-II**, 524–534 (August 1992).

8

EQUIPMENT AND CIRCUITS FOR DIGITAL MOBILE RADIO

This chapter describes equipment and circuit aspects of digital mobile radio systems. The basic structures of the base station and mobile station equipment and some detailed circuit implementations are shown. Some methods and circuits shown here as well as in other parts of this book might be patented. With the evolution of digital signal processors, some parts of mobile radio systems have been implemented using them. This topic is not covered in this book. Reader may refer to reference [25].

8.1 BASE STATION

Figure 8.1 shows a simplified block diagram of a base station for the Japanese digital cellular system in the NTT Mobile Communications Network Inc. [1]. A base station covers three-sectorized zones. Two-branch diversity reception is used. The antenna gains are 17 dBi and 14 dBi at 800 MHz for antenna aperture lengths of 5.4 m and 2.7 m, respectively. A transmit and receive duplexer is connected to one antenna, from which signals are transmitted.

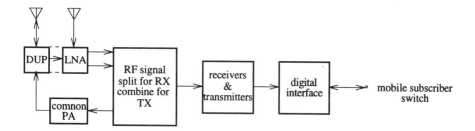

Figure 8.1. Block diagram of base station for a digital cellular system.

In order to achieve a low noise figure, an outdoor common receive low-noise amplifier is used, whose noise figure is less than 3 dB and gain 40 dB. The beam can be tilted by 0–5 or 3–11 degrees in the vertical plane by phase shifting the feeder circuits for the vertical array antenna.

The transmit common amplifier can handle a total power of 32 W, accommodating 16 RF signals ($\pi/4$ shifted QPSK modulation) and keeping the intermodulation at -60 dB. This performance was achieved by the feed-forward nonlinearity compensation technique, SAFF, described in Section 6.3.4. This common amplifier contributed to a drastic reduction in the size and cost of the base station transmitter: the bulky conventional signal combiner, which is made of many waveguide filters, is replaced with a very small microstripline combiner. This becomes possible because, due to the common amplifier, the signals are combined at a low level, making them free of transmitter intermodulation. The importance of compactness and low weight of the base station equipment cannot be overstated because of its flexibility for installation at different sites such as rooms in commercial buildings. In addition, channel assignment, such as dynamic channel assignment, becomes flexible since it is free from the restrictions of the conventional channel combining system.

The transmitter/receiver parts of the block diagram deal with modulation/demodulation, TDMA frame composition/decomposition, and so on, with output/input of RF signals and input/output of coded speech signals. The interface circuit covers multiplexed transmission of coded speech signals and other signals for control.

8.2 MOBILE STATION

A block diagram of a hand-held mobile station for the 800 MHz Japanese digital cellular system [2,3] is shown in Fig. 8.2. Antenna selection diversity is used. One antenna is a whip antenna used for transmission and reception via a duplexer. The other is an inverse F-type antenna embedded on the housing and is used for only reception. Antenna selection is controlled by the output of the demodulator. A 3-slot TDMA signal is pulse shaped, modulated with $\pi/4$ shifted QPSK, and amplified to a maximum output power of 0.8 W. The output power is automatically controlled. The envelope of the modulating signal is fed to the power amplifier for nonlinearity compensation purposes. The 11.2 kbps VSELP voice coder/decoder (Section 7.9.6) is implemented with a one-chip digital signal processor. VOX (voice-operated transmission) and intermittent reception are adopted, decreasing the power consumption. The battery life on each charging is about 30 hours in the waiting mode and 60 minutes in the speaking mode at maximum power without the VOX.

The receive circuit (RX) includes a RF low-noise amplifier, frequency downconverter, a channel selection band-pass filter, and an IF amplifier. A

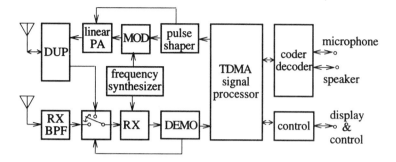

Figure 8.2. Block diagram of a mobile station for a digital cellular system.

carrier frequency is generated using a frequency synthesizer. Fast switching of less than 2 ms is achieved with a digital loop reset technique.

The two-receiver post-demodulation diversity system is adopted for the car-mounted and the shoulder-type mobile stations.

8.3 SUPERHETERODYNE AND DIRECT CONVERSION RECEIVERS

A block diagram of the superheterodyne receiver is shown in Fig. 8.3. The RF received signal is amplified and frequency downconverted into an IF signal. The gain of the low-noise amplifier should not be high so as to prevent intermodulation between the desired signal and others. The band-pass filter is used for channel selection. A steep roll-off transfer function is desired for the channel selection capability. An example of the square root of the Nyquist-I transfer function of the channel selection filter for the American TDMA digital cellular system is shown in Fig. 8.4. The delay as well as amplitude characteristics are taken into consideration for low distortion of the modulated signal waveform. A high gain is implemented for the IF amplifier since a channel is already selected and therefore no intermodulation is expected. The output of the IF amplifier is fed to an amplitude limiter or

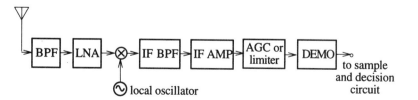

Figure 8.3. Block diagram of a superheterodyne receiver.

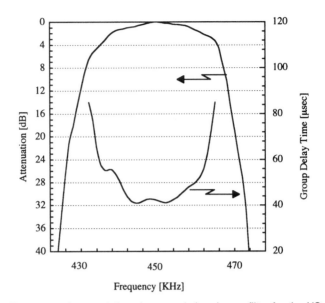

Figure 8.4. Frequency characteristics of a ceramic band-pass filter for the US digital cellular system. Courtesy of Murata Manufacturing Co. Ltd.

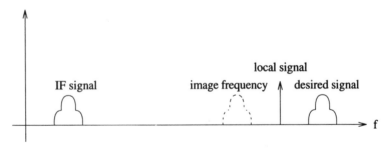

Figure 8.5. Image frequency for a heterodyne receiver.

an automatic gain control circuit followed by a demodulator. A further frequency conversion is sometimes made (double conversion system).

Advantages of the heterodyne receiver are high channel selectivity owing to moderate signal bandwidth relative to the IF frequency, and stable, high-gain amplification owing to the frequency difference between the input RF signal and the IF signal.

The disadvantage of the heterodyne receiver is the image frequency problem: undesired signal that is located at a frequency in mirror image relation to the desired signal with respect to the local frequency (Fig. 8.5), may be downconverted to the same IF frequency and not discriminated with

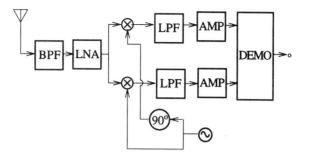

Figure 8.6. Block diagram of a direct conversion receiver.

the channel selection filter. This is known as the image frequency problem. To avoid this problem, a band-pass filter that attenuates the image frequency signal is employed before the frequency downconversion.

The direct conversion, or homodyne, receiver block diagram is shown in Fig. 8.6. The RF signal is downconverted into quadrature baseband signals using a pair of downconverters with a local signal whose frequency is the same as the carrier signal. The channel selection is made with the low-pass filters. Since the low-pass filters and the following circuits treat baseband signal, they are suited for implementation with integrated circuits. Moreover, we have no image frequency problem. However, we have new problems: (i) the $1/f$ noise appears in baseband frequency; (ii) dc-coupled circuits are required for baseband signals; and (iii) leaked local signals easily interfere with other receivers located close by.

The principle of demodulation for the direct conversion signal is based on the zero-IF complex or quadrature representation of modulated signals. Consider an FM signal, for example; the RF signal is represented as

$$f(t) = A_0 \cos[\omega_c t + \theta(t)]$$
$$= \text{Re}[A_0 e^{j\theta(t)} e^{j\omega_c t}],$$

where ω_c is the carrier frequency and $\theta(t) = k\int_{-\infty}^{t} s(t)\,dt$, where $s(t)$ is the transmit baseband signal, and k is the modulation constant. The demodulated quadrature signals become $x(t) = A_0 \cos\theta(t)$ and $y(t) = A_0 \sin\theta(t)$. The demodulator works on $x(t)$ and $y(t)$ as

$$d(t) = \frac{d}{dt}\theta(t) = \frac{d}{dt}\tan^{-1}\frac{y(t)}{x(t)} = \left(x(t)\frac{dy(t)}{dt} - y(t)\frac{x(t)}{dt}\right)\Big/ A_0^2 = ks(t) \quad (8.1)$$

8.4 TRANSMIT AND RECEIVE DUPLEXING

An antenna is used for signal transmission and reception via a duplexer. Figure 8.7 shows a block diagram of a duplexer. The center frequencies of

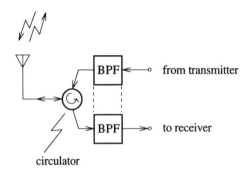

Figure 8.7. Antenna and duplexer.

the band-pass filters are chosen at the transmit and receive frequency bands, respectively. The circulator is a nonreciprocal circuit where the input signal is fed out to the next port in a rotational manner. The circulator is sometimes removed, connecting the band-pass filters in parallel to the antenna.

The same frequency is sometimes used for transmission and reception, that is, a half-duplex (press-to-talk) system and a time division duplex (TDD) system. Transmission and reception are controlled by pressing or releasing a control button in the press-to-talk system. Time is divided in turn for transmission and reception in the time division duplex or "ping-pong" system. In these systems, the duplexer is replaced with a switch, reducing the size and cost of the transmitter and receiver, since signal transmission and reception never take place at the same time.

8.5 FREQUENCY SYNTHESIZER [4]

For a system in which a number of frequency channels can be selected, a frequency synthesizer is indispensable. A block diagram of a frequency synthesizer is shown in Fig. 8.8. This is a PLL (phase-locked loop) circuit, where the VCO (voltage controlled oscillator) output signal is frequency divided and phase locked to a stable reference signal. The VCO output signal frequency f_0 is determined as

$$f_0 = Nf_r, \tag{8.2}$$

where N is the total number of frequency divisions and f_r is the reference signal frequency. If $f_r = 12.5\,\text{Hz}$, f_0 can be controlled by a step of 12.5 kHz; for example, from 800 MHz to 825 MHz by letting $N = 64\,000$ to $N = 66\,000$. The reference signal is generated by a stable quartz oscillator followed by a counter. Using temperature-dependent compensation technique, frequency stability of 1.5 ppm (part per million) is achieved. Frequency division is

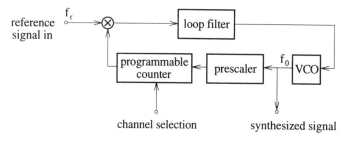

Figure 8.8. Frequency synthesizer.

implemented with the so-called prescaler and the programmable counter. The former is a fast counter with a fixed counting up number and the latter is a frequency-dividing programmable counter where the counting up number is controlled from the channel selection digital signal. The prescaler is used to cope with high frequencies at low power consumption. A GaAs circuit is suitable for this device.

Compactness, low power consumption, low noise, frequency stability, and fast channel switching are major technical issues for frequency synthesizers in mobile radio equipment. The compactness is governed by the VCO resonator and other electronic circuits. A high-Q resonator and a low-noise transistor are required for low-noise performance. Improvements in transistor technology, large-scale integrated circuits, and high-dielectric substrates for the resonator yield compactness and low-noise performance.

The prescaler is the main power-consuming circuit, since it operates at the highest frequency; the other circuits work at low frequencies, where low power-consuming CMOS (complementary metal oxide semiconductor) transistors can be used. With the improvements in semiconductor technology, the power consumption of the 800 MHz prescaler is reduced to 10 mW for GaAs circuits and 20 mW for more economical Si-bipolar transistor circuits. Intermittent operation of a synthesizer lowers the power consumption. To make this technique effective, the rise-up time must be short. For this purpose, a method termed SPILL (state preserving intermittently locked loop) is proposed, where the states of the programmable counter and the prescaler are stored just before intermittent power-down periods and are used for the initial states at the time of restarting operation.

The stability of the frequency synthesizer is determined by the stability of the reference signal. A high frequency stability is achieved at the base station, since the requirements for size, cost, and power consumption are less stringent than those at the mobile station. One technique is to adjust the frequency of the reference signal with reference to the received signal frequency transmitted from the base station. Using this kind of technique, a stability of less than 0.3 ppm is achieved.

Fast switching from one frequency to another is required at the moment of hand-off in a cellular system. This is more important for a TDMA system since a mobile station must switch its frequencies to monitor other channels (mobile-assisted hand-off; Section 9.1.2) in short time slots and come back again to the dedicated frequency and time slot. For this purpose, two synthesizers are prepared in a primitive method. A single synthesizer with a fast switching time of less than 1 ms is proposed. In this method, fast and slow counters are employed as phase detectors (comparators) and the expected dc voltage is added to the control voltage of the VCO. The fast and slow counters cover fine and coarse phase error detection, respectively.

8.6 TRANSMITTER CIRCUITS

8.6.1 Digital Signal Waveform Generator

In general, digital signal waveforms are generated by applying impulses to a transmit filter. In addition to this, a table look-up method can be used as shown in Fig. 8.9. The waveforms are calculated for the data sequences in advance and stored in ROM (read-only memory) and read out corresponding to the address (data sequence) entered into the shift register. The contents of memory are

$$s(n\,\Delta T) = \sum_{m=-M}^{M} a_m h(n\,\Delta T - mT)w(n\,\Delta T - mT), \qquad n = 0, 1, 2, \ldots, N-1,$$
$$(8.3)$$

where a_m are digital data, T is the symbol duration, ΔT is the sampling interval, $N = T/\Delta T$, $h(t)$ is the impulse response, which is truncated within

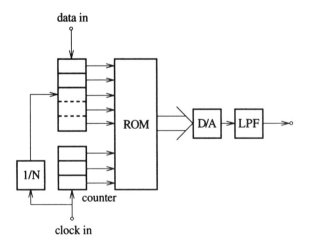

Figure 8.9. Digital signal waveform generator.

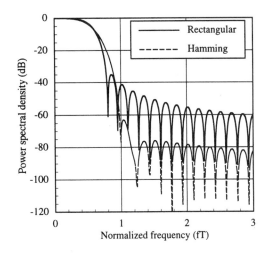

Figure 8.10. The effect of truncation and window on spectra $M = 3$.

$(2M + 1)\,T$, and $w(t)$ is the window function. For a 2-level polar signal, a_n is $+1$ or -1. For example, for a square root of the Nyquist-I pulse waveform with roll-off factor α and symbol frequency f_s, we have

$$h(t) = \frac{f_s}{\pi f_s t} \sin[(1 - \alpha)\,\pi f_s t] + \frac{f_s}{(\pi/4\alpha)^2 - (\pi f_s t)^2}$$
$$\times \left(\frac{\pi}{4\alpha} \cos[(1 + \alpha)\,\pi f_s t] + \pi f_s t \sin[(1 - \alpha)\,\pi f_s t] \right). \qquad (8.4)$$

For a rectangular window,

$$w(t) = \begin{cases} 1, & -MT \le t \le MT \\ 0, & \text{otherwise.} \end{cases} \qquad (8.5a)$$

For the generalized Hamming window,

$$w(t) = \begin{cases} \beta + (1 - \beta) \cos\left(\dfrac{\pi}{2}\dfrac{t}{MT} \right), & -MT \le t \le MT, \beta = 0 \sim 1 \\ 0, & \text{otherwise.} \end{cases} \qquad (8.5b)$$

In the above system, the required number of addresses for ROM is $N2^{M+1}$.

The effect of truncating the impulse response and the window function ($\beta = 0.56$) is shown in Fig. 8.10. The out-of band radiation is decreased by the introduction of the window function.

8.6.2 Modulator

8.6.2.1 FSK Modulator. A VCO (voltage controlled oscillator) is a typical frequency modulator. For a frequency modulator, stability of center frequency and modulation index is important. Figure 8.11 shows a frequency modulator with center frequency stabilization. The center frequency is locked to a reference signal. To reduce the effect of frequency modulation on the center frequency stabilization, the VCO output signal is frequency-divided, resulting in a very small modulation index. This circuit is appropriate for modulating signals without dc components such as analog voice signals and Manchester-coded signals.

Figure 8.12 shows an FSK modulator [6], with stabilization of both center frequency and modulation index using the properties of the family of FSK signals (i.e., MSK, tamed FM, and GMSK): the modulation index is 0.5 for these signals. As discussed in Section 6.2.1, discrete frequency components

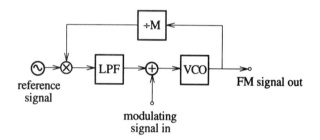

Figure 8.11. Center frequency stabilized FM modulator.

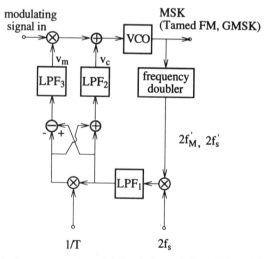

Figure 8.12. Center frequency and modulation index stabilized FM modulator for FSK signals with modulation index of 0.5.

of $2f'_M = 2(f'_c - \Delta f')$ and $2f'_s = 2(f'_c + \Delta f')$ are obtained by passing the signal into a frequency doubler, where f'_c is the center frequency and $\Delta f'$ is the frequency deviation, which should be $\Delta f' = 1/4T$ ($1/T = f_b$: bit rate frequency). f'_c and $\Delta f'$ are assumed to be deviated slightly from the desired frequencies f_c and $1/4T$. In Fig. 8.12, LPF$_1$ rejects the higher-order carrier frequencies and LPF$_2$ and LPF$_3$ pass the dc component. When the loops are locked, we can see that the center frequency control signal v_c is produced so that $f'_c + f_b/2 = f'_s \rightarrow f_s$ and the gain or (modulation index) control signal v_m is produced so that $\Delta f' \rightarrow 1/4T$.

An FM modulator with quadrature modulation is shown in Fig. 8.13. The principle of this modulator is described by the mathematical expression

$$s(t) = \cos[\omega_c t + \varphi(t)]$$
$$= \cos \varphi(t) \cos \omega_c t - \sin \varphi(t) \sin \omega_c t, \tag{8.6}$$

where ω_c is the carrier frequency and $\varphi(t)$ is the phase. The phase $\varphi(t)$ is given by

$$\varphi(t) = k \sum_{m=-M}^{M} a_m g(t - mT), \tag{8.7}$$

where k is the modulation constant and $g(t)$ is the phase impulse response,

$$g(t) = \int_{-\infty}^{t} h(x) \, dx,$$

and $h(t)$ is the frequency impulse response.

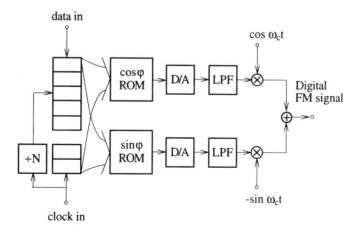

Figure 8.13. Digital FM signal generator with quadrature modulation.

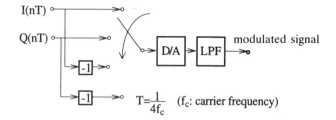

Figure 8.14. Digital signal processing quadrature modulator.

The center frequency and modulation index are kept stable with this circuit. However, balancing the quadrature modulation and spurious carrier frequency component becomes a problem in this system.

The above problems can be avoided by introducing digital signal processing to the modulator at an IF frequency. However, power-consuming digital multipliers and adders are required if the analog circuits are replaced with a digital signal processor. A simple method is proposed in which signal processing effort is minimum, shown in Fig. 8.14 [7].

A sampled quadrature-modulated signal is expressed as

$$s(nT) = I(nT)\cos(\omega_c nT) + Q(nT)\sin(\omega_c nT), \qquad n = 0, 1, 2, \ldots$$

If we choose $T = \pi/2\omega_c = 1/4f_c$ $(f_c = \omega_c/2\pi)$, we have

$$s(4nT) = I(4nT),$$
$$s[(4n + 1)T] = Q[(4n + 1)T],$$
$$s[(4n + 2)T] = -I[(4n + 2)T],$$

and

$$s[(4n + 3)T] = -Q[(4n + 3)T].$$

No multipliers or adders are used in this system.

8.6.3 Transmit Power Control

Transmit power control reduces the interference with other signals by preventing signal transmission at unnecessary power levels. The difficulty of circuit implementation for the transmit power control differs between the types of transmit power amplifiers employed: highly saturated (e.g., the class-C power) amplifiers or (quasi-)linear amplifiers. The input–output relations of the class C and the class AB amplifiers are shown in Fig. 8.15. We can see for the class C power amplifier that it is hard to manage the output

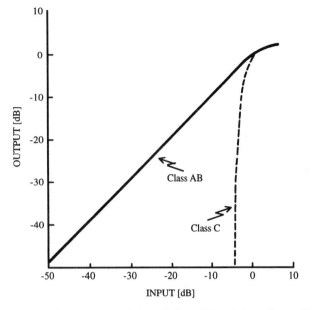

Figure 8.15. Input–output relation of class AB and class C amplifiers.

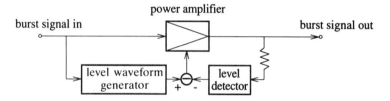

Figure 8.16. Output signal level waveform controlled amplifier.

power levels, since a very small change in the input signal level corresponds to a big change in the output power; we must introduce a feedback control technique. On the other hand, for a quasi-linear power amplifier we can merely use a variable attenuator at the input of the power amplifier (feed-forward control).

Transmit power control is necessary for burst signal transmission in a TDMA system. The output signal level should be gradually ramped up and down at the beginning and ending of the burst signal to prevent the spectrum from spreading at these times. Figure 8.16 shows a power amplifier with output signal waveform control for the above purpose. The class C amplifier is assumed. Such feedback control is unnecessary for a linear power amplifier. Hence, to avoid the feedback control circuit, a (quasi-)linear power amplifier is often adopted in TDMA systems, even when constant envelope modulation is used: the use of a linear power amplifier diminishes the advantage of constant envelope modulation described in Section 5.2.

The transmit power control should be implemented automatically, corresponding to fading channels. Feed-forward and feedback methods that achieve this result are known. The former is used for a system where the up-link and down-link are correlated such as in the time division duplex system using one frequency. The latter is used for a two-frequency duplex system where no correlation is expected between the down-link and up-link. In the feedback control method, received signal information at the receiver is reported to the transmitter.

In a fast-fading channel, the tracking speed of the automatic power control system is important. Prediction of signal level using past data is effective for this purpose. In predicting a signal level, prediction based on signal amplitude is not effective, since the amplitude is subjected to an acute notchlike fading characteristic. Instead, prediction based on in-phase and quadrature components of the received signal is more effective [8], since the signal components will never have such fading notches.

Different criteria for power control are known: one is to keep the received signal level constant, and the other is to keep signal to interference power ratio at a given value. For the latter, a distributed controlled system is analyzed in [23]. An automatic power control system with the latter criterion is proposed in [24] to make the system stable.

8.7 RECEIVER CIRCUITS

8.7.1 AGC Circuit

An AGC (automatic gain control) circuit is needed to cope with the wide dynamic range of the received signal in a mobile radio channel. The constant envelope modulation signal can be received with an amplitude limiter instead of an AGC circuit. Even with this kind of modulation, AGC is required if a channel equalizer is used since channel equalization is difficult if the distorted signal is amplitude limited.

A typical AGC circuit is shown in Fig. 8.17. A pilot signal may be used

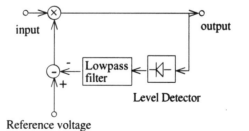

Figure 8.17. Feedback controlled automatic gain control circuit.

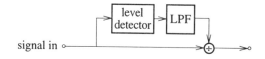

Figure 8.18. Feed-forward automatic gain control.

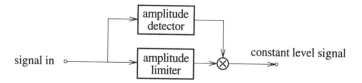

Figure 8.19. Feed-forward automatic gain control with amplitude limiter.

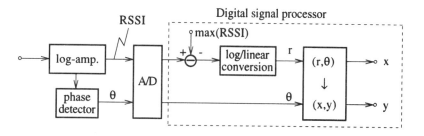

Figure 8.20. Digital signal processing receiver coping with wide dynamic range signal [9]. (Copyright © IEICE 1994.)

to allow level detection to be free from interference from signal level perturbations by modulation. The output signal level is feedback-controlled to a constant value. The bandwidth of the low-pass filter must be narrow so as not to track the modulated signal amplitude. In this system, tracking speed versus stability of the operation becomes an issue.

A feed-forward AGC is shown in Fig. 8.18. The average level is detected and the inverse of it is multiplied with the input signal to give a constant-level signal preserving the amplitude waveform of the modulated signal. Another version of feed-forward AGC is shown in Fig. 8.19. The received signal is amplitude limited and amplitude detected in parallel. The amplitude detector deals only with the signal amplitude due to modulation. The constant-level signal is restored at the output of the multiplier.

The circuit shown in Fig. 8.20 [9] is not the AGC circuit; however, it resolves the wide dynamic range problem. The input signal is applied to a log-amplifier, whose typical implementation is shown in Fig. 8.21. The

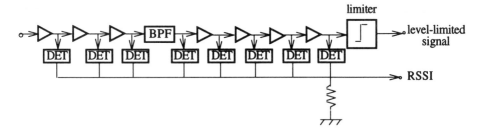

Figure 8.21. Log-amplifier with RSSI (received signal strength indicator) [9]. (Copyright © IEICE 1994.)

log-amplifier outputs the logarithm of the input signal level or RSSI (received signal strength indicator) and amplitude-limited signal. The amplitude-limited signal is fed to a phase detector. The detected phase θ and the RSSI signal are applied to the analog-to-digital converter. Since the signal level is compressed by the log-amplifier, the dynamic range of the analog-to-digital converter never comes in question. The maximum value of sampled RSSI signal in a block of the samples is found and is subtracted from the samples to normalize the received signal level. After this, the RSSI signal is log-to-linear converted. The amplitude and phase are translated into quadrature components for further signal processing such as demodulation.

8.7.2 Signal Processing with Logic Circuits

In this system, baseband pulse signals and angle-modulated signals are processed with logic circuits after conversion to 2-level logic signals.

A delay line is realized with a shift register as shown in Fig. 8.22. The time delay is accurate since it is controlled with a clock frequency and the number of shift register stages. The size of the shift register becomes large when the signal frequency is high or the required time delay is long.

Figure 8.23 shows a circuit that takes the moving average of the input signal. The output is obtained as an integer and is suitable for treatment by a microprocessor thereafter. The moving average is equivalent to a low-pass filter with transfer function $H(\omega) = \sin(\omega T)/\omega T$, where T is the duration of

Figure 8.22. Delay line consisting of shift register.

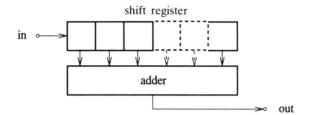

Figure 8.23. A moving-average circuit.

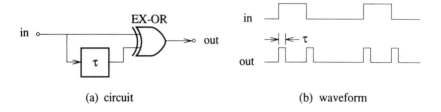

(a) circuit   (b) waveform

Figure 8.24. Time-differentiating and rectifying circuit.

the average, and is equivalent to integrate-and-dump filtering at the sampling instant (Section 2.1.7).

A time-differentiating and rectifying circuit is shown in Fig. 8.24(a) with schematic explanation of its operation (Fig. 8.24(b)). This circuit is useful in a clock recovery system.

A phase detector or comparator is shown in Fig. 8.25 together with the exclusive-OR output waveform. The moving-average circuit in Fig. 8.23 may be used for the low-pass filter (LPF). Another method for the LPF is a counter that counts the number of fast pulses during the high-level state of the EX-OR output signal for a period of the local frequency. The counted number is proportional to the phase difference between the signals. The phase detection characteristic curve is linear. This phase detector covers a phase difference of 0 to π. This circuit works as a frequency downconverter when the LPF cut-off frequency is properly selected. The quadrature phase detector is shown in Fig. 8.26. This circuits covers a phase difference of $-\pi$ to π.

A phase detector that detects only 2-state phase differences, that is, advance or delay, is shown in Fig. 8.27. This is a D-type flip-flop circuit. We need no low-pass filter for this phase detector.

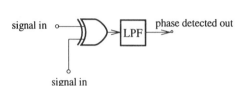

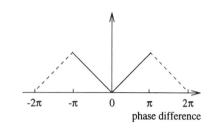

(a) circuit

(b) phase detection characteristic

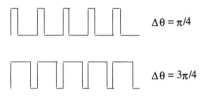

$\Delta\theta = \pi/4$

$\Delta\theta = 3\pi/4$

(c) EX-OR circuit output signal waveform.

Figure 8.25. Phase detector.

8.7.3 Demodulator

A coherent demodulator for MSK, GMSK, and tamed FM is shown in Fig. 8.28 [26]. It is assumed that the input signal is band-pass filtered and converted into a logic signal. Most of the circuits consist of logic circuits. Operation of this circuit may be understood by referring to previously described circuit elements and comparing this circuit with that in Fig. 6.11.

Coherent demodulators under fast-fading conditions have poorer floor error rate performance than noncoherent demodulators, for example, differential demodulators, although they give better performance under static or slow-fading conditions. A coherent demodulator which improves performance under fast fading is shown in Fig. 8.29 [10]. The demodulator works with dual modes of operation: when the fading frequency is slow and the RF signal level is low, it selects the conventional Costas loop carrier recovery system; otherwise, it selects the adaptive carrier tracking mode. In the adaptive carrier tracking mode, the carrier phase deviation due to fast fading is corrected if the deviation becomes larger than a given threshold value. It is shown that the adaptive carrier tracking mode of operation is equivalent to differential detection. It therefore shows improved floor error rate performance under fast-fading conditions.

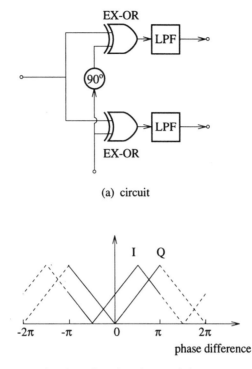

(a) circuit

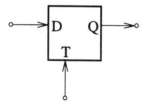

(b) phase detection characteristics

Figure 8.26. Quadrature phase detector.

Figure 8.27. Two-state phase detector.

A differential demodulator for QPSK or $\pi/4$ shifted QPSK can be constructed with a delay line and phase detectors at IF band. Instead of using the delay line, there is another method of detecting instantaneous phase relative to the asynchronized local carrier signal phase. This method is shown in Fig. 8.30. The logic circuit implemented demodulator is shown in Fig. 8.31. The quadrature demodulator consists of exclusive-OR circuits and counters. From the phase detection characteristics in Fig. 8.26, the instantaneous phase

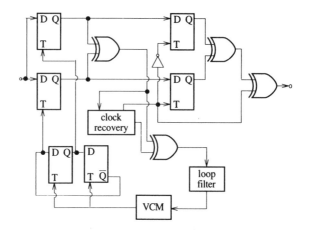

Figure 8.28. A coherent demodulator for MSK, GMSK, and tamed FM [26]. (Copyright © IEEE 1981.)

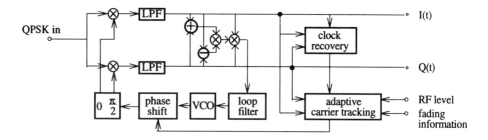

Figure 8.29. Block diagram of dual-mode carrier recovery system with dual mode of operation with Costas loop and adaptive carrier tracking.

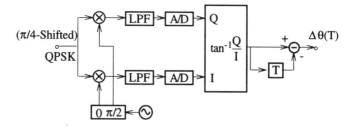

Figure 8.30. Differential detector of (π/4 shifted) QPSK.

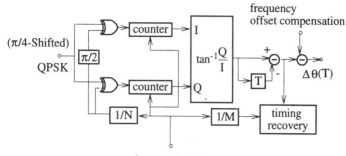

Figure 8.31. Logic circuit-implementated differential detector of (π/4 shifted) QPSK.

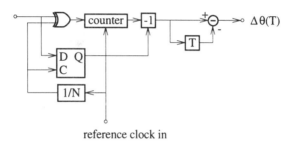

Figure 8.32. Differential detector using an EX-OR phase detector and 2-level phase detector.

θ can be found from the absolute value of θ and its sign. A differential detector based on this fact is shown in Fig. 8.32. The D-type flip-flop circuit detects the sign of the phase difference (i.e., 2-level phase) as described before. An all-digital differential detector including post-detection diversity is described in [11].

The carrier frequency offset worsens the error rate performance of the differential detector. The frequency offset $\Delta\omega$ results in an extra phase shift $\Delta\omega T$ in the demodulated signal $\Delta\theta(T)$. Thus, the phase error from signal phase shifts of $\pm\pi/4$ or $3\pi/4$ for $\pi/4$ shifted QPSK is used to compensate for the frequency offset [12].

A direct conversion demodulator for 2-level FM can be constructed as shown in Fig. 8.33 [13]. Limiters and a D-type flip-flop are used for demodulation. The principle of this demodulation is to detect the direction of rotation of the signal trajectory in the quadrature plane. The detection is made once per rotation of 2π. The detector cannot cope with signals with modulation index less than 2. Detection is made four times per 2π rotation with the demodulator shown in Fig. 8.34 [14]. Detection is made at each rise

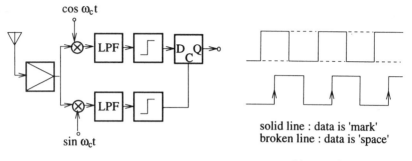

solid line : data is 'mark'
broken line : data is 'space'

(a) block diagram   (b) waveform

Figure 8.33. Direct conversion demodulator for 2-level FM.

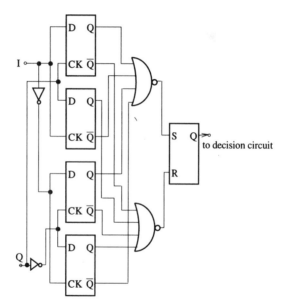

Figure 8.34. FSK demodulator.

up and down instant of the quadrature signals. Another demodulator is shown in Fig. 8.35 [15]. The principle of this circuit is detection of the rotating direction of the equivalent baseband frequency-doubled signal. Figure 8.36 shows another demodulator [16]. This circuit also operates on the frequency-doubled concept.

The group modulator, or its counterpart the group demodulator, modulates (demodulates) frequency division multiplexed signals (Fig. 8.37). It modulates/demodulates multiple signals at the same time. Thus, it becomes

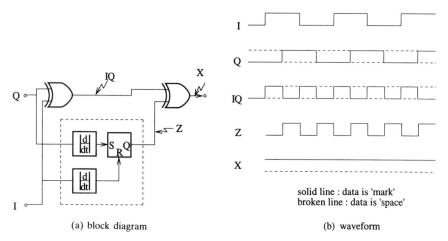

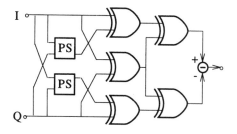

PS : phase shifter

Figure 8.36. FSK demodulator.

useful for a multicarrier transmission system or a base station transceiver for an FDMA system.

Group modulators and demodulators using the discrete Fourier transform technique are implemented as follows [21,22]. Assume that N complex baseband signals $x_k(t)$ are multiplexed in the frequency domain with equal spacing at frequencies of $\omega_k = -N/2 + 1 + k$ $(k = 0, 1, 2, \ldots, N-1)$. The z-transform of the modulator output signal becomes (Appendix 8.1)

$$Y(z) = \sum_{i=0}^{N-1} z^{-i} \exp\left(-j\frac{N-1}{N}i\pi\right)\left[\sum_{k=0}^{N-1} \exp\left(j\frac{ik}{N}2\pi\right)X_k(z^N)\right], \quad (8.8)$$

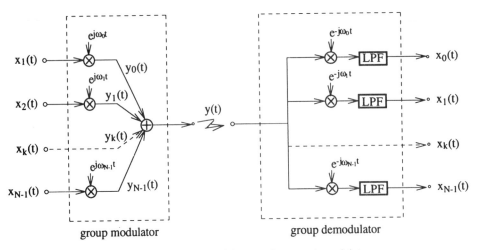

Figure 8.37. Group modulator and group demodulator.

where Δt is the sampling interval, $z = \exp(j\omega\Delta t)$, and

$$X_k = \sum_m x_k(mN\Delta t)\, e^{-j(N-1)m\pi}(z^{-N})^m \qquad (m = 0, 1, 2, \ldots). \qquad (8.9)$$

The term in the brackets in Eq. (8.8) is the discrete inverse Fourier transform. Thus, the group modulator is implemented as shown in Fig. 8.38(a).

The z-transform of a demodulator output signal becomes (Appendix 8.2)

$$X_k(z) = \sum_{i=0}^{N-1} Y_i(z^{-N})\, z^{-i} \exp\left(j\frac{N-1}{N} i\pi\right) \exp\left(-j\frac{ik}{N} 2\pi\right), \qquad (8.10)$$

where

$$Y_i(z^N) = \sum_m e^{j(N-1)m\pi}\, y(mN\Delta t + i\Delta t)(z^{-N})^m. \qquad (8.11)$$

Thus, the group demodulator is implemented with the discrete Fourier transform, as shown in Fig. 8.38(b).

8.8 COUNTERMEASURES AGAINST DC BLOCKING AND DC OFFSET

DC Blocking or ac coupling is desirable for preventing baseband signals from the dc offsets of the circuits. For signal transmissions through dc-blocked

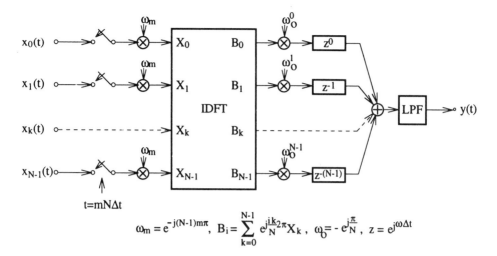

$$\omega_m = e^{-j(N-1)m\pi}, \quad B_i = \sum_{k=0}^{N-1} e^{j\frac{ik}{N}2\pi}X_k, \quad \omega_0 = -e^{j\frac{\pi}{N}}, \quad z = e^{j\omega\Delta t}$$

(a) modulator

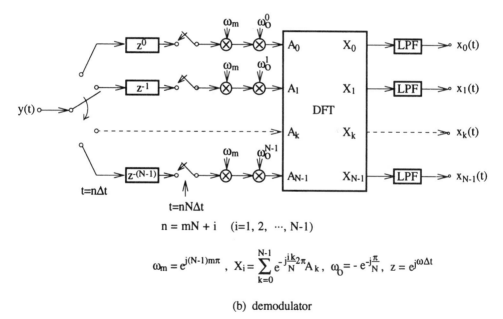

$$n = mN + i \quad (i=1, 2, \cdots, N-1)$$

$$\omega_m = e^{j(N-1)m\pi}, \quad X_i = \sum_{k=0}^{N-1} e^{-j\frac{ik}{N}2\pi}A_k, \quad \omega_0 = -e^{-j\frac{\pi}{N}}, \quad z = e^{j\omega\Delta t}$$

(b) demodulator

Figure 8.38. Group modulator and group demodulator implemented with discrete Fourier transform.

channels, waveforms without dc components are transmitted; for example, the Manchester codes are used, sacrificing spectrum performance.

Another approach is to use compensating techniques for waveforms distorted by dc-blocked channels. Figure 8.39 shows such a system for digital

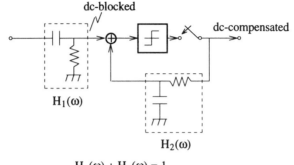

$$H_1(\omega) + H_2(\omega) = 1$$

Figure 8.39. DC-compensating circuit for digital signal.

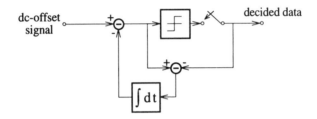

Figure 8.40. DC-offset control circuit.

signal transmission. The lost dc component is restored by applying the received data signals to a low-pass filter. It constitutes a decision feedback equalizer.

Another class of countermeasures against dc offsets is based on detecting the dc offset and subtracting it from the input dc-offset signal as shown in Fig. 8.40. The integrator can be replaced with a low-pass filter. Such a technique can be applied to the ALC (automatic level control) circuit as shown in Fig. 8.41. The combined dc- and level-offset control was implemented in a 4-level digital FM with limiter–discriminator system [17].

At a base station receiver for a TDMA system, dc-offset control is difficult, since it must be performed for each burst with different dc-offset values. A TDMA digital FM/limiter–discriminator detection system has such a problem, due to carrier frequency difference between the bursts transmitted from subscriber stations: the carrier frequency difference results in a dc offset. As a countermeasure against this problem, a method was proposed in [18] in which demodulated signals are subtracted from the signal delayed

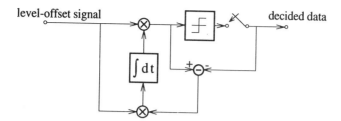

Figure 8.41. Automatic level-control circuit.

by one symbol duration. By means of this subtraction, the dc offset is removed at the symbol rate. This is equivalent to the partial response system with an impulse response $h(t) = \delta(t) - \delta(t - T_s)$, where T_s is the symbol duration. However, the error rate performance deteriorates with this method. The performance deterioration is recovered [19] using maximum-likelihood sequence estimation (Viterbi algorithm; Section 3.3.8).

For a burst signal receiver, dc- and level-offset compensation in Figs. 8.40 and 8.41 is operated using some given preamble bits [20].

APPENDIX 8.1 THE z-TRANSFORM REPRESENTATION OF THE GROUP MODULATOR

The kth modulated signal is expressed as

$$y_k(t) = x_k(t) e^{j\omega_k t}. \tag{A8.1}$$

The sampled signal at an interval of Δt becomes

$$y_k(n\Delta t) = x_k(t) e^{j\omega_k t} \sum_n \delta(t - n\Delta t).$$

Considering Eq. (2.5),

$$y_k(n\Delta t) = \sum_n x_k(n\Delta t) e^{j\omega_k n\Delta t} \delta(t - n\Delta t). \tag{A8.2}$$

Taking the Fourier transform of Eq. (A8.2), we have

$$Y_k(\omega) = \sum_n x_k(n\Delta t) e^{j\omega_k n\Delta t} e^{-j\omega n\Delta t}.$$

Using z-transform notation, we get

$$Y_k(z) = \sum_n x'_n z^{-n} \equiv X'_k(z), \tag{A8.3}$$

where

$$x'_n = x_k(n \Delta t) e^{j\omega_k n \Delta t} \quad \text{and} \quad z = e^{j\omega \Delta t}.$$

We interleave $X'_k(z)$ as

$$X'_k(z) = \sum_{i=0}^{N-1} X'_{ki}(z), \tag{A8.4}$$

where

$$X'_{ki}(z) = \sum_{n=mN+i} x_k(n \Delta t) e^{j\omega_k n \Delta t} z^{-n},$$
$$m = 0, 1, 2, \ldots; \; i = 0, 1, 2, \ldots, N-1. \tag{A8.5}$$

We assume equal frequency spacing as

$$\omega_k = \left(-\frac{N}{2} + k + \frac{1}{2} \right) \omega_s, \quad k = 0, 1, 2, \ldots, N-1, \tag{A8.6}$$

where $\omega_s = 2\pi/(N\Delta t)$. We then have

$$X'_{ki}(z) = z^{-i} \exp\left(-j\frac{N-1}{N} i\pi \right) \exp\left(j\frac{ik}{N} 2\pi \right) X''_{ki}(z^N), \tag{A8.7}$$

where

$$X''_{ki}(z^N) = \sum_m x_k(mN\Delta t + i\Delta t) e^{j(N-1)m\pi} (z^{-N})^m.$$

If we shift the time origin of $x_k(t)$, the term $i\Delta t$ is neglected and therefore $X''_{ki}(z^N)$ is independent of i. Thus we denote $X''_{ki}(z^N)$ by $X''_k(z^N)$. With the use of

$$Y(z) = \sum_{k=0}^{N-1} Y_k(z)$$

we have Eq. (8.8).

APPENDIX 8.2 THE z-TRANSFORM REPRESENTATION OF THE GROUP DEMODULATOR

The kth demodulated signal is expressed as

$$x_k(t) = y(t) e^{-j\omega_k t}.$$

(A8.8)

Taking the z-transform of the above equation, we have

$$X_k(z) = \sum_n y(n \Delta t) e^{-j\omega_k n \Delta t} z^{-n},$$

(A8.9)

where $z = e^{j\omega \Delta t}$.

Interleaving $y(n \Delta t)$ as $y(mN \Delta t + i \Delta t)$ $(i = 0, 1, 2, \ldots, N-1)$, we get

$$
\begin{aligned}
X_k(z) &= \sum_m \sum_{i=0}^{N-1} y(mN \Delta t + i \Delta t) \exp[j(N-1)m\pi] \exp\left(j\frac{N-1}{N} i\pi\right) \\
&\quad \times \exp\left(-j\frac{ik}{N} 2\pi\right) (z^{-N})^m z^{-i} \\
&= \sum_{i=0}^{N-1} Y_i(z^{-N}) z^{-i} \exp\left(j\frac{N-1}{N} i\pi\right) \exp\left(-j\frac{ik}{N} 2\pi\right),
\end{aligned}
$$

(A8.10)

where

$$Y_i(z^{-N}) = \sum_m y(mN \Delta t + i \Delta t) e^{j(N-1)m\pi}(z^{-N})^m.$$

(A8.11)

REFERENCES

1. S. Saitoh et al., "Basestation equipment; new technical report on digital mobile communication system," *NTT DoCoMo Technical Journal*, **1**, 33–38 (July 1993).

2. K. Murota et al., "Mobile station equipment," *NTT DoCoMo Technical Journal*, **1**, 43–46 (July 1993).

3. N. Tokuhiro et al., "Portable telephone for personal digital cellular system," *Proc. 43th Vehicular Technology Conference*, 718–721 (1993).

4. Y. Tarusawa and T. Nojima, "Frequency synthesizer; new technical report on fundamental technologies on mobile communications," *NTT DoCoMo Technical Journal*, **1**, 31–36 (January 1994).

5. S. Saito, Y. Taruzawa, and H. Suzuki, "State-preserving intermittently locked loop (SPILL) frequency synthesizer for portable radio," *IEEE Trans. Microwave Theory and Techniques*, **MTT-37**, 1898–1903 (December 1989).

6. C. B. Dekker, "The application of tamed frequency modulation to digital transmission via radio," *Proc. IEEE National Telecommunication Conference*, 55.3.1–55.3.7 (1979).

7. K. Kobayashi, Y. Matsumoto, T. Sakata, K. Seki, and S. Kato, "High-Speed QPSK/OQPSK burst modem VLSIC," *Proc. IEEE International Conference on Communications*, 1735–1739 (May 1993).

8. Y. Akaiwa and H. Koga, "Automatic power control for mobile communication channel," *Proc. International Symposium on Information Theory & its Applications*, **1**, 487–491 (November 1994).

9. K. Okanoue, A. Ushirokawa, H. Tomita, and Y. Furuya, "New MLSE receiver free from sample timing and input level controls," *Proc. Vehicular Technology Conference*, 408–411 (May 1993). Also in "A fast tracking adaptive MLSE for TDMA digital cellular systems," *IEICE Trans. Communications*, **E77-B**, 557–565 (May 1994).

10. S. Saito and H. Suzuki, "Fast carrier-tracking coherent detection with dual-mode carrier recovery circuit for digital land mobile radio transmission," *IEEE Journal on Selected Areas in Communications*, **7**, 130–139 (January 1989).

11. C. P. LaRosa and M. J. Carney, "A fully digital hardware detector for $\pi/4$ QPSK," *Proc. Vehicular Technology Conference*, 293–297 (May 1992).

12. M. Ikura, K. Ohno, and F. Adachi, "Baseband processing frequency-drift-compensation for QDPSK signal transmission," *Electronics Letters*, **27**, 1521–1523 (August 1991).

13. I. A. W. Vance, "Fully integrated radio paging receiver," *IEEE Proc.*, **129**, Part F, 2–6 (February 1982).

14. I. A. W. Vance, "Radio receiver for FSK signals," *UK Patent Application* GB 2 057 820 A, filed 4 September 1979.

15. Y. Akaiwa, "Demodulator for digital FM signals," *US Patent*, No. 4 651 107.

16. M. Hasegawa, M. Mimura, K. Takahashi, and M. Makimoto, "A direct conversion receiver employing a frequency multiplied digital phase-shifting demodulator," *Trans. IEICE*, **J76-C-I**, 462–469 (November 1993).

17. K. Kage, Y. Sasaki, M. Ichihara, and T. Sato, "The feasibility study of the Nyquist baseband filtered 4-level FM for digital mobile communications," *Proc. Vehicular Technology Conference*, 200–204 (May 1985).

18. Y. Nakamura and Y. Saito, "Discriminator with partial response detection of NRZ-FSK signals," *Trans. IEICE*, **J67-B**, 607–614 (June 1984).

19. Y. Akaiwa and T. Konishi, "An application of the Viterbi decoding to differential detection of frequency-discriminator demodulated FSK signal," *Proc. 4th International Symposium on Personal, Indoor and Mobile Radio Communications*, 210–213 (September 1993).

20. S. Sampei and K. Feher, "Adaptive dc-offset compensation algorithm for burst mode operated direct conversion receivers," *Proc. Vehicular Technology Conference*, 93–96 (May 1992).

21. M. G. Bellanger and J. I. Daguet, "TDM-FDM transmultiplexer: digital polyphase and FFT," *IEEE Trans. Communications*, **COM-22**, 1199–1205 (September 1974).

22. F. Takahata, M. Yasunaga, Y. Hirata, T. Ohsawa, and J. Namiki, "A PSK group

modem for satellite communications," *IEEE Journal on Selected Areas in Communications*, **SAC-5**, 648–661 (May 1987).

23. J. Zander, "Distributed cochannel interference control in cellular radio systems," *IEEE Trans. Vehicular Technology*, **41**, 305–311 (August 1992).

24. M. Almgren, H. Andersson, and K. Wallstedt, "Power control in a cellular system," *Proc. IEEE Vehicular Technology Conference*, 833–837 (June 1994).

25. Special issue on software radio, *IEEE Communications Magazine*, **33**, No. 5, (May 1995).

26. K. Murota and K. Hirade, "Transmission performance of GMSK modulation," *Trans. IECE*, **J64-B**, 1123–1130 (October 1981).

9

DIGITAL MOBILE RADIO
COMMUNICATION SYSTEMS

This chapter describes digital mobile radio communication systems. Some basic concepts are discussed first, followed by system descriptions. At the time of writing, the situation on digital mobile system is changing rapidly and new systems are appearing.

For further description of mobile radio communication systems, reader may refer to references [2] and [59]–[68].

9.1 FUNDAMENTAL CONCEPTS

9.1.1 Traffic Theory

Traffic theory plays an important role in analyzing and designing the performance of information transmission systems including public switched telephone systems, mobile radio communication systems, and others. The carried traffic under a given service quality or grade of service is discussed under conditions such as the number of channels provided, channel assignment strategies, and characteristics of information. This section concentrates on arguments in telephone systems.

The traffic volume is measured as the overall call holding time of a system during a measuring time period. The traffic density is defined as the traffic volume divided by the measuring time. The unit of traffic density is called an erlang, after A. K. Erlang.

9.1.1.1 Call Originating Rates. The calls are characterized by rate of origination and by holding time. For calls arriving at random, the originating rates are calculated as follows. Consider a time span of t and divide it into n short time intervals $\Delta t(t = n\Delta t)$, where Δt is short enough that only one call can occur in that period. Denoting call arrival rate by λ; then the probability that a call originates in the interval is $\lambda \Delta t$. Since random calls originate independently, the probability that k calls originate during the time t becomes

$$
\begin{aligned}
v_k(t) &= \lim_{\Delta t \to 0} \binom{n}{k} (\lambda \Delta t)^k (1 - \lambda \Delta t)^{n-k} \\
&= \frac{(\lambda t)^k}{k!} \lim_{n \to \infty} \left(1 - \lambda \frac{t}{n}\right)^n,
\end{aligned}
$$

where

$$
\binom{n}{k} = \frac{n!}{(n-k)!\,k!}.
$$

Using $\lim_{x \to 0}(1 + x)^{1/x} = e$, we have

$$
v_k(t) = \frac{(\lambda t)^k}{k!} e^{-\lambda t}. \tag{9.1}
$$

The above is termed the *Poisson arrival process*.

9.1.1.2 Holding Time and Terminating Rates.

For telephone conversation, the distribution function of holding time is well approximated by the exponential holding time, that is,

$$
H(t) = \begin{cases} 1 - e^{-\mu t} & (t \geq 0) \\ 0 & (t < 0). \end{cases} \tag{9.2}
$$

The average holding time is given by

$$
h = \int_0^\infty t \frac{d}{dt} H(t)\,dt = \frac{1}{\mu}. \tag{9.3}
$$

The call termination rate is calculated as follows. Conditional probability that a call being held for time τ will terminate in the following time t is given as

$$
H(t|\tau) = \frac{H(\tau + t) - H(\tau)}{1 - H(\tau)}.
$$

Using Eq. (9.2) we have

$$
H(t|\tau) = 1 - e^{-\mu t}. \tag{9.4}
$$

This equation means that the call terminating probability never depends on the past held time, τ. This property of the exponential distribution serves to simplify the treatment of the traffic theory.

For a short time Δt, Eq. (9.4) becomes

$$H(\Delta t \mid \tau) \approx \mu \Delta t \qquad (\Delta t \ll 1/\mu) \tag{9.5}$$

The termination rate becomes μ.

9.1.1.3 State Formula and Erlang's Loss Formula.

Consider a communication system with n channels. The state of the system is described by the number of channels that are held. The state changes according to the arrival and termination of calls. The state transition diagram is shown in Fig. 9.1.

At equilibrium, the number of transitions coming in and going out for a state is the same. The equations of state are given as

$$\lambda P_0 = \mu P_1,$$
$$(\lambda + \mu) P_1 = \lambda P_0 + 2\mu P_2,$$
$$\vdots$$
$$(\lambda + r\mu) P_r = \lambda P_{r-1} + (r+1)\mu P_{r+1}, \tag{9.6}$$
$$\vdots$$
$$n\mu P_n = \lambda P_{n-1},$$

and

$$P_0 + P_1 + P_2 + \cdots + P_n = 1, \tag{9.7}$$

where P_r denotes the probability that r channels are held. Solving Eq. (9.6) under Eq. (9.7), we get

$$P_r = \frac{a^r/r!}{\displaystyle\sum_{x=0}^{n} (a^x/x!)}, \tag{9.8}$$

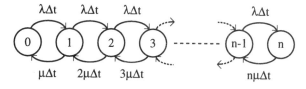

Figure 9.1. State transition diagram of an n-channel communication system.

where $a = \lambda/\mu = \lambda h$ is the applied traffic. The call blocking probability B is given by P_n as

$$B = \frac{a^n/n!}{\displaystyle\sum_{x=0}^{n}(a^x/x!)}.$$ (9.9)

The average carried traffic a_c is given as

$$a_c = \sum_{x=0}^{n} xP_x$$
$$= a(1-B).$$ (9.10)

The efficiency of the channel usage is then

$$\eta = \frac{a_c}{n} = \frac{a(1-B)}{n}.$$ (9.11)

The efficiency increases as the number of channels increases (Fig. 9.2). This is called the *trunking effect*.

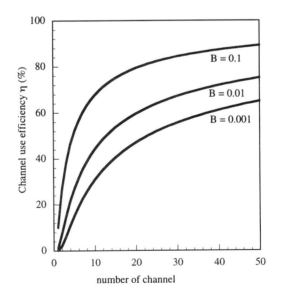

Figure 9.2. Channel efficiency vs. number of channels.

9.1.2 The Cellular Concept

In a cellular system, a service area is covered with many small zones or cells (Fig. 9.3). A zone is illuminated by a base station. With this system, the system capacity, that is, the number of subscribers that can be accommodated, can be increased through reuse of radio channels at different cells, where cochannel interference can be neglected. Thus, a cell group is repeated to cover the total service area. In addition, a lower transmit power is possible because of the smaller coverage of a base station compared to a single cell system with a large coverage area. On the other hand, the cellular system needs sophisticated procedures for control of location registration for roaming subscribers and call hand-offs between neighboring zones for continuation of a conversation by subscribers moving over zone boundaries. These functions of a cellular system became possible with modern network technologies that include electronic switching and improved radio communication techniques such as frequency-synthesizing transceivers and digital communication.

The technologies supporting a cellular system are classified into two groups: One is related to system control, such as location registration, paging, and hand-offs. The other is related to spectrum efficiency, which is discussed in the following. The spectrum efficiency of a cellular system is given as [1],

$$\eta = \frac{1}{s} \frac{a_c}{2\Delta w n_g n_z},\qquad(9.12)$$

where a_c is the traffic carried per base station, s is the area size for each cell, $2\Delta w$ is the spectrum bandwidth per pair of channels, n_g is the number of cells in one group, or the cell cluster size, and n_z is the number of channels per cell. The system bandwidth is $2\Delta w n_g n_z$. A smaller cell size can accommodate a higher density of subscribers and hence, the total capacity of the system increases; however, a larger number of base stations must be

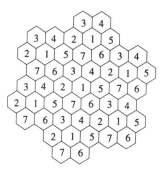

Figure 9.3. A zone layout for cellular systems.

deployed. A smaller cell group means a shorter reuse distance. The reuse distance is affected by how rapidly the radio waves attenuate with distance and the magnitude of the required protection ratio for cochannel interference.

The term a_c/n_z reflects how much traffic is carried for a given number of channels. When we assume the Erlang B formula (Eq. 9.9), a_c is given as a function of the number of channels and the blocking probability. A larger number of channels per bandwidth is obtained if we use a narrow-band signal. However, a narrow-band signal generally requires a higher protection ratio, resulting in a larger n_g. Thus, finding the optimum bandwidth per channel is not easy.

When we assume a hexagonal cell layout, the number of cells in one group is given as [2]

$$n_g = \frac{1}{3}\left(\frac{D}{R}\right)^2 = i^2 + j^2 + ij \qquad (i, j = 1, 2, 3, \ldots), \tag{9.13}$$

where R is the cell radius and D is the distance from the center of a cell to the center of the nearest cell reusing the same channel. Determining the cell group number is a difficult task, since it depends on the wave propagation characteristics, including shadowing, zone shape, correlation between desired and interfering signals, and protection ratio [3].

9.1.2.1 Omni Cell.

A cell or zone is the area covered by a base station antenna. When an omnidirectional antenna is used, a zone is covered circularly around the base station. When we plan to place base stations to cover a whole service area by repeating the cell cluster, strategies such as triangular, rectangular, and hexagonal layouts are used. The hexagonal layout is optimal in the sense that the cell cluster size is minimum under the condition of given cochannel interference.

9.1.2.2 Sector Cell.

If a base station employs N antennas, each of which covers a different direction in horizontal space, N-sector cells are produced around the base station, as shown schematically in Fig. 9.4. Different channels are used between the sectored cells. The sector cell layout has advantages in view of efficiency of deploying base stations. An omnicell is split into N-sector cells, each has a cell sizes that is $1/N$ of the omnicell. Thus, a smaller cell is implemented with little effort—that is, the number of base station sites is not increased. Indeed, this technique was introduced to cope with traffic increase in urban areas.

The sector cell layout is classified into two systems, back-to-back and parallel-beam systems, as shown in Fig. 9.5, where hatched cells use the same channels. Cochannel interference is different in different directions due to the directivity of the sector antenna. Using this fact, an efficient parallel-beam

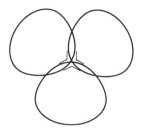

Figure 9.4. Three-sectored cell.

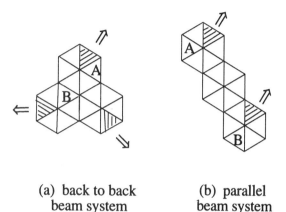

(a) back to back
     beam system

(b) parallel
     beam system

Figure 9.5. Different layout for 6-sectored cell system.

sector cell layout was proposed [4] for which the reuse distance differed in the horizontal and vertical directions. An example of a 5-site reuse system is shown in Fig. 9.6. In this figure the blacked out sectors, which use the same channels, can be placed closer on the vertical axis than on the horizontal axis, assuming horizontal directional beam antennas. A 3-sectored 5-site layout is claimed under the condition that the signal to interference power ratio (CIR) is 13 dB.

9.1.2.3 Beam Tilting. A base station antenna consists of many vertically placed radiating elements, showing directivity in the vertical plane. Beam tilting is a technique in which the antenna directivity is tilted downward to reduce interference to other cells [5]. This technique is also effective in reducing frequency-selective fading by decreasing the multipath echo signal level [6].

9.1.2.4 Reuse Partitioning. In a cell the signal to interference ratio (CIR) is dependent on the location of the mobile station: if the mobile is close to

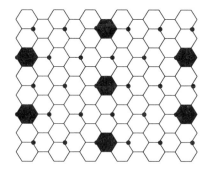

Figure 9.6. Efficient 5-site 3-sectored cell layout.

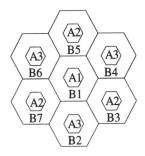

Figure 9.7. Reuse partitioning.

the base station, the CIR is high. This leads to the reuse partitioning concept [7]. The reuse factor is reduced, to be, for example, three in the inner part of the cell shown in Fig. 9.7.

Let us assume the reuse is partitioned into two groups where the reuse factor is N_A and N_B. Consider a total system consisting of S channels of which a part P is assigned to the N_A group. The number of channels assigned to a cell is then

$$C = \left(\frac{P}{N_A} + \frac{1-P}{N_B} \right) S. \qquad (9.14)$$

The equivalent reuse factor N_{eq} is then

$$N_{eq} = \frac{S}{C}$$

$$= \frac{N_A N_B}{P N_B + (1-P) N_A},$$

$$= f N_A + (1-f) N_B \qquad (9.15)$$

where

$$f = \frac{PN_B}{PN_B + (1-P)N_A}.$$

The equivalent reuse factor changes depending on P from N_A ($P = 1$) to N_B ($P = 0$).

9.1.2.5 *Cell Layout for Nonuniform Traffic Distribution.*

Traffic density is hardly uniform in all the service areas. It sometimes shows a so-called bell-shaped distribution with its peak in the urban areas. To cope with this, small cell sizes are used in this area and cell sizes are gradually increased as the density decreases.

Another case is where traffic density is locally increased in some multiple small areas. In this case it is not economic to deploy multiple small cells all over the service area. They are used only at the increased traffic areas and the total area is covered with large (umbrella) cells (Fig. 9.8).

9.1.2.6 *Microcell System.*

As mentioned earlier, a system with smaller cell size yields a higher spectrum efficiency (Eq. 9.12). Conventional cells have a radius larger than 1 km. Cellular systems with smaller cell radii, on the order of several hundred meters, are called "microcell" systems; the term "picocell" is even used for cell sizes on the order tens of meters. As the cell size is reduced, the base station antenna height will be lowered to, for example, lamp-post height and therefore buildings, roads, and other obstructions will seriously affect the radio wave propagation. The cell shape is then heavily deformed [8] and is barely representable by circular or hexagonal regions. In this situation, it becomes hard to estimate the radio wave propagation and hence channel reuse planning becomes difficult. In addition, the cells must be highly overlapped to reduce dead-spot areas where service is unavailable because no radio signals reach it. Consequently, the cell cluster size must be increased, resulting in low spectrum efficiency. Furthermore, hand-overs occur more often for a fast-moving terminal. Solutions to these problems will be discussed in the following.

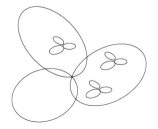

Figure 9.8. A small-cell overlaid with large-cell system.

9.1.2.7 Microcell and Macrocell Overlaid System.

A microcell and macrocell overlaid system [9,10] yields some advantages, such as paging and location registration procedures being carried out through macrocell systems, and fast-moving terminals being connected to the macrocells to lessen the hand-off frequency. If the same standards for radio channels (air interface) are used between the microcell and macrocell systems, modifications to mobile terminals can be kept minimal: implementation of automatic power control function is highly desirable to limit unnecessarily high output power for communication through microcells. Several methods for channel assignment in the microcell and macrocell overlaid system are known: (i) channels are divided between the microcell and macrocell systems; (ii) channels are assigned for a microcell system that are not used in the macrocell system; and (iii) the same channels are used at the same time between the microcell and macrocell systems. The last strategy, due to Kinoshita, Tsuchiya, and Ohnuki [11], offers the highest spectrum efficiency. This strategy is possible by increasing the transmit power of the microcell system to overcome the interference from the macrocell system. A rough estimate of the necessary increase in the microcell transmit power is around 5 dB, assuming a cell radius of 3 km for the macrocell and 640 m for the microcell [12]. Since the transmit power of the microcell system is much lower than that for macrocell system, only interference from the macrocell to microcell is significant and interference in the reverse direction can be neglected.

9.1.2.8 Hand-Off.

When a mobile terminal in communication crosses the cell boundary and moves into another cell, the call should be connected through the new cell base station. This is called the (intercell) hand-off or hand-over. The hand-off is initiated when the radio signal quality drops to a certain threshold level. The base station reports the hand-off request to its radio control station, which in turn orders some other base stations to measure and report the signal level of the channel under consideration. The radio channel control station selects the base station that reports the highest received signal level and that has an idle channel(s). If there is no base station that can accommodate the traffic, the hand-off fails and the call is terminated (dropped).

The actual hand-off system is more sophisticated. For example, the hand-off is carried out not with one threshold but with two, to make a hysteresis loop control. To explain this, assume that the signal levels at the boundary region of two coverage areas are as illustrated in Fig. 9.9. When the received signal level at base station #1 drops below the threshold level (a), a hand-off to base station #2 is triggered. Although the received signal level at base station #2 drops below the threshold level (a), the hand-off to base station #1 is not triggered as long as the signal level does not drop below the threshold level b (<a). If we do not use this technique, hand-off is carried out many times during a short time period; this is not desirable

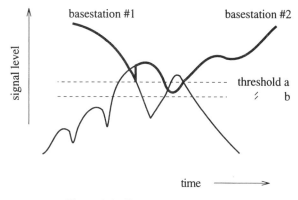

Figure 9.9. Illustration of hand-off.

in view of the increase in network control procedures and the interruptions to the voice signal.

In a TDMA system, a mobile station takes part in the hand-off procedure by selecting the base station to which the call is handed (the mobile-assisted hand-off). This becomes possible because the mobile station, in communicating with the current base station, can monitor signals from the candidate base stations for hand-off at the TDMA time slots which are not dedicated to the mobile station. With this decentralized technique, the hand-off procedure load in the control station is decreased.

Since a forced termination due to a hand-off failure is serious for users, it is important to minimize the probability of a hand-off failure. To this end, schemes such as reserving some channels for hand-off and introducing a queuing system are considered.

9.1.2.9 *Location Registration and Paging.* A mobile station may move all around the service area. When a call arrives for a mobile station, it is not efficient to page the mobile station from all the base stations, since the paging traffic becomes large. In a system, the service area is divided into multiple paging areas as shown schematically in Fig. 9.10. A mobile station is paged from a paging area. To know which paging area is responsible to a mobile station, each mobile station is registered to a paging area. This location registration is carried out when a mobile station receives a paging area identification code that is different from the one for the last location registration. If the paging area size becomes smaller, the location registration traffic increases. Therefore, the paging area sizes should be decided by compromising the traffic for paging and for location registration.

The location information is stored in a file, the home location register, at a switching center. When a mobile station moves into the territory of another switching center, the location information of the mobile station is

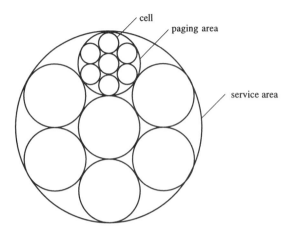

Figure 9.10. Cell, paging area, and service area.

recorded in a file, the visitor location register, and is also sent to the home switching center to record this fact.

Even in an analog cellular systems, a digital data transmission system is necessary for system control such as paging, location registration, and call set-up and release. In this system, the data rate is slow because digital data are transmitted through an analog voice channel. This places a burden on the system control. A high data rate transmission capability of a digital cellular system can relieve the problem.

As the paging area size becomes small, the frequency of location registration grows. In addition, location registration is made backward and forward frequently between the boundaries of location registration areas. In order to prevent such a phenomenon, the overlaid location registration area is adopted in the NTT digital cellular system [13]. In this system, boundaries of the location registration areas in a layer are placed on the center of those in the other layers.

A technique was proposed in [14] for autonomous creation of location registrations and paging areas. A fully distributed location registration strategy is discussed in [15].

9.1.3 Multiple Access

The communication media, the radio channels in mobile communication, are used by multiple users. Multiple access methods are classified into random or contention access schemes and controlled access schemes. The controlled access schemes include the demand assignment system, the polling system, and the token-passing system. In the demand assignment system, a channel is assigned on demand. This scheme is used for telephone systems. The others are used for data communication systems. In a polling system, a central station polls subscriber stations in turn and only a polled station can get access to the channel. In a token-passing system, a subscriber station that has received a token has the right to access the channel. The token is passed to the next station when a transmission ends.

9.1.3.1 *Random Access System.*

Random access is used for data communications or in the reservation phase of an on-demand channel assignment system. In a random access system, collision may occur between signals transmitted from different stations. If the stations in conflict retransmit immediately after a collision, the collision continues forever. To avoid the succession of collisions, the signal is retransmitted after a back-off time, which is randomly selected from a given time interval.

In a random access data transmission system, the data are transmitted in a form of a so-called, "packet," which is a block of data consisting of a flag pattern and fields for control data, address, information data, and checksum for error detection (Fig. 9.11). In terms of switching technologies, a conventional telephone system is called a circuit-switched system, where the channel is held throughout a conversation; packet data transmission is called a packet-switched system. In a packet transmission system, the average throughput and average transmission delay are important measures of performance. The throughput, denoted by S, is defined as the rate of effective channel usage per unit time. The delay, denoted by D, is the average time between generation and reception of a packet. These performance measures are discussed in the following [16].

Let us assume random arrivals of packets. The probability that k packets arrive during a packet time length T is given by (Eq. 9.1),

$$P_k(T) = \frac{(\lambda T)^k}{k!} e^{-\lambda T}, \tag{9.16}$$

where λ is packet arrival rate per unit time, including packet retransmissions. Let $G = \lambda T$ denote the average number of packets to be transmitted.

The ALOHA system is a famous random access system. It has two versions: pure ALOHA and slotted ALOHA. In the pure ALOHA system, when a packet is generated a station sends the packet immediately. In the slotted ALOHA system, packet transmissions are synchronized to time slots, which are known throughout the system.

01111110	Address	Control	Data	Checksum	01111110
(flag)	8	8		6	(bits)

Figure 9.11. Frame format for a packet.

The throughput S is given by the product of G and the probability that a packet experiences no collisions. For the pure ALOHA system, the probability for a packet to meet no collision is given by $P_0(2T)$. The throughput $S(<G)$ is then

$$S = GP_0(2T)$$
$$= Ge^{-2G}. \qquad (9.17)$$

The average number of packet retransmissions for a packet to be successfully transmitted is $G/S - 1$. The average transmission delay is then

$$D = T + 2\tau + (e^{2G} - 1)(T + 2\tau + B), \qquad (9.18)$$

where τ is the round-trip time and B is the average back-off time at collision.

For the slotted ALOHA system, the probability that a packet meets no collision is given by $P_0(T) = e^{-G}$, which is smaller than that for the pure ALOHA system for a fixed G, $e^{-G} > e^{-2G}$. The average throughput in this case is

$$S = Ge^{-G}. \qquad (9.19)$$

A packet waits on average $T/2$ until the next slot starts. Let us assume a collided packet waits for a random number of time slots between 0 to $K - 1$; then the average back-off time B is $(K-1)T/2$. Thus, the average transmission delay is

$$D = 1.5T + 2\tau + (e^G - 1)[(K + 1)T/2 + 2\tau]. \qquad (9.20)$$

9.1.3.2 CSMA System. In this system, stations measure the carrier level of the channel to monitor whether the channel is busy or not. If the channel is sensed to be idle, the station sends a packet, and otherwise it waits for another chance. The carrier sense is not complete in detecting whether the channel is busy or not, since there is a delay for signals transmitted from other stations. The CSMA schemes are classified into three subgroups: nonpersistent CSMA, 1-persistent CSMA, and p-persistent CSMA. The nonpersistent system obeys the back-off algorithm if the channel is sensed busy. In the 1-persistent system, the packet is sent immediately after the time when the channel is idle. In the p-persistent system, even when the channel is sensed busy, the packet is transmitted with a probability of p or waits for t time units with a probability of $1 - p$ and senses the carrier again. The case $p = 1$ is the 1-persistent system.

The CSMA is improved with collision detection (CSMA/CD). In this system, carrier sensing is carried out even during signal transmission. If a

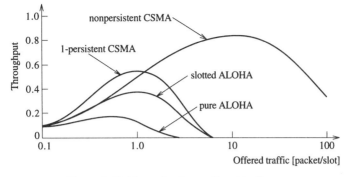

Figure 9.12. Throughput vs. offered traffic.

collision is detected, transmission is stopped so as not to waste the channel. Carrier sensing performs well in cable transmission systems. However, it may become uncertain in wireless systems, since the radio waves are not guaranteed to reach all the stations involved in the current communication, because of radio wave propagation uncertainty. The throughput vs. offered traffic load is shown in Fig. 9.12 [17].

9.1.4 Channel Assignment

In some mobile telephone systems, a radio channel is assigned for a call on demand. In a cellular system, the channel assignment for calls in each cell is important since it directly affects the efficiency of channel reuse between different cells. The channel assignment schemes are divided into three categories: fixed channel assignment, dynamic channel assignment, and hybrid channel assignment. In the fixed channel assignment, channels are fixed for each cell and the channels are never used by the other cells in the same cell cluster. In the dynamic channel assignment, channels are dynamically used between cells depending on the situation. In the hybrid channel assignment, some channels are fixed for each cell and others are used dynamically.

The fixed channel assignment scheme requires channel assignment planning in advance according to cell layout design. The cell layout design and the channel assignment are probably the most important technical matters in implementing an actual cellular system. This is so because radio propagation in practice is not uniform in contrast to the idealized situation that we often assume in theoretical analyses.

The dynamic channel assignment systems were known long ago. The first available, mean square, the nearest neighbor, and the channel borrowing methods appeared in reference [2]. In recent years, the dynamic channel assignment schemes have received attention as microcell systems have became a hot topic: Their aspect of adaptive or decentralized operation is recognized as necessary to channel assignment in microcell systems, where radio wave propagation is too irregular to allow estimation of interference between cells. Under this condition, we should not design the cell layout and the channel assignment in advance.

9.1.4.1 *First Available.*

In the first available or the random assignment scheme, when a call arrives at a cell, the cell base station searches for an available channel in a given order or at random. The examination of the channel is simply done by monitoring the received signal levels: if the level is lower than a threshold level, the channel can be used. Thus, channels are selected from all the channels at each cell with its own decision changing depending on the channel usage status in the interfering cells. Channels are assigned autonomously in a decentralized manner. Consequently, the term channel "assignment" would be better replaced with the concept of channel "selection." Thus, we are free from channel assignment planning. Not all of the dynamic channel assignments are autonomous or decentralized. The performance of the first available system is very different with the different schemes for searching for a channel, viz., random search and ordered search. The random search system has a lower channel reuse performance than the ordered search system since no systematic channel reuse is obtained. A version of the first available system with ordered search is known as the ARP (autonomous reuse partitioning) system and will be discussed later.

9.1.4.2 *Channel Segregation.*

The channel segregation method was proposed by Akaiwa and Furuya in 1985 [18,19]. It was the first method where the learning concept was introduced to the channel assignment system. In this system, each cell learns, through interference and channel usage between other cells, to acquire some favorite channels. Channel reuse is established autonomously. The term "channel segregation" comes from the analogy with ecological territories of animals. The territories (channel usage in each cell) are established through several contests (interference) between the owners (cells). This, termed "segregation," can be understood as a system where resources are shared in a decentralized manner by many users. The channel segregation system is described as follows. Each base station has a table in which priority function of channels are stored; a higher priority is given to a channel with a higher value of the priority function.

1. When a call arrives at a base station, the channel with the highest priority is chosen from the channels that are not used in the base station.
2. The channel is tested for whether it can be used for the new call. Specifically, the signal level of that channel is measured at the base station and/or at the mobile station. If the measured signal level is

below a given threshold level, the channel is decided idle, otherwise
it is decided busy.

3. If the channel is decided idle, the priority of the channel is increased
   and the channel is used.

4. If the channel is decided busy, its priority function is decreased and
   the next highest priority channel is set to be tested.

5. Step (1) to (4) are repeated. If there is no available channel, the call
   is blocked.

The estimated carrier to interference level instead of the received signal
level may be a better criterion for the test of a candidate channel. It is
important that the above procedure is made at each base station independently. There is neither information exchange nor central control information
between base stations.

The most interesting question is whether the channels segregate automatically between cells. To investigate this a computer simulation experiment is
done (see Fig. 9.13). Random calls are originated in each cell. As a time
proceeds, each cell acquires a different channel, in other words, channel
segregation is established. We consider a one-dimensional system, where cells
are partially overlapped as shown in Fig. 9.13(a). Cells #1 and #4 never
overlap, that is, no interference occurs between them. The learning process
in terms of the priority function and channel usage frequency is shown in
Fig. 9.13(b). The interfering cells capture different channels and cells #1 and
#4 capture the same channel. Thus, the channel reuse is automatically
established. The channel segregation is also confirmed in two-dimensional
systems. Owing to the structured channel reuse pattern, channel segregation
shows a better blocking probability performance than does the random
assignment system.

Channel segregation was applied for the TDMA/FDMA system in [20].
In a TDMA/FDMA system, time-division as well as frequency-division
channels are prepared. If we apply the random channel assignment to a
TDMA/FDMA system, the problem occurs that a call can be blocked even
when a radio channel is idle. This happens when all channels on a carrier
frequency are used and the idle channel is on another carrier frequency; this
occurs because of the inability of a base station transceiver to switch its carrier
frequency time-slot by time-slot so that it therefore cannot access the idle
channel. The probability of the occurrence of this inaccessible channel
problem is high when time- and frequency-division channels are used at
random. The computer simulation results of a TDMA/FDMA system with
channel segregation show that the call blocking probability due to the
inaccessible channel problem is decreased. This is the result of the bunching
effect, where channels to be used in a base station tend to gather on the same
frequency (Fig. 9.14).

The microcell and macrocell overlaid system is another example where

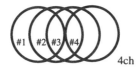

(a) Partially overlapped cells

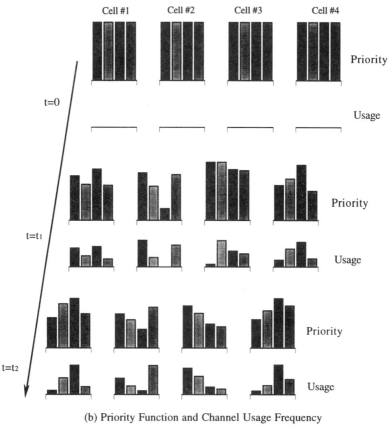

(b) Priority Function and Channel Usage Frequency

Figure 9.13. Channel segregation.

channel segregation has been applied [12]. The microcell systems use dynamically the same channels that are assigned to the macrocell system, even when those channels are being used in the macrocell system as described in Section 9.1.2. On learning the cochannel interference from the macrocell system, the microcell system tends to use the channels which show the least interference from the macrocells. Figure 9.15 shows computer simulation

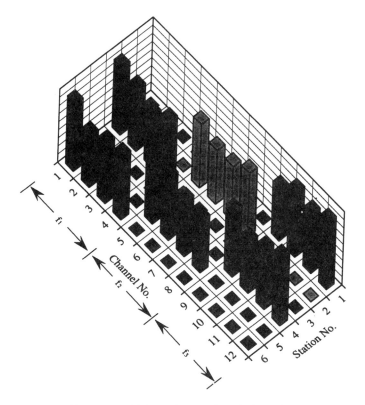

Figure 9.14. Relative frequencies of channel usage.

results for channel usage frequencies in the microcell system. Microcells A1–A8 (B1–B8) in macrocell #9 (#3) uses channels that are assigned for macrocells #1 (#5) and #2 (#7).

When multiple transmitters/receivers are located close to each other, a problem called *intermodulation* occurs: interfering signals are produced due to nonlinearity of the transceiver circuits. The frequencies of the these interfering signals are given by the algebraic sum of the carrier frequencies and their harmonics. We must carefully select a set of carrier frequencies that will not be interfered with by the intermodulated signals. Channel segregation can successfully avoid the intermodulation interference by automatically selecting an appropriate set of carrier frequencies [21].

9.1.4.3 *Autonomous Reuse Partitioning.* The autonomous reuse partitioning (ARP) dynamic channel assignment method due to Kanai [22] automatically establishes the reuse partitioning. The steps in this method are simple:

1. Channels are ordered identically throughout all base stations.

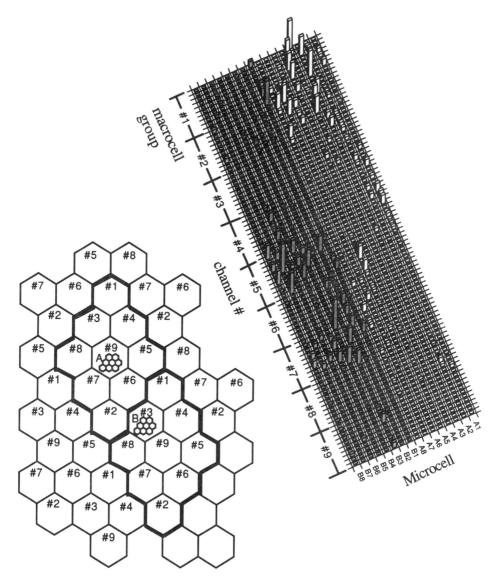

Figure 9.15. Channel usage frequencies of microcells underlaid with macrocell.

2. When a call arrives at a base station, the channels are tested for whether they can be used or not in that order. If a channel shows CIR (carrier to interference ratio) above a given threshold value in both the forward and backward link, the channel is determined idle. The first available channel is used.

3. If there is no available channel, the call is blocked.

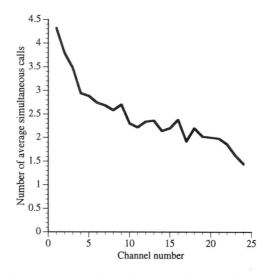

Figure 9.16. Average number of simultaneous calls versus channel number with autonomous reuse partitioning method.

This is a fully decentralized system. The evidence for the establishment of automatic reuse partioning is shown in Fig. 9.16. In this figure, computer simulation result of the average number of simultaneous calls versus channel number is shown. We can see that the lower-numbered channels are more densely used throughout the service area: this means that the reuse distance for the channel is shorter than that for a channel with a higher number channel. Even though the reuse distance is short for a low-numbered channel, the CIR value is similar to that for high number channel, as shown in Fig. 9.17.

Self-organized reuse partitioning (SORP) [23] was proposed for decreasing the channel sensing times in ARP. In this method, the average power for each channel is calculated at each base station using the signal levels at the base stations. The channels are sensed from those showing an average power level which is closest to the mobile transmit power. Another method for decreasing the number of channels sensed with ARP was proposed in [24], where the starting channel for carrier sensing is adoptively selected depending on the received signal level at the base station. At each base station the average received signal level is calculated and stored for each channel: consider channel j, then the average received signal level $s(j)$ is calculated as

$$s(j) \leftarrow \frac{ns(j) + P}{n + 1}, \qquad (9.21)$$

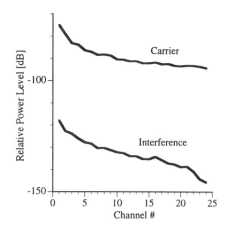

Figure 9.17. Average carrier and interference signal level versus channel number with autonomous reuse partitioning method.

where n is the number allocation of that channel so far and P is the received signal power in dB at the current location. The starting channel is the channel that has the average received signal level larger than the current signal level by a given margin.

Self-organized reuse partitioning is combined with channel segregation in [25].

Investigation must be continued into clarifying the performance of the decentralized dynamic channel assignment schemes.

9.1.4.4 Performance of Some Channel Allocation Methods.
In comparing different channel assignment methods, we can consider such performance criteria for channel allocation methods as call blocking probability, cochannel interference, number of hand-offs and forced terminations, average number of carrier sense times, and others. These performance measures depend on the channel assignment method as well as the assumed conditions, such as number of channels, radio propagation conditions including fading, power control, intra/intercell hand-offs, distribution of mobile stations, and so on.

Figure 9.18 shows the performance comparisons between fixed, random, channel segregation, ARP channel assignment, and SORP. A two-dimensional cell layout with uniform distribution of mobile stations is assumed. Other parameter are listed in Table 9.1. SORP shows the best performance under the assumed conditions.

9.1.4.5 Packet Reservation Multiple Access (PRMA).
This method due to Goodman, Valenzuela, Gayliard, and Ramamurthi [26] is closely related to the reservation ALOHA method. The channel is divided into some

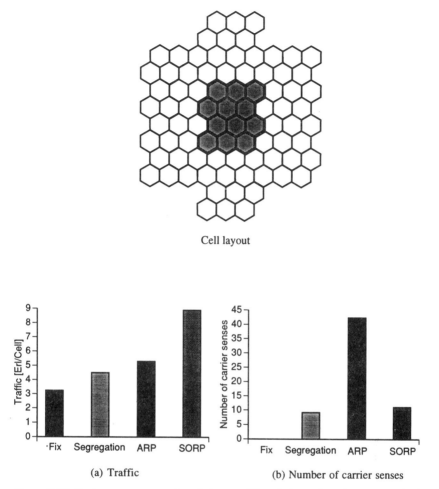

Cell layout

(a) Traffic

(b) Number of carrier senses

Figure 9.18. Performance comparison between different channel assignment methods. Blocking probability = 1% and interference probability = 1%.

number of time slots to form a frame. Voice signals are transmitted only at talkspurts in the form of a packet, so a voice activity detector is employed at the voice terminal. In order to avoid collision between a voice packet and other packets, a time slot is reserved by a voice terminal for each talkspurt period. At the beginning of a talkspurt, the voice terminal makes a reservation for a time slot through a random access slot which is not reserved. The status "reserved" or "available" is broadcast from the base station. The speech signal waits until the reservation become successful. If the reservation is not successful within 32 ms then the terminal discards the speech packet. This distinguishes PRMA from reservation ALOHA.

TABLE 9.1 Simulation parameter for Fig. 9.19

Required blocking probability	0.03
Required interference probability	0.03
Power control	Yes
Required CIR	15 dB
Number of channels	84
Path loss figure	3.5
Hand-off	No
Reuse factor for fixed channel assignment	12
Margin of CIR	
Fixed assignment	0 dB
Segregation	10.0 dB
ARP	10.5 dB
SORP	4.1 dB
Terminal mobility	Stationary
Shadowing	Log-normal (6 dB standard deviation)

Since speech signals have a talkspurt rate of around 40%, the radio channel can be used effectively by statistically multiplexing speech signals in the PRMA system: The number of served radio channels can be less than the number of speech terminals in conversation. The statistical multiplexing technique in a TDMA system is known as DSI (digital speech interpolation) [27], which requires complex control; on the other hand, the DSI effect is easily embedded in the PRMA system.

In applying the PRMA method to digital cellular systems, a method combining the PRMA and channel segregation was proposed in [28].

9.1.4.6 Other Channel Allocation Methods. For other methods, refer to references [29]–[33].

9.1.5 FDMA, TDMA, and CDMA

Many channels must be prepared to accommodating a number of users in a circuit-switched systems. We have three ways to produce channels using a given bandwidth: frequency, time, and code division multiplexing. When we consider the up-link (subscriber to base station), the prepared channels are accessed by many users (multiple access). Thus, we have FDMA (frequency division multiple access), TDMA (time division multiple access), and CDMA (code division multiple access) systems. The radio transmission system differs considerably between these multiple access systems, so that which multiple access system is selected is the subject of heated debate.

FDMA is used for analog mobile radio systems. This system is implemented with minimum signal processing. The number of channels given with this system is decided by the carrier frequency spacing. Since the channels are split (orthogonally) into frequency divisions, a stable carrier frequency source and a channel selection filter with a sharp roll-off transfer function are required. A narrower spectrum bandwidth of the modulated carrier signal is advantageous. The modulated signal bandwidth per user for digital cellular systems becomes very narrow owing to the evolution of low-bit-rate voice coding and spectrum-efficient linear modulation. Furthermore, a higher frequency band tends to be used. In this situation, a guard band between channels given for discrimination in frequency division becomes relatively crucial so as not to decrease the spectrum efficiency. Although FDMA systems have the advantage that they are immune against frequency-selective fading owing to the narrow spectrum, they have not been favored in recent arguments about digital mobile communication systems. This is because FDMA systems are not superior to TDMA or CDMA systems in other aspects described later.

TDMA systems have become feasible for digital communication systems, where voice signals can be converted into digital signals and therefore the multiplexing in time-division is an easy task. Channels are given periodic time slots in a period time called a frame. Since the channels are orthogonal in time, it is easy to select a channel: we merely open a gate at the wanted time slot. Furthermore, no interference between the time-divided channels occurs even if they are subjected to a nonlinear channel as long as a single carrier is assumed. The number of base station transceivers is decreased by a factor of $1/N$, where N is the number of time division multiplexed channels on a carrier.

Since N channels are multiplexed, the total transmission data rate becomes N times the rate of a single channel. The increase of the data rate requires N times the bandwidth and N times the peak power as compared to an FDMA system: however, the bandwidth per channel and the average power are the same. A high data rate may cause intersymbol interference in a frequency-selective fading channel.

In a TDMA system, time slot synchronization in a frame becomes important since, if the synchronization is lost, the channels may collide with each other. In order to establish the synchronization the base station periodically sends a frame timing signal which subscriber stations use for synchronization. When subscriber stations are located in a coverage area with different distances from the base station, the signal propagation delay difference becomes a problem. A solution to problem is to adjust the subscriber timings by sending a control signal from the base station: At the initial stage of the synchronization procedure, the subscriber station sends a burst signal which is short enough to avoid collisions even when the subscriber is located at the most distant place in the predetermined coverage area. The base station then measures the timing error for that station and feeds back the information to that station. To avoid the effects of residual errors in this timing control system, guard times are placed between time slots. This guard time and the preamble signal, which bears the fixed signal, are the overheads that reduce the spectrum efficiency of TDMA systems.

CDMA systems use different spectrum-spreading codes to discriminate channels from each other. Direct sequence (DS) and frequency hopping (FH) are known as spread-spectrum systems. Consider first the DS system. If we use orthogonal codes such as Walsh functions (Section 2.1.5), N channels are orthogonally multiplexed on the same frequency and at the same time using N times the bandwidth of a channel before spreading. A CDMA system enjoys the benefits of SS (spread-spectrum) communications: a high time resolution and wide bandwidth. Furthermore, no management of frequencies or time slots is required for this system.

Since CDMA systems are based on the orthogonality between spectrum-spreading codes, the degree of correlation becomes crucial for actual systems. A residual cross correlation between codes, and timing differences between codes, reduce the orthogonality. When applied to mobile radio communications, the near–far problem accelerates the loss of orthogonality. For this system, automatic transmit power control, which keeps the received signal level constant at the base station, is necessary.

FH systems are considerably different from DS systems in terms of the above argument. For a slow FH system, channel orthogonality is similar to that of a FDMA system, due to the difference in frequencies. FH systems can randomize the cochannel interference and fading effects.

9.1.6   Inter-Base-Station Synchronization

Inter-base-station synchronization becomes important for TDMA systems and even for FDMA systems that use time-division transmit and receive duplexing (TDD). If the frame timing is different between two neighboring base stations, the overlapped time slots cannot be used by either station. In a TDD system, if the timing of the transmit and receive time periods are different between two neighboring stations, the base station-to-base station interference becomes serious [34]. A strong interfering signal from a base station is received at another base station, since they are placed in higher positions.

There are two approaches to inter-base station synchronization. One is a central control system and the other is a distributed control system. In the former approach, a reference timing signal is broadcast from a central station. For example, a switching station sends the signal over a wireline channel or the timing signal from GPS (Global Positioning System) [35] is used. Using the wireline system, the delay difference between wireline channel is large and changes as the line is switched to another line. The approach using GPS is limited to systems where the GPS receiver is available and its cost contribution is negligible.

As a result of the above assessment, distributed control is preferable to central control for some systems. For this purpose, Akaiwa, Andoh, and Kohama [36] proposed a method in which the inter-base station synchronization is established autonomously with a decentralized control. In this system, each base station measures the timing errors between it and the other base stations and calculates an average error to adjust its timing. The synchronization is established by iterating this procedure. Let us discuss it more specifically.

The timing error between base stations i and j is defined as

$$\Delta T_{ij} = T_j(n) - T_i(n) + 2t_0, \tag{9.22}$$

where $T_i(n)$ and $T_j(n)$ are timings at the nth iteration for stations i and j, respectively, and t_0 is a constant time whose meaning is explained later. The average timing error at base station i is defined as

$$\Delta T_i = \frac{\displaystyle\sum_{j=1, j\neq i}^{N} P_{ij}\Delta T_{ij}}{\displaystyle\sum_{j=1, j\neq i}^{N} P_{ij}}, \tag{9.23}$$

where P_{ij}, the weighting factor for averaging, is the level of received signal transmitted from station j at base station i. Weighting with received signal level emphasizes the timing error of a base station with a higher received signal level. The advantage of this weighting is shown later by computer simulation.

The new timing of base station i is updated as

$$T_i(n + 1) = T_i(n) + \varepsilon\Delta T_i, \tag{9.24}$$

where ε is a constant for the iteration process.

Computer simulation of the above system is made with assumed models. A power–distance relation is assumed as

$$P(r) = P_0 r^{-4},$$

where $P(r)$ is received signal power at distance r, and P_0 is a constant proportional to the transmit power. The same transmit powers for base stations are assumed if not stated. For an intuitive understanding of the calculated value, we assume a transmission bit rate of 512 kbps. Timing errors are normalized with the bit duration T_b ($= 1/512$ ms).

For the simplest model, five stations are placed at the same place and $t_0 = 0$ is used. The synchronization process is shown in Fig. 9.19. The initial timings are chosen arbitrarily. It can be confirmed that the timings synchronize to a phase keeping the average timing a constant throughout the synchronization process. The second model is shown in Fig. 9.20, where five base stations are placed linearly with equal distance. The result with $t_0 = 0$ is shown in Fig. 9.21. Although base stations synchronize with each other,

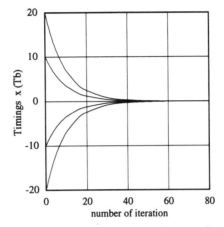

Figure 9.19. Synchronization process. Five base stations are located at the same place.

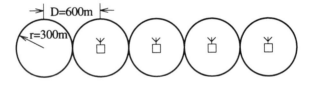

Figure 9.20. Five base stations are linearly placed with equal distance.

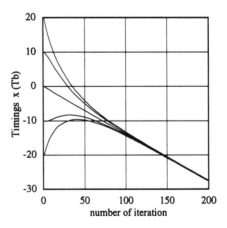

Figure 9.21. Synchronization process for the model in Fig. 9.20.

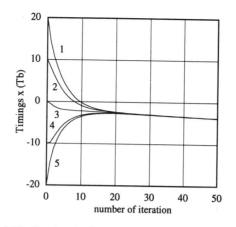

Figure 9.22. Synchronization process. $t_0 = 0.51 T_b$ is assumed.

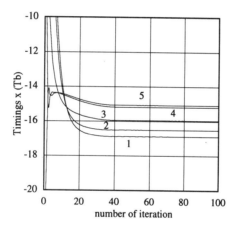

Figure 9.23. Synchronization process. Power is increased 100-fold at base station #5.

their timings are linearly delayed as time proceeds. This can be understood by the fact that, being synchronized, every base station finds its timing is advanced because of the radio wave propagation time and tries to delay its timing. We then assume $t_0 = 0.51 T_b$, which corresponds to the propagation delay for a distance of 300 m. The results are shown in Fig. 9.22. The timings synchronize and stay on a fixed phase.

With the same configuration, Fig. 9.23 shows the results when the power of base station #5 is increased one hundred times. At first glance, we get the impression that synchronization is not good. However, it is good: base station #3 delayed its timing relative to that for base station #5 so as to synchronize to the strong signal from the base station. If base station #3 synchronized to the base station in the absolute timing, the signals would collide with each other due to the time delay. Base station #2 delays its timing for the same reason. The timing delay of base station #1 relative to base station #2 is small, since the received signal level from base station #5 is not significant at its location due to propagation loss. In the above situation, absolute timing synchronization is not good, but the relative timing synchronization is good. This effect is due to the weighting factor introduced in our algorithm. The same effect is confirmed for a model in which base stations are unequally spaced. The relative synchronization characteristics of our algorithm become effective for microcell systems, where radio wave propagation is highly irregular.

The inter-base station synchronization is confirmed in a two-dimensional cell model. A theoretical analysis [37] of the synchronizing process is described in Appendix 9.1.

To prevent the phenomenon of linear drifting of the autonomous synchronization method, one technique is for each base station to control the timing drift to be zero after timings are within a given range [38]. Another technique [34] is to use a few timing reference stations, to which other base stations synchronize autonomously. Another method [39] uses a correction factor t_0, which is determined by measuring the distance between base stations in advance by, for example, reading a map or using the GPS. A further technique [40] is to calculate the time delay by reporting the timing error with each base station using a common radio control channel.

The residual timing error depends on the selection of the weighting factor for averaging and on the value of the iteration coefficient.

9.2 DIGITAL TRANSMISSION IN ANALOG MOBILE COMMUNICATION SYSTEMS

In analog mobile communication systems, digital signals must be transmitted for system control purposes. Table 9.2 [41] shows a summary of digital transmission technologies used in analog mobile communication systems such as cellular systems for NTT, US (AMPS), the Nordic (NMT), and the two-way radio system (MCA) in Japan. The data transmission speed is slow. Major parts of circuits including an FM modulator and a limiter–discriminator are used in common between the digital transmission and the analog FM voice signal transmission. Digital data are transmitted through analog voice channels, where the dc component is blocked. Therefore, Manchester coding or subcarrier systems are used.

9.3 PAGING SYSTEMS

A paging system is a one-way (wireline to mobile) digital transmission system. Initially, tone-only systems appeared, where an audio tone is generated when

TABLE 9.2 Digital transmission technologies in some analog mobile communication systems [41] (Copyright © IEICE 1985)

	Mobile Telephone System			Trunked System
	NTT	AMPS	NMT-450	MCA[a]
Bit rate (bps)	300	10 000	1200	1200
Code	Split phase	Split phase	NRZ	NRZ
Modulation	FSK	FSK	Subcarrier FSK[b]	Subcarrier FSK[b]
Maximum frequency deviation (kHz)	±4.5	±8	±3.5	±3.5
Error correction	BCH (recycled transmission)	BCH (recycled transmission)	Hagelberger	Hagelberger
Channel spacing (kHz)	25	60	25	25
Frequency band (MHz)	800	800	450	800

[a]Press-to-talk system for private communication.
[b]Mark frequency = 1200 Hz; space frequency = 1800 Hz.

a terminal is called. The next generation systems transmit numerical signals to be displayed at a receiver. Even messages can be sent in modern paging systems. Message paging systems will grow with evolution of the electronic-mail services. The message communication hardly interrupts the person being called, and a vibrator mode of a call alerting gives no interference to other people.

The pager terminals have became compact, lightweight, and slim. Direct conversion systems are employed for some card-type receivers to reduce the size of the terminal.

The POCSAG (Post Office Code Standardization Advisory Group) systems are widely known [42]. The specification is summarized in Table 9.3 and signal format is shown in Fig. 9.24. For a tone-only service, 230 000 subscribers can be accommodated per channel.

To cope with increasing numbers of users roaming within Europe, the ETSI (European Telecommunication Standards Institute) is planning a paging system called ERMES (European Radio Messaging System) [43]. The specification is shown in Table 9.4. The channel data rate is increased to 6250 bps and multilevel FM is employed to obtain high-speed transmission within the channel spacing of 25 kHz. Simultaneous transmission from multiple base stations is used. In this system, signals from different base stations should be time-synchronized within 50 ms. ERMES provides tone, numeric, alphanumeric, and transparent data paging services. Supplementary services, such as roaming, message repetition, legitimization, urgent message

TABLE 9.3 POCSAG paging system

Frequency	280 MHz band
Channel spacing	25 kHz
Modulation	NRZ-FSK
Channel speed	512 bps
Error correction code	BCH (31,21)

TABLE 9.4 ERMES specification

Code word type	Shortened cyclic code (30,18)
Address capacity	32 million for each country
Frequency band	169.425–169.8 MHz
Number of RF channels	16
Channel spacing	25 kHz
Modulation	4PAM/FM (4-level FSK)
System bit rate	6250 bps

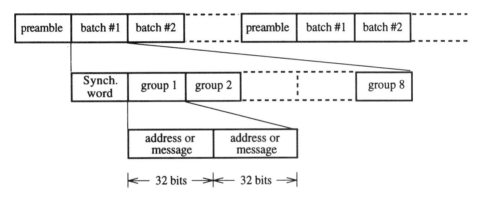

Figure 9.24. Signal format for POCSAG paging system.

indication, and so on, are planned. Pager terminals must be able to scan through 16 RF channels.

In Japan an advanced radio paging system is standardized based on a system called FLEX* (* registered trade mark of Motorola). Three data rates of 1600, 3200, and 6400 bps can be selected to adapt to the service requirements: a data rate is chosen by compromising the traffic density and the coverage area size; a high data rate service is used for a high subscriber density area with a shortened maximum service range. Four-level FSK is used for data rates of 6400 and 3200 bps and 2-level FSK for data rates of 3200 and 1600 bps. A time-diversity system where the same data are sent up to four-times is adopted to obtain long service ranges. A code word consists of BCH (31,21) and a parity bit. The flexible operation of the system is performed through sending a frame information signal that tells data rates, number of repeated transmissions, and so on. A variety of services such as tone-only, numeric, alphabetic, and message and data transmission are efficiently accommodated through the use of different fields (i.e., address, vector and message fields). The vector field data indicates the start and end positions of the message for a called terminal.

9.4 TWO-WAY DIGITAL MOBILE RADIO

So-called two-way mobile radio is a press-to-talk private radio system for applications such as car dispatch services and police communications. The first application of digital voice transmission in mobile radio communications was the two-way radio system. Since digital mobile radio technologies were not well developed at that time, ADM (adaptive delta modulation) voice coding and digital FM were used [44]. A 150 MHz digital mobile radio equipment using 16 kbps ADM voice coding and 4-level FM at 25 kHz channel spacing was reported [45,46]. The block diagram of the radio equipment is shown in Fig. 9.25. The RF, IF, and modulator/demodulator circuits are used for both analog and digital modes of communication.

A digital mobile radio for police service was reported in [47]. Using efficient voice coding of 8 kbps and constant envelope digital modulation, this system is operated at 12.5 kHz channel spacing with an error rate of 5×10^{-2}.

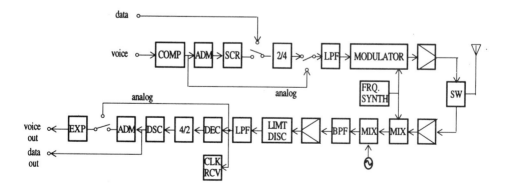

Figure 9.25. Block diagram of two-way analog/digital mobile radio using ADM voice coding and 4-level FM.

9.5 MOBILE DATA SERVICE SYSTEMS

Mobile radio communication services such as dispatch traffic are improved with the introduction of digital data transmission systems: Transmission of text and other data messages can replace or assist voice communication. These kinds of digital mobile communication systems have been in commercially use.

9.5.1 MOBITEX

MOBITEX is a mobile communication system [48,77] developed by the Swedish Telecommunications Administration, where digital data as well as analog voice signals are transmitted through trunked frequency channels. The base stations operate in duplex, while the mobile units operate in two-frequency simplex or duplex. The channel spacing is 25 kHz. The digital modulation is FFSK with a data rate of 1200 kbps. A BCH (15,10) code for data coding and ARQ are used in the radio channel.

9.5.2 Teleterminal System

Teleterminal System is a public land mobile radio data communication system developed in Japan [42,49]. Data signals are transmitted in trunked packet-switched channels. A summary of the system specifications is shown in Table 9.5. The Reed–Solomon code (15,11,5) and ARQ are adopted for error control in the radio channel. The multiple access method is a polling system, where a polled terminal can transmit data packets.

9.5.3 Mobile Data Systems in Analog Cellular Systems

Since cellular systems are connected to the wireline public switched telephone network (PSTN), it is desired to extend data services in the PSTN to existing

TABLE 9.5 Teleterminal System specification

	Mobile	Base
Frequency	800 MHz band	
Mode of radio transmission	Two-frequency simplex	Two-frequency duplex
Modulation	FSK	
Channel spacing	25 kHz	
Channel speed	9600 bps	
Transmit power	5 W	20 W
Transmit/receive spacing	55 MHz	

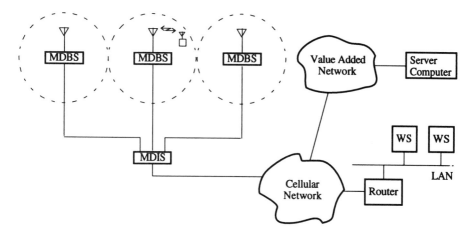

MDBS : Mobile Data Base System
MDIS : Mobile Data Intermediate System

Figure 9.26. CDPD network.

analog cellular mobile systems. To this end the CDLC (cellular data link control) system [50] and the CDPD (cellular digital packet data) system [51] are proposed.

In the CDLC system, the CCITT V26 modem is used for a baseband voice channel spanning from a PSTN to an analog FM cellular system. BCH (16,8) and Reed–Solomon (72,68) codes are selected for the forward channel depending on the circuit conditions, and the Golay (23,12) code is used in the backward channel. The protocol is a full-duplex with layer 2 protocol based on HDLC standards.

In the CDPD system, data signals are transmitted as packets through a full-duplex radio channel with channel data rates of 19.2 kbps. The CDPD network is shown in Fig. 9.26. The mobile data base stations are probably co-sited with analog FM cellular base stations. The CDPD system detects the cellular system channels that are not used for voice communication and uses them for data transmission. When a voice communication starts in the channel, it stops the data communication in that channel and switches it to another idle channel (channel hopping). The channel hopping is controlled by the mobile data base station broadcasting the channel hopping signal to the data terminals.

The CDPD protocol supports the physical layer and the data link layer. Any of the upper-layer protocols can be used: the TCP/IP (Transmission Control Protocol/Internet Protocol) and the OSI (Open, System Interconnection) [16] for computer networks. Through the use of the value-added networks, various data services will be provided for mobile data users.

9.6   DIGITAL CORDLESS TELEPHONE

Analog cordless telephones are widely used all over the world. Although mobility is limited to a distance of several tens of meters, many people enjoy the wireless communication. The penetration is limited to in-house use. The digital cordless telephone system is expected to be widely used in business and public areas as well as at home, owing to high capacity, high security due to digital scrambling, and improved data service capability. Considering many different applications of digital cordless telephone, the performance is limited by the air interface. In this section, proposed digital cordless telephone systems are described.

9.6.1   CT-2

The CT-2 (second-generation cordless telephone) was developed in the United Kingdom. The specification is summarized in Table 9.6 together with those for other systems. Time division transmit and receive duplexing (TDD) on a frequency was adopted. Since one block of frequency band is sufficient for this system, in contrast to the two-frequency duplexing system, the spectrum allocation becomes easier. In addition, diversity transmission is realized by employing circuits for the diversity system only at the base station due to high correlation between the up-link and down-link channels on the same frequency. Furthermore, direct communication between mobile terminals is easily made.

The FDMA system is adopted with transmission rates of 32 kbps. The voice coding is 32 kbps ADPCM. No channel coding is used. These concepts are based on emphasizing low system cost and high voice quality.

A public service known as the Telepoint with CT-2 was launched first in the United Kingdom. This system is limited to calls originating from the mobile station: a paging system is used for calling a mobile station. This fact

TABLE 9.6   Digital cordless telephone specification

	CT-2	DECT	PHS
Frequency band (MHz)	900	1900	1900
Access system	FDMA/TDD	TDMA/TDD	TDMA/TDD
Channels/carrier	1	12	4
Voice coding	32 kbps ADPCM	32 kbps ADPCM	32 kbps ADPCM
Modulation	GMSK	GMSK	$\pi/4$ QPSK
Channel speed (kbps)	72	1152	384
Carrier spacing (kHz)	100	1728	300
Frame length (ms)	2	10	5
Transmit power (mW)			
Peak	20	250	80
Average	10	10	10

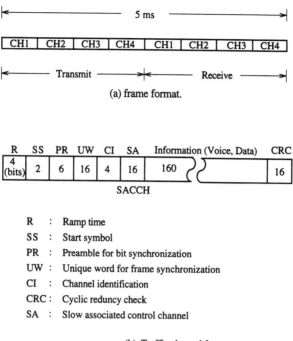

(a) frame format.

R : Ramp time
SS : Start symbol
PR : Preamble for bit synchronization
UW : Unique word for frame synchronization
CI : Channel identification
CRC : Cyclic reduncy check
SA : Slow associated control channel

(b) Traffic channel format.

Figure 9.27. PHS signals.

and the smaller service area than with cellular phone systems are the probable reasons for lack of success of the Telepoint service in the United Kingdom.

9.6.2 DECT

DECT (Digital European Cordless Telecommunications) was standardized by ETSI (European Telecommunication Standard Institute). It is a 12-channel-per-carrier TDMA/TDD system with channel speed of 1152 kbps. It uses frequencies in the 1900 MHz band. The other major specifications are the same as those for the CT-2 system. Rather high channel data rates will characterize the services provided by the DECT system.

9.6.3 PHS

The PHS (Personal Handy System) was standardized in Japan. Initially it was called PHP (Personal Handy Phone). It is a 4-channel per carrier TDMA/TDD system. The modulation is $\pi/4$ shifted QPSK. The frame length is 5 ms. The other major specifications are similar to those for DECT. The frame structure is shown in Fig. 9.27.

The peak transmit power is 80 mW, resulting in an expected zone radius of 100–200 m. Functions for interference detection before channel selection as well as during communication are required as part of the standardized functions. For starting communication, a radio channel which is proven not to interfere with other zones by performing the interference detection is selected at each base station. During communication, if interference is detected, the channel is switched to another channel (intracell hand-off).

9.7 DIGITAL MOBILE TELEPHONE SYSTEMS

The fast growth of digital mobile radio communication technologies is strongly driven by research and development activities for digital mobile telephone systems. A pioneering experimental TDMA system was reported in 1982 [53], recognizing the advantages of a TDMA approach. However, the field test results showed difficulty in high-speed TDMA digital transmission due to the fast frequency-selective fading effects. The first Nordic Seminar on Digital Land Mobile Radio Communications in 1985 astonished digital mobile radio communications researchers, or at least the author of this book, by presenting so many ambitious proposals, some of which affected the later pan-European (GSM) digital cellular systems. The pan-European organization for the GSM system standardization was launched. It was not long before efforts for standardization of digital mobile telephone systems started in the United States and Japan. In this section, the GSM, the American digital cellular, and the (Japanese) personal digital cellular systems are described in brief. For details refer to [52] and [68].

9.7.1 The GSM System

The European countries set up a committee called GSM (Group Special Mobile) for making a standard pan-European digital mobile telephone system that could be introduced in the 1990s. Criteria for the GSM (later called Global Systems for Mobile Communication) systems, are such as

- Spectral efficiency
- Subjective voice quality
- Mobile cost
- Hand-portable feasibility
- Base station costs
- Coexistence with the existing systems

A frequency band (890–915 MHz for mobile, 935–960 MHz for base) which had not been used for analog systems in many European countries was assigned.

Many systems were proposed as candidates for the GSM system. It is worth recalling them here. Table 9.7 summarizes the proposed systems. All the multiple access methods—FDMA, TDMA, and CDMA—are included. An automatic equalizer is assumed to be used in the TDMA systems. Through the efforts of many people, the GSM system was standardized. Table 9.8 [42] summarizes the GSM system specifications.

A TDMA system was selected by a comparison between analog vs. digital, FDMA vs. TDMA, and narrow-band vs. wide-band TDMA. The RPE-LTP (regular pulse excited predictive coding with long-term prediction) with a voice coding rate of 13 kbps was adopted. The block length for voice coding is 20 ms. The coded voice signal is protected with a convolutional code resulting in a gross rate of 22.8 kbps. For digital modulation, DPM was tentatively chosen; later, GMSK with baseband filtering, where the 3 dB bandwidth normalized by the bit rate $B_b T$ is 0.3, was selected. This digital modulation scheme is a constant envelope modulation that is different from the linear modulation that is used in the US and the Japanese mobile telephone systems. It is not clear whether linear modulation was discussed as a candidate method. The proposal for linear modulation for mobile communication appeared in 1987 [54].

The TDMA frame structure is shown in Fig. 9.28. A TDMA frame includes 8 time slots. The time slots for the up-link are intentionally delayed relative to those for the down-link (base to mobile) for use of a switch instead of a duplexing filter for transmit–receive duplexing purposes. The transmission rates become 270.833 kbps, accommodating 8 channel-coded time slots. This transmission rate cannot be achieved for a mobile telephone system unless an automatic equalizer is used. As a standardized performance of the equalizer, the (average) delay spread of 16 μs must be coped with. Standardization for the equalizer is not necessary; the choice of equalizer is left for the equipment manufacturers. Viterbi or decision feedback equalizers are used for this purpose. The training signals of length of 26 bits are used for equalizer training as shown in Table 9.9. The autocorrelation function of the training signal is shown in Fig. 9.29. This is calculated by taking correlation between the training signal code and the central 16 bits of the code, assuming -1 for 0. The autocorrelation function shows a sharp peak at the center and a null within ± 5 symbol time. Thus, with this code a channel impulse response with a time duration within ± 5 symbols results.

The GSM transceiver must be frequency-agile for two purposes: (i) for a mobile station to monitor one channel on different frequencies during a frame length of 4.6 ms including the transmit and reception periods; and (ii) to randomize cochannel interference. The base station hops at every frame slot, resulting in a hopping rate of 217 hops/s.

Channels are logically divided into traffic channels (TCH) and control channels. The TCH is an information-bearing channel. The control channels include the following:

TABLE 9.7 Proposed systems for the GSM system

System	Company	Multiple Access	Modulation	Bit Rate (kbps)	Channel Spacing (kHz)
S900D	Bosch/Ant/Matra	TDMA	4-level FSK	128	250
DMS90	Ericsson	TDMA	GMSK	340	300
SFH900	LCT/TRT	TDMA/CDMA	GMSK	200	150
CD900	SEL/AEG/ATR/SAT/ ITALTEL	TDMA/CDMA	QPSK	4000	6000
MATS-D	TeKaDe/TRT		QAM	1218	1250
			GTFM	19.5	25

TABLE 9.8 GSM system summary

Frequency band	890–915 MHz (mobile to base)
	935–960 MHz (base to mobile)
Channel spacing	200 kHz (interleaved)
Number of frequency channels	124
Multiple access	8 channel-TDMA
	Time-slot length 577 μs (156.25 bit)
Modulation	GMSK ($B_b T = 0.3$)
Channel speed	270.833 kbps
Voice coding	RPE-LTP (13 kbps, 22.8 kbps with FEC)
User data speed	9.6, 4.8, 2.4 kbps
Automatic equalizer	Capable of the maximum delay spread of 16 μs
Frequency hopping	217 hops/s
Base station coverage (radius)	0.5–35 km
Transmit power	0.8–20 W (mobile), 2.5–320 W (base)

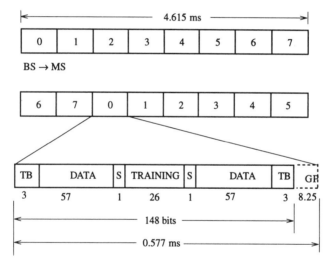

Figure 9.28. GSM system frame format.

- BCCH (broadcast control channels) are for broadcasting information about the system to all mobile stations.
- FCCH (frequency correction channels) are for mobile stations to adjust their frequency to a stable frequency of a base station.
- PCH (paging channel) is for paging mobile stations.
- RACH (random access channels) are up-link channels, where mobile stations make random access to the channels, in order to send control signals (e.g., channel demand for call origination at mobile stations).

TABLE 9.9 Training signal codes of the GSM system

Training Sequence No.	Values
1	00100101110000100010010111
2	00101101110111100010110111
3	01000011101110100100001110
4	01000111101101000100011110
5	00011010111001000001101011
6	01001110101100000100111010
7	10100111110110001010011111
8	11101111000100101110111100

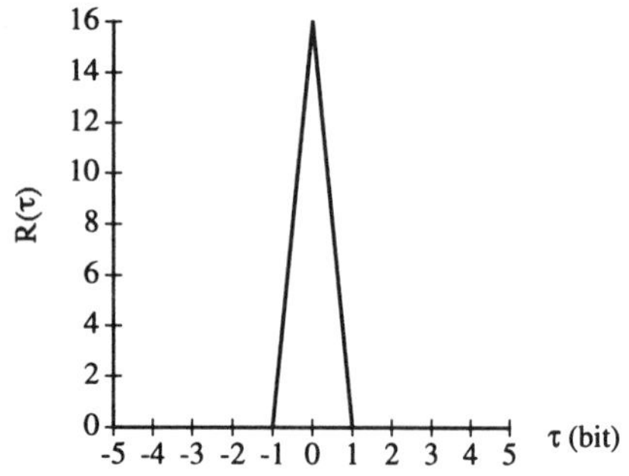

Figure 9.29. Autocorrelation function of a training signal for GSM systems.

- SACCH (slow associated control channels) are channels for transmitting slow control signals together with information signals.
- FACCH (fast associated control channels) steal traffic time slots to be used for fast control requirements such as hand-off.

The GSM network configuration is shown in Fig. 9.30. Some base station controllers are connected to a mobile switching center. Base transceiver stations are connected to the base station controller, where use of radio channel is controlled for paging, hand-off, and so on. A mobile station can be used between different people by use of the subscriber identity card. Mobile switching centers are equipped with file systems such as home location register, visiting location register, and equipment identity register. Many data services as well as voice service are provided with the standardized protocols, which are beyond the scope of this book.

9.7.2 Digital Cellular Systems in North America

In the United States the shortage of capacity of the existing analog system (AMPS) was expected with the increase in subscribers, especially in big cities

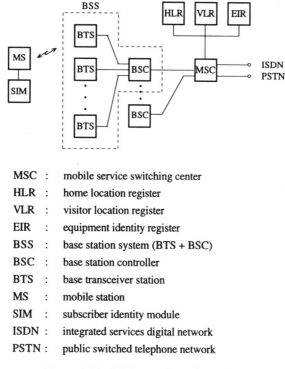

MSC : mobile service switching center
HLR : home location register
VLR : visitor location register
EIR : equipment identity register
BSS : base station system (BTS + BSC)
BSC : base station controller
BTS : base transceiver station
MS : mobile station
SIM : subscriber identity module
ISDN : integrated services digital network
PSTN : public switched telephone network

Figure 9.30. GSM network configuration.

such as New York, Chicago, and Los Angeles. To cope with this, efforts toward a higher-capacity system were begun. A condition was assumed that no new frequency spectrum was assigned to this system, but that the AMPS system should be gradually replaced with the new system. The TIA (Telecommunication Industry Association), which was the main body for the new system standardization, decided to adopt a TDMA system in 1990. In 1991 the TIA issued the standard system (IS-54). A dual-mode (analog and digital) system was considered owing to the requirement for gradual replacement.

Table 9.10 summarizes the system specifications [42]. The interleaved channel spacing of 30 kHz is the same as that for the AMPS system for ease of the dual mode of operation. A 3-channel TDMA with six time slots system is adopted. The frame structure is shown in Fig. 9.31. A voice signal uses slots 0 and 3 or 1 and 4 or 2 and 5. If half-rate voice coding is adopted, it becomes a 6-channel TDMA system. For voice coding VSELP is used with 7.95 kbps and 13.5 kbps without and with error protection, respectively. The digital modulation is $\pi/4$ shifted QPSK, which was the linear modulation first introduced into cellular systems. A maximum delay difference of $40\,\mu s$ in

TABLE 9.10 American digital cellular (IS-54) system summary

Frequency band	869–894 MHz (BSS to MS)
	824–849 MHz (MS to BSS)
Channel spacing	30 kHz (interleaved)
Number of frequency channels	832
Multiple access	3-channel/6-time slot TDMA
Time-slot length	40/6 ms
Modulation	$\pi/4$ shifted QPSK
Channel speed	48.6 kbps
Voice coding	VSELP (7.95 kbps, 13 kbps with FEC)
Automatic equalizer	Capable of maximum delay difference of 40 μs
Base station coverage (radius)	0.5–40 km
Transmit power	0.6–4 W (mobile)

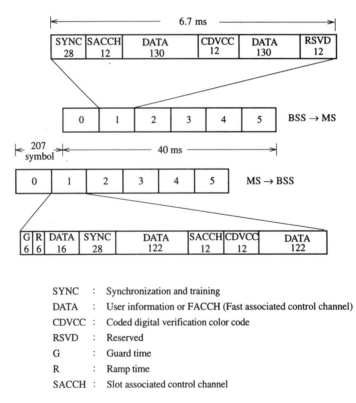

SYNC	:	Synchronization and training
DATA	:	User information or FACCH (Fast associated control channel)
CDVCC	:	Coded digital verification color code
RSVD	:	Reserved
G	:	Guard time
R	:	Ramp time
SACCH	:	Slot associated control channel

Figure 9.31. Frame structure for the American digital cellular (IS-54) system.

TABLE 9.11 CDMA digital cellular system (IS-95) summary

Spread spectrum system	Direct sequence
Chip rate	1.2288 MHz
Data rate	9600 bps
Spreading factor	128
Modulation	QPSK
Bandwidth	1.23 MHz
Voice coding	QCELP (8 kbps)
Error correction	Convolutional coding/Viterbi decoding
Automatic transmit power control at mobile station	Dynamic range of 85 dB, Step size of 0.5–1 dB

radio channels must be handled with an automatic equalizer. A long slot length of 6.7 ms requires a high tracking speed for the equalizer, since the channel characteristics change significantly during the burst time length under fast fading.

The IS-54 system has three times the capacity of the AMPS system, if the reuse factor is the same between the two systems. Discussion of the reuse factor is difficult since it is dependent on a required protection ratio for cochannel interference, the probability that the signal to noise ratio drops below a given threshold, and the voice quality under cochannel interference and fading conditions. For comparison of voice quality between different voice coding systems including an analog FM system, the test of opinion of many users must be carried out in an actual communication channel environment.

After the standardization of the IS-54 TDMA system, another candidate system using a CDMA architecture [55] was proposed by Qualcom, Inc. This CDMA system was also adopted as another standard (IS-94) system. The specifications are summarized in Table 9.11. It is a direct-sequence spread-spectrum system. The multiple access system is direct-sequence CDMA, which is the first spread-spectrum system standardized for cellular systems. The spreading factor is 128. The spreading code is a combination of PN (pseudo noise) codes and Walsh functions. By using the Walsh functions with length of 64 bits, 64 orthogonal channels are provided for a base station. By using different PN codes between base stations, the same frequency is used for each base station, that is, the reuse factor or cell group is 1. A spectrum efficiency of around 10 times that of the AMPS system is claimed. For this result, in addition to the one-frequency reuse, automatic transmit power control, soft hand-off or macrodiversity, voice-activated transmission, and sectorization are assumed. Automatic transmit power control, to maintain received signal level within a given range at a base station, is necessary for a spectrum-efficient CDMA system to cope with the near–far effect. Soft hand-off means that the hand-off is gradually made as in macrodiversity between base stations. This technique is effective under shadowing conditions. Some channels must be reserved for soft hand-off purposes.

Voice-activated transmission, where the carrier signal is transmitted only when voice signal is detected, decreases the interchannel interference by the ratio of silence time period to the whole speech period.

In this system, sectored cells are also considered. The effect of sectorization results in an N times increase in spectrum efficiency by reducing the size of the service area to $1/N$, where N is the number of sectors.

Frequency guard bands must be reserved between different frequency carriers, which reduces the spectrum efficiency.

Thus, the high capacity of the CDMA system is maintained through many advanced techniques. Validation of the claimed capacity of the CDMA system is difficult because all the mobile stations must in principle take part in the experiment. In contrast, interchannel interference experiment for TDMA or FDMA systems can be carried out using a small number of mobile stations that interfere with a considered channel.

9.7.3 Digital Cellular Systems in Japan

The Japanese digital cellular standard was made by the Research & Development Center for Radio Systems (RCR) in April 1991. A summary of the specifications is shown in Table 9.1.2 [56]. New frequency bands are assigned at 800/900 MHz and 1.5 GHz. The carrier spacing of 25 kHz is a little narrower than that for the US IS-54 system of 30 kHz. This is achieved with voice coding at a lower rate of 11.2 kbps. The modulation and multiple access system are the same as those for the US IS-54 system. The half-rate voice coding was already standardized: the PSI-CELP (Section 7.9.6) at a coding rate of 3.45 kbps and 5.6 kbps without and with error protection was adopted.

TABLE 9.12 Specification summary for the Japanese digital cellular system

Frequency band	810–826 MHz (BSS → MS)
	940–956 MHz (MS → BSS)
	1477–1489 MHz (BSS → MS)
	1429–1441 MHz (MS → BSS)
	1501–1513 MHz (BSS → MS)
	1453–1465 MHz (MS → BSS)
Carrier spacing	25 kHz (interleaved)
Modulation	$\pi/4$ shifted QPSK
Multiple Access	3-channel TDMA (full rate)
	6-channel TDMA (half rate)
Channel speed	42 kbps
Voice coding	VCELP (6.7 kbps, 11.2 kbps with error protection)
	PSI-CELP (3.45 kbps, 5.6 kbps with error protection)
Automatic equalizer	Optional
Diversity	Optional

The automatic equalizer and diversity system are optional. Since a diversity receiver shows immunity against frequency-selective fading as well as flat fading, the automatic equalizer might be unnecessary as long as the delay spread is not too high. Under Rayleigh fading, the diversity system requires an average carrier signal ratio (CIR) of 11 dB for a bit error rate of 2×10^{-2}, which is the threshold value for the system; without the diversity system, a CIR of 16 dB is required.

9.8 WIRELESS LOCAL AREA NETWORK

Local area networks (LAN) are widespread through networking of down-sized workstations or personal computers. Wireless LANs gives mobility to terminals, limited to where a power line outlet is placed. If the power source of the terminal is a battery, even portable terminals are available. Low-speed systems with data rate of 1200–9600 bps are used at shops for POS (point of sale) terminals and at factories for various forms of delivery control. High-speed systems with data rate of over several hundred kbps are used for personal computers or workstation systems. WaveLAN was developed by NCR for banking services and has a data rate of 2 Mbps at 900 MHz. ALTAIR is developed by Motorola for "Ethernet," which is a widespread wireline LAN, and has a data rate of 10 Mbps at the 18 GHz band.

IEEE (The Institute of Electrical and Electronics Engineers) established a committee (IEEE 802.11) to develop a standard for wireless LANs. The short- and long-term solutions are given. The short-term system adopts a data rate of 1–2 Mbps at 902–928 MHz. This spectrum band, called ISM (Industry, Science and Medical) band, can be used without license provided spread-spectrum communications are used. The long-term system adopts data rates higher than 10 Mbps. A bandwidth of 70–140 MHz at a frequency less than 5 GHz is expected.

Multimedia (voice, data and image) communications is an important issue for modern communications, including wireless LANS. ATM (asynchronous transfer mode) is the basic model assumed for providing multimedia services in wireline networks. It is certain that future mobile and wireless networks will also be required to interface with fixed multimedia networks, possibly carrying ATM-type traffic. The flexible ATM air interfaces could meet the varying service and bit rate needs and offer simpler internetworking with ATM networks and terminals. An ATM-based wireless LAN architecture has been proposed [57] where a dynamic reservation TDM/TDMA channel format is assumed for medium access control (MAC).

Frequency allocation and standardization for a new system are difficult and time-consuming tasks. Therefore, it is convenient for a wireless LAN to be implemented on the digital radio channels that are already assigned to other radio communication systems, for example, digital cordless telephone systems. Such a wireless data communication system is proposed [58]

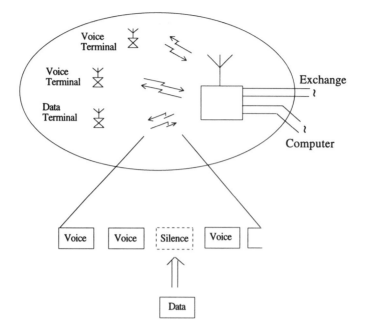

Figure 9.32. Voice and data integrated wireless local area network.

in which data communication is integrated into the digital voice channels. The data signal is transmitted only during silent periods of voice communication as well as through the idle channels in the form of packets. In this system, of course, the radio signal for voice must be transmitted only when voice signal appears (voice-activated transmission). The system model is shown in Fig. 9.32. The central issue for this system is how to avoid interference from data transmission on voice communication. In this system the base station controls the data transmissions by broadcasting a control signal in the down-link: when the base station finds a silent period in a voice channel, it broadcasts an idle signal to permit data transmission for any data terminals. The collision of voice and data signals may occur when the voice terminal restarts its transmission. In the proposed system, the voice terminal sends a short packet to indicate the restarting of transmission. The voice restart packet is either received or collides with a data packet. In both cases, the base station inhibits data transmission. The voice terminal does not care about the control from the base station. The success of the system depends on the collision detection capability. A CRC (cyclic redundancy check) can be used for detecting a collision. The collision may not be detected due to the near–far problem in mobile radio communication systems; signals are received at very different levels and the signal with the highest level is received, capturing the other signals. Computer simulation on the collision

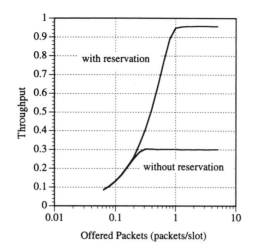

Figure 9.33. Throughput performance of an integrated voice and data wireless local area network.

detection capability for a $\pi/4$ shifted QPSK system with a packet length of several hundred bits shows that a collision can be completely detected if the received signal levels are kept at ± 7 dB; out of this range, a signal is captured by the other signal. Automatic power control must be incorporated in the proposed system.

We cannot differentiate a collision between data packet and data packet or between data packet and the voice restart packet, the base station inhibits data transmission after every collision. This causes a little deterioration in throughput performance. Techniques are proposed to raise the throughput performance by intentional use of the capture effect. One method is to send the voice restart packet with enough transmit power to capture a data packet; then collisions are only between data packets and therefore the base station does not need to inhibit data transmission after a collision. Another method is to increase transmit power for every voice restart packet and for randomly selected data packets. Although the base station inhibits data transmission after a collision, the collision probability between two data packets is decreased by one-half. The other method is to introduce a reservation scheme for data transmissions. This is very effective in TDMA systems, where the time slot length is much shorter than the average data packet length. The performance is shown in Fig. 9.33.

9.9 MOBILE SATELLITE SYSTEMS

In mobile satellite systems, mobile radio communication is made via transceivers aboard satellites. The wide coverage area of a mobile satellite system differentiates it from terrestrial mobile communication systems. This property of the mobile satellite system makes it suitable for communication services in wide areas where traffic density is sparse.

A satellite may be geostationary in orbit above the equator at a height of 36 000 km. This kind of satellite system is appropriate for fixed stations, since steering of antennas is not needed even for a directive antenna. The round-trip time delay between an earth station and the satellite is 300 ms, which cannot be neglected in full-duplex voice communication. An LEO (low earth orbit) satellite orbits the earth at a height as low as 1000 km. Since it appears and disappears periodically in a period of the orbit, many satellites must be prepared to make communication available at any time. The LEO satellite system is suitable for mobile radio communications because of low propagation loss and short time delays due to short distances.

INMARSAT (the International Maritime Satellite Organization), established in 1979, started communication services for ships and later for aircraft and land mobiles. Analog and digital voice, facsimile, and data services are provided through geostationary satellites. A geostationary satellite system, Omni-TRACS, is in service for slow data and positioning. American Mobile Satellite Corp. in the United States, Telesat Mobile Inc. and AUSSAT in Australia are planning mobile communication services.

Several LEO satellite systems are planned. The "Iridium" system is proposed by Motorola Inc. This system consists of 66 satellites on seven polar-orbits at a height of 780 km. (The number of the satellite was initially planned for 77. Incidentally iridium, a rare metal, has the atomic number of 77.) Digital voice and data communications are served at a rate of 2400 bps. For a portable terminal, an average transmit power of 300 mW and a peak power of 7 W are assumed. Transmitted signals from a portable terminal are received at a satellite and relayed to other satellites to be transmitted to a terrestrial gateway that is connected with the public telephone network.

The Orbcomm by Orbital Science Corp. uses 20 satellites in orbits at a height of 970 km. The Globalstar by Loral Qualcom Satellite Service uses 48 satellites. This system is complementary to terrestrial mobile radio system: a portable terminal communicates to the satellite system in out of the service area of the terrestrial system using a dual-mode transceiver.

The GPS (Global Positioning System) being operated by the US defense agency consists of 24 satellites in LEO at a height of 20 000 km. The satellites send spread-spectrum positioning signals using two codes: Precision (P) code and Clear and Acquisition (C/A) code. The latter is open to the public. A GPS terminal receives and decodes these codes for time and orbital information and, using this information, decides its position as the crossing point of spheres of signals radiated from three satellites. If four satellites are considered, height as well as position can be determined.

9.10 PERSONAL COMMUNICATION AND FPLMTS

Many concepts have been proposed for future telecommunication systems. Personal communication or UPT (universal personal telecommunication) is commonly referred concept. In a CCITT meeting in 1991, UPT was defined as the following:

> UPT enables access to telecommunication services while allowing personal mobility. It enables each UPT user to participate in a user-defined set of subscribed services and to initiate and receive calls on the basis of a personal, network-transparent UPT number across multiple networks on any fixed or mobile terminal, irrespective of geographical location, limited only by terminal and network capabilities and restrictions imposed by the network operator.

To enhance the personal mobility, a personal terminal should be able to be used anywhere in a communication network (terminal mobility). The terminal mobility is maximized with mobile radio communications and many mobile radio communication systems are under study for personal communication services. In Europe UMTS (Universal Mobile Telecommunication System) is being standardized in ETSI. The multiple access method is a key issue to this system. Both TDMA and CDMA alternatives are under investigation within the RACE (Research and Development in Advanced Communications Technologies in Europe) program. In the United States, seven task-groups have been established for a standardized personal communication service (PCS) system. For further description, readers may refer to [69–76].

FPLMTS (Future Public Land Mobile Telecommunication System) is being standardized in Task Group 8/1 of CCIR (International Radio Communications Consultative Committee). This is a worldwide standardized system. The FPLMTS is an evolutional mobile communication system that can provide a variety of services of quality comparable to that of fixed communications by accessing public networks, with use by many kinds of mobile terminal, which are supported with standardized radio interfaces. Comprehensive studies are being made.

APPENDIX 9.1 THEORETICAL ANALYSIS OF AUTONOMOUS INTER-BASE-STATION SYNCHRONIZATION SYSTEM

We rewrite Eq. (9.22) as

$$\Delta T_{ij} = [T_j(n) - d_{ij} + t_0] - [T_i(n) - t_0], \qquad (A9.1)$$

where d_{ij} is the propagation time delay from base station j to base station i. Substituting Eq. (A9.1) into Eq. (9.23), we get

$$\Delta T_i = \frac{1}{P_i} \sum_{j=1, j \neq i}^{N} p_{ij} T_j(n) - T_i(n) + M_i, \tag{A9.2}$$

where

$$M_i = 2t_0 - \frac{1}{P_i} \sum_{j=1, j \neq i}^{N} p_{ij} d_{ij} \tag{A9.3}$$

and

$$P_i = \sum_{j=1, j \neq i}^{N} p_{ij}. \tag{A9.4}$$

Equation (9.24) then becomes

$$T_i(n+1) = (1 - \varepsilon) T_i(n) + \frac{\varepsilon}{P_i} \sum_{j=1, j \neq i}^{N} p_{ij} T_j(n) + \varepsilon M_i. \tag{A9.5}$$

Same Received Power Case

We first consider the simplest case where the received signal levels, p_{ij}, are all the same. In this case we have

$$\Delta T_i = \frac{1}{N-1} \sum_{j=1, j \neq i}^{N} T_j(n) - T_i(n) + M_i, \tag{A9.6}$$

$$M_i = 2t_0 - \frac{1}{N-1} \sum_{j=1, j \neq i}^{N} d_{ij}, \tag{A9.7}$$

$$T_i(n+1) = (1 - \varepsilon) T_i(n) + \frac{\varepsilon}{N-1} \sum_{j=1, j \neq i}^{N} T_j(n) + \varepsilon M_i. \tag{A9.8}$$

Making use of Eq. (A9.8), we obtain

$$\sum_{j=1, j \neq i}^{N} T_j(n) = T_1(n) + \cdots + T_{i-1}(n) + T_{i+1}(n) + \cdots + T_N(n)$$

$$= \left(1 - \varepsilon + \frac{\varepsilon(N-2)}{N-1}\right) \frac{N-1}{\varepsilon} [T_i(n) - (1 - \varepsilon) T_i(n-1) - \varepsilon M_i]$$

$$+ \varepsilon T_i(n-1) + \varepsilon \sum_{j=1, j \neq i}^{N} M_i.$$

Substituting the above equation into Eq. (A9.8), we get

$$T_i(n+1) = \frac{2N-2-\varepsilon N}{N-1} T_i(n) + \frac{1-N+\varepsilon N}{N-1} T_i(n-1) + M, \quad (A9.9)$$

where

$$M = \frac{\varepsilon^2}{N-1} \sum_{j=1}^{N} M_j.$$

Solving the difference equation (A9.9), we have

$$T_i(n) = \frac{1}{N} \sum_{j=1}^{N} T_j(0) + \frac{N-1}{N} M_i - \frac{N-1-\varepsilon N}{\varepsilon N} \left(\frac{N-1-\varepsilon N}{N-1}\right)^{n-1} [T_i(1) - T_i(0)]$$
$$+ M\frac{N-1}{\varepsilon N} \left\{ n - \frac{N-1}{\varepsilon N}\left[1 - \left(\frac{N-1-\varepsilon N}{N-1}\right)^n\right]\right\}. \quad (A9.10)$$

Equation (A9.10) shows the timing convergence path for base station i. The last term in Eq. (A9.10) will cause timing drift. We should let $M=0$ by adjusting t_0 to avoid timing drift.

If a system is stable and no timing drift, by letting $n \to \infty$ in Eq. (A9.10), we get the stable value of the timing of base station i expressed as

$$T_i(\infty) = \frac{1}{N} \sum_{j=1}^{N} T_j(0) + \frac{N-1}{N} M_i. \quad (A9.11)$$

The residual timing error between base stations i and j measured at a location which is the midpoint of these two base stations is

$$T_i(\infty) - T_j(\infty) = \frac{N-1}{N} (M_i - M_j). \quad (A9.12)$$

Using Eq. (A9.7) in Eq. (A9.12), we can get the timing error between base station i and base station j as

$$T_i(\infty) - T_j(\infty) = \frac{1}{N}\left(\sum_{k=1,k\neq j}^{N} d_{jk} - \sum_{k=1,k\neq i}^{N} d_{ik} \right). \quad (A9.13)$$

Different Received Power Case

Taking the z-transform of Eq. (A9.5), we get a set of equations that may be written as single vector equation shown as

$$At = b, \quad (A9.14)$$

where

$$
A = \begin{bmatrix}
z - 1 + \varepsilon & -\dfrac{\varepsilon}{P_1}p_{12} & \cdots & -\dfrac{\varepsilon}{P_1}p_{1N} \\[2ex]
-\dfrac{\varepsilon}{P_2}p_{21} & z - 1 + \varepsilon & \cdots & -\dfrac{\varepsilon}{P_2}p_{2N} \\[2ex]
& & & \\[1ex]
-\dfrac{\varepsilon}{P_N}p_{N1} & & \cdots & z - 1 + \varepsilon
\end{bmatrix},
$$

$$
t = \left[\frac{T_1(z)}{z}, \frac{T_2(z)}{z}, \cdots \frac{T_N(z)}{z} \right]^{\mathrm{T}},
$$

$$
b = \left[T_1(0) + \frac{\varepsilon M_1}{z - 1}, T_2(0) + \frac{\varepsilon M_2}{z - 1}, \ldots T_N(0) + \frac{\varepsilon M_N}{z - 1} \right]^{\mathrm{T}},
$$

and $T_i(z)$ is the z-transform of $T_i(n)$; $T_i(0) = T_i(n)$ at $t = 0$; the superscript T represents the transpose operation. We can get a closed-form expression of the timing of base station i by solving Eq. (A9.14). For $N = 3$, the solution can be shown as

$$
\frac{T_i(z)}{z} = \frac{a_{i1}}{(z - 1)^2} + \frac{a_{i2}}{z - 1} + \frac{a_{i3}}{(z - \alpha_1)(z - \alpha_2)} \qquad (i = 1, 2, 3), \quad \text{(A9.15)}
$$

where a_{i1}, a_{i2}, and a_{i3} are coefficients; α_1 and α_2 are two roots. The first term of Eq. (A9.15) represents timing drift. We should let $a_{i1} = 0$ by adjusting t_0 to avoid timing drift. The second term represents the stable value to which the system converges. The residual timing error is determined by this term. The third term determines the system stability and convergence speed. The coefficient a_{i2} can be obtained as

$$
a_{i2} = \frac{2M_i P_1 P_2 P_3 + \displaystyle\sum_{k=1,k\neq i}^{3} M_k p_{ik} \displaystyle\prod_{k=1,k\neq i}^{3} P_k}{2P_1 P_2 P_3 + p_{12}p_{23}p_{31} + p_{13}p_{21}p_{32}} + C_3, \qquad \text{(A9.16)}
$$

where C_3 is a constant related to initial timing, p_{mq}, P_m, and M_m $(m, q = 1, 2, 3; m \neq q)$ but it is independent of i. Under general path loss conditions, it is quite reasonable to assume that the terms $p_{12}p_{23}p_{31}$ and $p_{13}p_{21}p_{32}$ are very small in comparison with $P_1 P_2 P_3$. Neglecting $p_{12}p_{23}p_{31}$ and $p_{13}p_{21}p_{32}$, we have

$$
a_{i2} \approx M_i + \frac{1}{2} \sum_{k=1,k\neq i}^{3} M_k \frac{p_{ik}}{P_i} + C_3 \qquad (i = 1, 2, 3). \qquad \text{(A9.17)}
$$

For $N \geq 4$, it is practically impossible to obtain a solution. By observing the above equation we assume the expression as

$$T_i(\infty) \approx M_i + h \sum_{k=1, k \neq i}^{N} M_k \frac{p_{ik}}{P_i} + C_N, \tag{A9.18}$$

where C_N is a constant and is independent of i; h is a coefficient to be determined. Comparing Eq. (A9.18) with computer simulation results, we can find a suitable value of $h = (\sqrt{N} - 1)/2$. The residual timing error between base stations i and j measured at the midpoint of these two base stations is

$$T_i(\infty) - T_j(\infty) \approx M_i - M_j + \frac{\sqrt{N} - 1}{2} \left(\sum_{k=1, k \neq i}^{N} M_k \frac{p_{ik}}{P_i} - \sum_{k=1, k \neq j}^{N} M_k \frac{p_{jk}}{P_j} \right)$$

Substituting M_i from Eq. (A9.3) and using Eq. (A9.4), yields

$$T_i(\infty) - T_j(\infty) \approx \sum_{k=1, k \neq j}^{N} \frac{p_{jk} d_{jk}}{P_j} - \sum_{k=1, k \neq i}^{N} \frac{p_{ik} d_{ik}}{P_j} + \frac{\sqrt{N} - 1}{2}$$

$$\times \left(\sum_{k=1, k \neq j}^{N} \sum_{m=1, m \neq k}^{N} \frac{p_{km} d_{km}}{P_k} \frac{p_{jk}}{P_j} - \sum_{k=1, k \neq i}^{N} \sum_{m=1, m \neq k}^{N} \frac{p_{km} d_{km}}{P_k} \frac{p_{ik}}{P_i} \right). $$

$$\tag{A9.19}$$

Equation (A9.19) shows that the residual timing error depends on weighting factors besides the propagation delay. As a result, the residual timing error for a specific system of size N is not a fixed value. For a $k \times k$ ($k = 2$–14) base station system, the average timing standard deviation obtained with Eq. (A9.18) almost agrees with that obtained by computer simulation.

REFERENCES

1. Y. Nagata and Y. Akaiwa, "Analysis for spectrum efficiency in single cell trunked and cellular mobile radio," *IEEE Trans. Vehicular Technology*, **VT-36**, 100–113 (August 1987).

2. W. C. Jakes (ed.), *Microwave Mobile Communications*, Wiley, New York, 1974.

3. J. E. Stjernvall, "Calculation of capacity and cochannel interference in a cellular system," *Proc. First Nordic Seminar on Digital Land Mobile Radio Communication*, 209–217 (1985).

4. T. Kanai, "Channel assignment for sector cell layout," *Trans. IEICE*, **J73-B-II**, 595–601 (November 1990) [in Japanese].

5. Y. Yamada, Y. Ebine, and K. Tsunekawa, "Base and mobile station antennas for land mobile radio systems," *Trans. IEICE*, **E74**, 1547–1555 (June 1991).

6. F. Tong and Y. Akaiwa, "Effect of beam tilting on bit-rate selection in mobile multipath channel," *Proc. 3rd International Conference on Universal Personal Communications*, 225–229 (1994).

7. S. W. Halpern, "Reuse partioning in cellular systems," *Proc. IEEE Vehicular Technology Conference*, 322–327 (May 1983).

8. J. J. Mikulski, "DynaT*A*C cellular portable telephone system experience in the U.S. and the UK," *IEEE Communication Magazine*, **24**, 40–46 (February 1986).

9. L. Lathin, "Radio network structures for high traffic density areas," *Proc. Third Nordic Seminar on Digital Land Mobile Radio Communication*, No. 14.10 (1988).

10. J. Worsham and J. Avery, "A cellular band personal communications system," *Proc. Second International Conference on Universal Personal Communications*, 254–257 (1993).

11. Y. Kinoshita, Tsuchiya, and Ohnuki, "Common air interface between wide-area cordless telephone and urban cellular radio: frequency channel doubly reused cellular systems," *Trans. IEICE*, **J76-B-II**, 487–495 (June 1993).

12. H. Furukawa and Y. Akaiwa, "A microcell overlaid with umbrella cell system," *Proc. IEEE Vehicular Technology Conference*, 1455–1459 (June 1994).

13. N. Nakajima, "Evolution of cell layout techniques—toward microcell systems," *NTT DoCoMo Technical Journal*, **1**, 21–29 (October 1993).

14. M. Taketsugu and Y. Ohteru, "Holonic location registration/paging procedure in microcellular systems," *IEICE Trans. Fundamentals*, **E75-A**, 1652–1659 (December 1992).

15. J. Z. Wang, "A fully distributed location registration strategy for universal personal communication systems," *IEEE Journal on Selected Areas in Communications*, **11**, 850–860 (August 1993).

16. A. S. Tanenbaum, *Computer Networks*, Prentice-Hall, Englewood Cliffs, N.J. (1981).

17. J. L. Hammond and P. J. O'Reilly, *Performance Analysis of Local Computer Networks*, Addison-Wesley, Reading, Mass. (1986).

18. Y. Akaiwa, Y. Furuya, and K. Kobasyshi, "Method of determining optimal transmission channel in multi-station communications system," *United States Patent*, No.4 747 101.

19. Y. Furuya and Y. Akaiwa, "Channel segregation, a distributed adaptive channel allocation scheme for mobile communication systems," *Proc. Second Nordic Seminar on Digital Land Mobile Radio Communication*, Stockholm (October 1986); also in *IEICE Trans.*, **E-74**, 1531–1537 (June 1991).

20. Y. Akaiwa and H. Andoh, "Channel Segregation—a self-organized dynamic channel allocation method: application to TDMA/FDMA microcellular system," *IEEE Journal on Selected Areas in Communications*, **11**, 949–954 (August 1993).

21. Y. Akaiwa and H. Furukawa, "Application of channel segregation for automatic channel selection free from intermodulation interference," *Proc. Seventh IEEE*

International Symposium on Personal, Indoor and Mobile Radio Communication, 1235–1236 (October 1996).

22. T. Kanai, "Autonomous reuse partioning in cellular systems," *Proc. IEEE Vehicular Technology Conference*, 782–785 (May 1992).

23. H. Furukawa and Y. Akaiwa, "Self-organized reuse partioning a dynamic channel assignment method in cellular system," *Proc. IEEE Vehicular Technology Conference*, 524–527 (May 1993).

24. N. Kataoka, M. Miyabe, and T. Fujino, "A distributed dynamic channel assignment scheme using information of received signal level," *IEICE Technical Report*, **RCS-93-70**, 1–7 (November 1993).

25. H. Furukawa and Y. Akaiwa, "Self-organized reuse partioning (SORP), a distributed dynamic channel assignment method," *IEICE Technical Report*, RCS-92-126, 61–66 (January 1993).

26. D. J. Goodman, R. A. Valenzuela, K. T. Gayliard, and B. Ramamurthi, "Packet reservation multiple access for local wireless communications," *IEEE Trans. Communications*, **COM-37**, 885–890 (August 1989).

27. K. Feher (ed.), *Advanced Digital Communications*, Prentice-Hall, Englewood Cliffs, N.J. (1987).

28. M. Frullone, G. Riva, P. Grazioso, and C. Carciofi, "Self-adaptive channel allocation strategies in cellular environments with PRMA," *Proc. IEEE Vehicular Technology Conference*, 819–823 (June 1994).

29. R. Beck and H. Panzer, "Strategies for handover and dynamic channel allocation in micro-cellular mobile radio systems," *Proc. IEEE Vehicular Technology Conference*, 178–185 (May 1989). H. Furukawa and Y. Akaiwa, "A self-organized reuse-partitioning dynamic channel assignment scheme with quality based power control," *Proc. Seventh International Symposium on Personal, Indoor and Mobile Radio Communications*, 562–566 (1995).

30. M. Yokayama, "Decentralization and distribution in network control of mobile radio communications," *Trans. IEICE*, **E73**, 1579–1586 (October 1990).

31. I. Katzela and M. Naghshineh, "Channel assignment schemes for cellular mobile telecommunication systems: a comprehensive survey," *IEEE Personal Communications*, 10–31 (June 1996).

32. N. Amitay, "Distributed switching and control with fast resource assignment/handoff for personal communications system," *IEEE Journal on Selected Areas in Communications*, **11**, 842–849 (August 1993).

33. J. C.-I. Chung, "Performance issues and algorithms for dynamic channel assignment," *IEEE Journal on Selected Areas in Communications*, **11**, 955–963 (August 1993).

34. J. C.-I. Chuang, "Autonomous time synchronization among radio ports in wireless personal communications," *Proc. IEEE Vehicular Technology Conference*, 700–705 (May 1993).

35. Information on GPS is available at the Web site: http://www.utexas.edu/depts/grg/gcraft/notes/gps/gps.html

36. Y. Akaiwa, H. Andoh, and T. Kohama, "Autonomous decentralized inter-base-station synchronization for TDMA microcellular systems," *Proc. IEEE Vehicular Technology Conference*, 257–262 (May 1991).

37. T. Fangwei and Y. Akaiwa, "Theoretical analysis of autonomous inter-basestation synchronization," *Proc. International Conference on Universal Personal Communications*, 878–882 (November 1995).

38. Y. Akaiwa and H. Andoh, "Improved scheme of autonomous inter-basestation synchronization," *Proc. Fall National Convention of IEICE*, **B-251** (1991).

39. H. Kazama, S. Nitta, M. Morikura, and S. Kato, "Semi-autonomous synchronization among base stations for TDMA-TDD communication systems," *IEICE Trans. Communications*, **E77-B**, 862–867 (July 1994).

40. X. Lagrange and P. Godlewski, "Autonomous inter base station synchronization via a common broadcast control channel," *Proc. IEEE Vehicular Technology Conference*, 1050–1054 (June 1994).

41. M. Kuwabara (ed.), *Car Telephone*, IEICE, Tokyo (1985).

42. Y. Yasuda (ed.), *Mobile Communications in ISDN Era*, Ohm, Tokyo (1992).

43. A. Alshamali and R. Macario, "Technical features of the planned European radio messaging system — ERMES," *IEEE Vehicular Technology Society News*, 22–25 (August 1994).

44. T. Nakanishi, E. Murata, K. Honma, and Y. Rikou, "Digital voice processing land mobile radio," *Proc. 1980 International Conference on Security through Science and Engineering*, 58–62 (September 1980).

45. M. Ikoma, K. Kimura, N. Saegusa, Y. Akaiwa, and I. Takase, "Narrow-band digital mobile radio equipment," *Proc. IEEE International Conference on Communications*, 23.3.1–23.3.5 (1981).

46. T. Hiyama, A. Yotsutani, K. Kage, and M. Ichihara, "4-level FM digital mobile radio equipment," *NEC Research and Development*, No. 71, 20–26 (October 1983).

47. M. Nakajima, T. Watanabe, S. Saka, H. Nogami, and T. Karasawa, A narrow band digital portable radio using 8 kbps speech CODEC, *Proc. Fall National Convention of IEICE*, **B-273** (1990).

48. E. Lycksell, "MOBITEX, a new radio communication system for dispatch traffic," *TELE*, 68–75 (January 1983).

49. T. Miyamoto, H. Tatsumi, and H. Orikasa, "Tele-Terminal System (mobile data communications)," *Proc. First International Workshop on Mobile Multimedia Communications*, A.1.5-1–A.1.5-8 (December 1993).

50. T. Harris, "Data services over cellular radio," in *Cellular Radio Systems*, D. M. Balston and R. C. V. Macario (eds.), chapter 12, Artech House, Boston (1993).

51. N. J. Muller, *Wireless Data Networking*, Artech House, Boston (1995).

52. D. M. Balston and R. C. V. Macario (eds.), *Cellular Radio Systems*, Artech House, Boston (1993).

53. K. Kinoshita, M. Hata, and K. Hirade, "Digital mobile telephone system using TD/FDMA schemes," *IEEE Trans. Vehicular Technology*, **VT-31**, 153–157 (November 1982).

54. Y. Akaiwa and Y. Nagata, "A linear modulation scheme for spectrum efficient digital mobile telephone systems," *Proc. International Conference on Digital Land Mobile Radio Communications, Venice*, 218–226, (1987).

55. K. S. Gilhousen, I. M. Jacobs, R. Padovani, A. J. Viterbi, L. A. Weaver, Jr., and C. E. Wheatley III, "On the capacity of a cellular CDMA system," *IEEE Trans. Vehicular Technology*, **VT-40**, 303–312 (May 1991).

56. N. Kakajima, "Japanese Digital Cellular Radio," in *Cellular Radio Systems*, D. M. Balston and R. C. U. Macario (eds.), chapter 10, Artech House, Boston (1993).

57. D. Raychaudhuri and N. D. Wilson, "ATM-based transport architecture for multiservices wireless personal communication networks," *IEEE Journal on Selected Areas in Communications*, **12**, 1401–1414 (October 1994).

58. Y. Akaiwa, T. Nomura, and S. Minami, "An integrated voice and data radio access system," *Proc. 42nd IEEE Vehicular Technology Conference*, 255–258 (1992).

59. W. C. Y. Lee, *Mobile Cellular Telecommunications*, 2nd ed., McGraw-Hill, New York (1995).

60. W. C. Y. Lee, *Mobile Communications Design Fundamentals*, 2nd ed., Wiley, New York (1993).

61. G. L. Stüber, *Principles of Mobile Communication*, Kluwer Academic Publishers, Boston (1996).

62. T. S. Rappaport, *Wireless Communications, Principles and Practice*, Prentice-Hall PTR, Englewood Cliffs, N.J. (1996).

63. R. Steele (ed.), *Mobile Radio Communications*, Pentech Press, London, and IEEE Press, New York (1992).

64. M. D. Yacoub, *Foundations of Mobile Radio Engineering*, CRC Press, Boca Raton, Fla. (1993).

65. G. C. Hess, *Land-Mobile Radio System Engineering*, Artech House, Boston (1993).

66. D. M. Balston and R. C. V. Macario (eds.), *Cellular Radio Systems*, Artech House, Boston (1993).

67. K. Pahlavan and A. H. Levesque, *Wireless Information Networks*, Wiley, New York (1995).

68. J. D. Gibson (ed.), *The Mobile Communications Handbook*, CRC Press, Boca Raton and IEEE Press, New York (1996).

69. J. Jayapalan and M. Burke, "Cellular data services architecture and signaling," *IEEE Personal Communications*, **1**, No. 2, 44–55 (1994).

70. R. Steele, "The evolution of personal communications," *IEEE Personal Communications*, **1**, No. 2, 6–11 (1994).

71. Special issue on PCS, *IEEE Personal Communications*, **1**, No. 4 (1994).

72. Special issue on Standards, *IEEE Communications Magazine*, **32**, No. 1 (January 1994).

73. Special issue on PCS, *IEEE Communications Magazine*, **30**, No. 12 (December 1992).

74. Special issue on European radio propagation and subsystems research for the evolution of mobile communications, *IEEE Communications Magazine*, **34**, No. 2 (February 1996).

75. Special issue on the European path toward UMTS, *IEEE Personal Communications*, **2**, No. 1 (February 1995).

76. D. C. Cox, "Wireless personal communications: what is it," *IEEE Personal Communications*, **2**, No. 2, 20–35 (April 1995).

77. M. Khan and J. Kilpatrick, "MOBITEX and mobile data standards," *IEEE Communications Magazine*, **33**, No. 3, 96–101 (March 1995).

INDEX

425